The Science of Reading

.

THE
Science
OF
Reading

INFORMATION, MEDIA,

AND MIND IN

MODERN AMERICA

Adrian Johns

The University of Chicago Press

Chicago and London

The University of Chicago Press, Chicago 60637
The University of Chicago Press, Ltd., London
© 2023 by The University of Chicago
Published 2023
Printed in the United States of America

32 31 30 29 28 27 26 25 24 23   1 2 3 4 5

ISBN-13: 978-0-226-82148-1 (cloth)
ISBN-13: 978-0-226-82149-8 (e-book)
DOI: https://doi.org/10.7208/chicago/9780226821498.001.0001

Library of Congress Cataloging-in-Publication Data

Names: Johns, Adrian, author.
Title: The science of reading : information, media, and mind in
    modern America / Adrian Johns.
Description: Chicago ; London : The University of Chicago Press,
    2023. | Includes bibliographical references and index.
Identifiers: LCCN 2022026310 | ISBN 9780226821481 (cloth) |
    ISBN 9780226821498 (ebook)
Subjects: LCSH: Reading, Psychology of—Research—United
    States—History. | Reading—Research—History.
Classification: LCC BF456.R2 J65 2023 | DDC 418/.4—dc23/
    eng/20220715
LC record available at https://lccn.loc.gov/2022026310

♾ This paper meets the requirements of ANSI/NISO Z39.48-1992
(Permanence of Paper).

For Zoe

CONTENTS

ACKNOWLEDGMENTS

This book was largely written during 2020–21 under the constraints of the COVID-19 pandemic. But I have been thinking about the themes that it addresses for some years—decades, in fact. I initially started reflecting on the science of reading back in the mid-1990s, when I was a research fellow at Downing College in Cambridge (a research fellowship is roughly equivalent to a postdoctoral fellowship in the United States). That gave rise to a chapter on "the physiology of reading" in my first book, *The Nature of the Book* (1998). The period that that book dealt with was the seventeenth century, however, long before modern scientific research enterprises came into being. Since then I have returned to the topic every so often, wondering about how people in later times conceived of the practice and impact of reading and how their conceptions affected cultural experiences. For the most part, those reflections took the form of talks rather than publications. They did range increasingly widely, however, and eventually they extended into the modern era. Once or twice I even dared to mention neuroscience. The notion of writing in earnest about the topic came to seem a real possibility, if still a remote one.

For the stimulus to develop all those piecemeal ideas into a book I must thank Paula McDowell of NYU. In the spring of 2020, just days before New York locked down, Paula ran a conference on the theme of Marshall McLuhan as a reader, and she invited me to give a keynote speech. It was a gamble on her part given that I had been professionally under the radar since the death of my wife in 2016. But it worked out: I found myself bitten by the research bug once again. I followed the science of reading down a rabbit hole, reaching back from McLuhan's rendezvous with Samuel Renshaw (for which, see chapter 7) to, at one end, James McKeen Cattell and Edmund Burke Huey, and, at the other, the strange contemporary technology of "machine reading." So my first gratitude is to Paula for making possible my own version of Berlioz's *Lélio*—a return to intellectual life.

It so happened that that conference happened just before I began a period of research leave. This sabbatical, which ran from July 2020 to June 2021, occurred thanks to a grant from the National Endowment for the Humanities (FEL-267420-20) and additional support from the University of Chicago. I am extremely grateful to both. Without the freedom to concentrate wholly on this project without the distractions of teaching, administration, and pandemic-avoidance, it could never have been substantially begun, let alone completed. As it was, I was able to retreat from the world and submerge myself in the extensive and often odd literature of the science of reading.

The first tentative version of this book was completed on the day of the presidential election in November 2020. Karen Darling at the University of Chicago Press then took it in hand. She saw its potential immediately. That it eventually appeared in this form is a tribute to her editorial prowess. She and I have worked together for some years on the Press's science.culture series, and it would be impossible to imagine a better collaborator. I am also very grateful to both the outside readers—I have no idea who they were—to whom she sent the manuscript. They not only agreed to read through the whole thing but also provided a number of extraordinarily acute observations, both on what the text said and on what it neglected to say. The final version has been improved immeasurably thanks to their insights.

Other than that, my debts are mainly general ones, for conversations undertaken over many years. It would be possible and warranted to list hundreds of such occasions. I have been at the University of Chicago since 2001, and the community there has been a constant source of inspiration. I am especially thankful to Bob Richards, who is a truly extraordinary human being, generous and humane to an extent that few can match. James Evans has been a constant friend and coworker on allied topics, especially those to do with the information sciences. And although they won't recognize their contributions, I want to signal my appreciation for important critical insights I gleaned from conversations with Mauricio Tenorio, Michael Rossi, Susan Goldin-Meadow, and David Nirenberg. Debby Davis, an elementary-school teacher at the University of Chicago Laboratory Schools, in fact inspired a lot of the conceptualization and planning of this book, although I don't think she realized that she was doing so at the time. Her creativity, intelligence, acute critical insight, professionalism, and kindness combined to make me see the interest and importance of themes that I would surely have missed otherwise. The students at Chicago are frighteningly smart, dedicated, and imaginative, and I have learned an immense amount from them too; I must mention here Michael Castelle, Jillian Foley, Eric Gourevitch, Carmine Grimaldi, Alex Moffett, Parysa Mostajir, Justin Niermeier-Dohoney, Elizabeth Sartell, Adam Shapiro, Joy Wattawa, and Tali Winkler, but the reality is that such a short list represents an injustice: it is really the collective that is so extraordinary.

For advice on particular subjects I am thankful to Michael Sokal (for Cattell), Andrew Abbott (for library sciences), and Jamie Cohen-Cole (for psycholinguistics and the cognitive revolution). Jane Dailey provided some crucial help about issues of African American education in the South. Norman Stahl and Douglas Hartman provided important assistance on Edmund Burke Huey at a late stage. And Nate Bolt and Edwin Hutchins were very helpful on contemporary issues of cognition, computing, and display technologies. Beyond those, I shall merely mention a few individuals who have had especially extensive influence in a more general sense. Steven Shapin and Dan Kevles are two senior figures who have become my guiding lights over many years now; I do not think I would have dared take on this project but for their support. Conversations with Ann Blair, Alex Csiszar, and Lisa Gitelman over the years have been as influential on the project as their work has been on the information-history field in general. Nancy Barnes was a greatly supportive pen pal during much of the writing. Although Bob Brain and I did not happen to speak during the writing of this book, his work on modernism and psychophysics has long fascinated me—not only for its content, but for the curious, intellectually generous way in which it is pursued. And Olena Marshall provided sharp comments when needed, and, much more significantly, put up with me throughout.

Most of the materials I used in writing *The Science of Reading* are printed texts of one kind or another. A thorough exploration of the archival sources for the science of reading could not be undertaken because of the pandemic. (It is worth noting here that they are extensive—future researchers will find a lot to fascinate them.) But I am greatly indebted to the Special Collections Department at the University of Chicago Library, the University Archives at Ohio State University, and the Archives at the Cummings Center for the History of Psychology in Akron, Ohio, for permission to use original materials—and even to see original instruments—in their care. Other archives and collections supplied images and are acknowledged as appropriate in picture credits.

The text of this book is almost entirely new. But along the way, early thoughts on the broad topic have appeared in print here and there. In particular, some parts of chapter 1 were aired in "The Physiology of Reading," in *Books and the Sciences in History*, ed. N. Jardine and M. Frasca-Spada (Cambridge: Cambridge University Press, 2000), 291–314; and the story about Samuel Renshaw and Marshall McLuhan that forms part of chapter 7 was published in "Watching Readers Reading," *Textual Practice* 35, no. 9 (September 2021): 1429–52. Otherwise nothing you read here has appeared before.

Researching and writing a book is, for me at least, an all-consuming activity. So I am particularly grateful to my family, David, Lizzie, Zoe, and Benjamin, for tolerating my absence of mind during this strange time.

The last thing I will say is that producing this book has been a helter-skelter intellectual adventure. I found myself pursuing at great speed topics that were new to me, in a discipline (psychology) about which I knew little, for a period outside my usual zone of expertise, and in a country not my own. I heartily recommend this kind of recklessness: it can be quite exhilarating. But I shall certainly have made mistakes, and I want to stress that, however embarrassing they may be, they are my own responsibility.

Introduction

The Mysterious Art of Reading

Reading is a good thing. We like to believe that it is a fundamental element of any modern, enlightened, and free society. We may even think of it as *the* fundamental element. It has long been standard to identify the emergence of contemporary virtues like democracy, secularism, science, and tolerance with the spread of literacy that occurred in the wake of Johannes Gutenberg's invention of printing in the fifteenth century. And of course we maintain that the ability to read successfully is functionally essential for anyone who wishes to become a fully actualized, participating citizen in the modern world. Almost nobody nowadays would argue that reading is anything but a beneficial and intrinsically meritorious practice for everyone. If there is one practice that unites the most elevated moral reflections on modernity with the most quotidian of everyday experiences, reading is it.

All of us who are literate—and it is worth remembering for a moment that many even in the developed world are not—have, of course, learned to become so. Reading, as one of its first scientific investigators pointed out, is not natural. No nonhuman creature has ever done it, as far as we know. And yet, "this habit," as Edmund Burke Huey marveled in 1908, "has become the most striking and important artificial activity to which the human race has ever been moulded." Huey was surely right in that arresting realization. And the questions that forced themselves upon his mind in consequence of it were surely the appropriate ones too. Since reading is unnatural, he asked, "What are the unusual conditions and functionings that are enforced upon the organism in reading? Just what, indeed, do we do, with eye and mind and brain and nerves, when we read?"[1] Apparently simple, these questions are in fact deep and complex; and they are extremely difficult to answer. They require not only sophisticated psychological and physiological concepts but stances on such matters as the mind-body relationship and the nature of knowledge itself. All of science and philosophy, we might almost say, are implicit in them. That is surely why, Huey observed, in ancient times reading was accounted "one of the most mysterious of the arts," and why its operation was still accounted "almost as good as a miracle" even in his own day. And yet, starting in about 1870, generations of scientists did take on Huey's questions. *The Science of Reading* is about the rise and fall—and subsequent rise again—of the enterprise these scientists created to answer them.

Huey posed those questions at the beginning of what was the first major book in this new science to be published in America. *The Psychology and Pedagogy of Reading* first appeared in 1908 and proved to have extraordinary longevity. It was reissued several times in the next two decades, it was reprinted again by MIT Press in 1968 as a classic of cognitive science, and it enjoyed a new edition as recently as 2009. The volume is valuable as a gateway into the subject of this book, not only because of its prominence in the field, which is unrivaled, but also because Huey was remarkably and explicitly reflective about the cultural concerns that underpinned his new science and gave it its purpose. I shall say more about this in chapters 1 and 2, but for now it is useful simply to call to mind the historical distance that separates readers today from those whom Huey investigated in the late nineteenth and early twentieth centuries. Although the questions that he posed in his research were in one sense naturalistic—that is, they were questions about the properties of readers considered as human beings in general, independent of time and place—Huey was well aware that what made those questions meaningful were contemporary contexts both large and small. He was writing in the era of the first mass education and the first mass democracy. Industrialization and the Gilded Age had given rise to giant capitalist institutions that transformed perceptions of society and people's places in it. Telegraphy and telephony were transforming communications, and radio would soon do so even more. The mass-circulation newspaper was changing how people thought about themselves, their privacy, and that

oddly numinous entity "the public." Optimism about social and technological progress was tempered with anxieties about decadence, degeneration, addiction, atavism, and other perils. And Darwinism—social as
well as natural—suggested powerful ways to understand and master the
dynamics of all these processes, for good and ill. As we shall see, Huey had
all these hopes and fears very much in mind when he made his remarks
about reading's marvelous and mysterious power. They played a signal
part in motivating his pursuit of a scientific approach to the practice.

One aim of this book is to explain the origin, development, and consequences of the science of reading that Huey and his peers inaugurated.
In that light, its approach is thoroughly, and, I hope, convincingly, historical. Yet it is also worth considering that the questions that excited
researchers in Huey's time do have their echoes in our own age, just over
a century later. We too have our optimistic hopes and our existential anxieties, many of which have to do with new communications systems and
the problems of large-scale capitalist institutions.[2] The economic and social inequalities of 2020s society, notoriously, are greater than they have
been at any time since Huey's, and it is possible that the moral and political instability arising from the conjunction of communications technologies and social strains may prove as great. True, we now talk about
our situation in rather different terms than Huey used to address his. We
invoke *information technology*, *surveillance capitalism*, and *attention*, and
we worry about what happens in and to our brains as they are exposed
to the firehose blast of multichannel, polysensory information that char-
acterizes twenty-first century life. Those are concepts and technologies
quite different from Huey's. But when we ask how we can educate the
next generation so they may live full lives in this environment, and nobody seems to have a definitive answer, our concerns are not so far removed from his generation's. And in many ways our capacity to pose and
tackle such questions is indebted to that generation's work. Moreover,
as we shall see, the science of reading that evolved from that time is in
fact responsible for central aspects of the very experience that inspires
our own anxious questioning. The story told in this book therefore does
not end with the ascendancy of the science of reading in the 1930s, 1940s,
and 1950s, nor even with its eclipse—temporary, as it turned out—in the
1960s and 1970s. It extends into the present. One point is to cast light on
the ways in which we think about equivalent problems today. Although

the science of reading that Huey and his fellows brought into being does not provide answers for us in any simple way, considering it historically does help us appreciate our own questions and their meanings in a better light. And a history of the science of reading need not be so rigorously self-denying as to shy away from profound questions about how and why we now think, wonder, and fear as we do.

History, Reading, and Science

A historical appreciation of the science of reading carries enormous implications for knowledge of all kinds, and for history in particular. The reason is simple but worth stating explicitly. It is empirically obvious that readers' responses to a given printed work can and do vary across wide ranges of possibility. Indeed, that truism is perhaps most evident precisely when readers themselves claim to be observing the literal meaning of a text. Nobody argues more vehemently over literal meanings than scriptural fundamentalists, after all. Yet what is obvious is often disregarded as a matter of everyday practice. Any understanding of past cultures, after all—indeed, any understanding worth its name of other cultures in our own time—tends to rest on a tacitly shared assumption that our own readings of textual sources, if not precisely the same as those of the people we are interested in, are at least commensurable with them. The differences are assumed to be comprehensible and explicable, and in fact a field like intellectual history could be defined in large part as the attempt to achieve such comprehension. But the basis of any such enterprise includes, in the end, a tacit agreement to set the implications of such differences aside. As professional scholars, we act as if reading itself—the intimate, uninterrupted heart of the practice, when one comes face-to-face with a page of text in the solitude of a study or the hush of a library—is fundamentally *one thing*. The history uncovered in this book calls that concurrence into doubt.

The paradox here is one that historians, anthropologists, and other researchers have struggled with for some time now. On the one hand, print clearly has cultural consequences extending beyond the local—indeed, extending beyond the local is its very point. But on the other, for all that printed works take on relatively constant forms, their uses obviously vary widely with local contexts. For several decades now, historians have sought to account for such variety by revealing the history of reading practices alongside those of authorship, manufacture, and circulation.[3] These, more than texts considered in the abstract, were said to be the motors of change or the foundations of cultural stability. And there were ways to bring them to light, perhaps most directly through analyzing the traces left in printed books themselves by readers who took notes, doodled, folded back pages, and the like. Even without such direct evidence, a sensitive historian could reveal a lot about the different contexts in which reading had taken place and make suggestions about how those contexts had affected the received meanings of genres and particular, even literary, works.[4] The work produced in this growing field of "the history of reading" is often very sophisticated indeed, if occasionally a little dryly microscopic. It has succeeded in refocusing attention onto the complex ways in which the writing, printing, circulation, and reading of books (and other informational objects) all combine to give rise to cultural processes. But the enterprise has the potential to generate something distinctly more ambitious, and less congenial to prevailing modes of thought, than its current achievements amount to. Taken at face value, the historical study of reading could—and perhaps should—trigger an epistemic crisis in the humanistic social sciences, and indeed in the nonhumanistic ones too.

In one sense—which will become clearer later—the history of reading plays a similar role vis-à-vis traditional historical writing to that which cognitive science has played vis-à-vis behaviorist psychology. It focuses on the dynamic, active character of reading as a practice and highlights the degrees of freedom that readers enjoy in arriving at their interpretations. Reading, it argues, is a creative skill. Exercising that skill is work, and work that eludes policing by states and similar institutions because it is partly carried out inside our bodies and brains. Where the cognitive scientists of the 1950s and 1960s charged that behaviorists were intellectual authoritarians who downplayed the active, exploratory character of learning, the historians of reading of the 1980s and 1990s leveled a comparable charge against social historians who classified cultures in terms of the circulations of texts that presumptively had deterministic effects on their readers. There was something Romantically emancipatory about restoring to view the autonomy of readers who broke the bounds of those categories. To focus on women, ethnic minorities, censors, or scientists as readers was to contribute to an enriched and nonreductive understanding of how an interlaced society was sustained in practice.

Perhaps the most striking formula expressing this kind of perspective was one coined by the French cultural theorist Michel de Certeau, who, in a much-cited article originally published in 1980, referred to reading as "poaching." De Certeau was interested in the active, unruly, appropriative character of everyday life in general—he was writing partly in opposition to Michel Foucault's concept of disciplining—and for that the practice of reading struck him as the exemplary case.[5] His own background was an intellectually extraordinary combination of Freudian psychoanalysis, Lacanian theory, Jesuit theology, mystical studies, classics, and Renaissance history. But his notion of reading as fugitive, active, and uncontainable struck a powerful chord in the United States, where this background was not widely appreciated. There, followers welcomed it in the light of a cognitive reconfiguration in the human and social sciences that had been under way since about 1960, the nature of which will be a theme in the later chapters of this book. They saw de Certeau as encouraging them to identify the diverse ways in which local cultures could escape subsumption in the overweening political systems of modernity by eluding the simple kinds of cause-and-effect models that attributed compulsive power to media. To that extent, de Certeau's views aligned with those of other influential thinkers of the time, including Pierre Bourdieu, who had pointed to the varied ways in which French readers made use of tabloid newspapers, and Richard Hoggart, whose pioneering book *The Uses of Literacy* (1957) had made a similar point about working-class citizens in Britain.[6] All three insisted that culture—and therefore politics—could not plausibly be accounted for in terms of the "impact" of mass print on passive populations. Readers were active, questing, and cussed subjects who took up printed objects and put them to their own uses. And those uses were best explained historically.

As appealing as the metaphor of poaching proved to be, however, nobody seriously suggested that reading was simply and uninhibitedly anarchic, let alone recommended that it *should* be so. Not least, such a notion would have had a hard time accounting for one of the most remarkable things about the practice, namely its propensity to have a radically transformative effect on an individual reader. Anyone who read habitually had experienced this kind of momentary power, and by common consent the experience was striking enough to cast doubt on any notion that readers were always in conscious control. Instead, the implication was to focus on the constructive interplay between an active reader and an object such as a book. That interplay always occurred in some specific time and place. Historians of reading therefore became very attuned to the material and formal characteristics of informational objects, and to what they implied in specific contexts—to matters of typography, paper, binding, and the like, and to settings such as coffeehouses and monasteries. Moreover, they would approach these characteristics in light of a revolution of sorts that occurred at the same time in the ascetic world of bibliography. The principal figure in that revolution was D. F. (Don) McKenzie, a charismatic New Zealander, whose trio of lectures published under the significant title *Bibliography and the Sociology of Texts* challenged a scientistic norm that had reigned in that desiccated discipline for almost a century.[7] McKenzie and those he inspired—who included the present writer—brought to notice how the material, tactile nature of such objects affected their readings. They paid attention to whether a reader of John Locke's *Essay Concerning Human Understanding*, say, was using an octavo or a folio, a paperback or a hardback, a microfiche or an audiobook. Looking at graffiti differed, to them, from deciphering checkout receipts or turning doctors' prescriptions into medicaments.

In the generation since McKenzie and his first interlocutors, historians of reading have addressed all of these questions with great dedication and skill. But in practice there remains a distinct limit, a non plus ultra, to the enterprise. That limit is a matter of convention, not logic. For all that its advocates argue that reading needs to be understood historically, and that particular cases of reading must be explained in terms of the physical objects and social contexts at hand, they have been reluctant to pursue the approach to its fullest implications. In their work there remains, in effect, a shared thing, a common practical experience, that reading *is*—and that thing is also, ultimately, what reading *was*. It is, of course, quite understandable that they would be reluctant to confront such an assumption explicitly. It is reasonable to assume that a fundamental commonality at this level undergirds the very possibility of historical knowledge—for that matter, perhaps all knowledge. Denying it would look like epistemic nihilism. That is, it would amount to a professed claim to undermine the viability of knowledge itself in a way that might all too predictably be decried as ridiculous, self-refuting, or childishly solipsistic. Yet, understandable as it is, there is surely a failure of nerve here. The implication of the history of reading, I would contend, is indeed this radical, and if so then it should be faced up to. The question is how to acknowledge this implication in full and embrace it, while still maintaining the possibility—more, the necessity—of historical investigation and interpretation. We need no less in order to furnish defensible accounts of how cultures arise, subsist, and change.

An answer to that question lies in the history of science. In particular, it lies in the history of the science of reading itself. It is by exploring this history that we come to understand not just *that* reading has differed across cultures, but *how*, *why*, and *to what effect*. It confirms the radical scope of the history of reading while also bounding it. If we historicize the science of reading, we come to see not only that the historians of reading are right, and not only that the implications of their work are greater than they themselves have felt ready to acknowledge, but also that there are ways to preserve and revive the historian's craft in the face of those implications. How exactly to do that is what is contended for, and perhaps exemplified, in this book. But the first lesson that we stand to learn when we bring together the history of reading and that of its science is this: that there is even more depth to the history of reading than its practitioners believe. Reading is historical "all the way down." At least, we have to start with that position. We cannot begin by assuming that there is a single, shared practice, common to human nature as such, that underlies all of the different applications that historians (and, in their different ways, psychologists, sociologists, and anthropologists) have brought to light. The scientists of reading themselves did begin that way, of course. They commenced with the assumption that a certain reading propensity was hard-wired into human beings; but they discovered in the end that it was not so. The Holy Grail did not exist. In that sense, their enterprise failed by virtue of succeeding. But thanks to them, that distinctly troubling idea may itself be treated as historical—as, in a real sense, an achievement. And when it is, we stand to realize that we can break what may otherwise seem an endless and rather vicious cycle. By attending to what may be called the epistemic constitutions of reading, we can hope to attain the kind of enriched understanding that has been in prospect for a while now, in principle at least, but for which we have been too timid to reach out.

It is important to add one more point, however. Not just any kind of approach to the history of science will serve this purpose. Since the 1970s, researchers keen to understand the sources of science's authority have emphasized the practical, emergent, and communicative character of scientific work, and sensitivity to these kinds of issues will be essential for the task at hand now. In that light, incidentally, it is interesting to note how strong the parallels have been between the enterprises of the history of reading and the history of science in the last generation. Both have stressed explorations of practice, material culture, and local context in their bids to understand the making of meanings. The connections to cognitive research, too, are real, if often not immediately evident. David Bloor, for example, designer of the "strong program" in the sociology of knowledge, had been an acolyte of F. C. Bartlett, a British psychologist who collaborated in the prewar science of reading. Much could no doubt be said on this theme. But for the moment it is sufficient to point out that in applying the history of science to the practices of reading, it turns out that we are not venturing so very far afield.

Given that it has such ambitious objectives, making clear what this book is not is almost as important as explaining what it is. It is not a history of literacy, for a start. There are several such histories, and they are all very earnest and important. It is also not—or not directly, at least—a history of the "reading public," that focus of so much critical attention in the 1920s and beyond. The figures who play large parts in this story were not, in general, the grand visionaries of publicity and democracy that we tend now to remember, like John Dewey and Walter Lippmann. Dewey and Lippmann do make an appearance, to be sure, but they are here supporting actors. More central are the active researchers—the people who built laboratory instruments and designed innovative experiments to investigate what happened when an individual read a page of words or a community encountered propaganda. Nor, third, is this a history of the *teaching* of reading. Here too there are many books that curious readers may track down, produced over the decades—if, rather dispiritingly, more often from the network of metaeducational institutions (that is, programs that teach teachers) than from teachers drawing upon their own experiences. I shall have much to say about pedagogy, to be sure, but it is not my primary focus. Indeed, one point of the book is to disengage, where appropriate, research on reading from the strident arguments about teaching that have always threatened to overwhelm it.

Finally, this book is not a comprehensive history of experimental or educational psychology. I do suspect—but am not competent to prove—that the conventional divisions of psychology are a little too neatly drawn in existing disciplinary histories. In researching this topic I have found myself traversing what are often portrayed as fairly hard boundaries, particularly that between behaviorism and cognitive science. Sometimes these boundaries loomed very large for the people I have been pursuing, but sometimes they did not. In part, this is simply because boundaries between disciplines are always shifting and have to be sustained by careful and diligent work. They are often defined precisely in order to bolster one side or another in a debate.[8] But I have come to feel that there may be more to it in this case. In fact, one of the most remarkable aspects of the science of reading was the frequency with which scientists in other psychological fields appropriated and redeployed its experiments and technologies in order to tackle fundamental problems in their own areas. When they did so, the origin of those experiments and devices in examinations of reading would often be forgotten. In light of this tendency, I suspect that the history of psychology may need to be rethought in terms of a timeline for which experiments on reading—rather than, say, concepts of association or faculty, or convictions about stimulus and response—were fundamental. That is a task that lies far beyond my own reach, however. Suffice it to say that the origins of the science of reading are those conventionally identified for experimental psychology too, and that the fields have overlapped for much of their subsequent history. And since about 1910, when *educational* psychology emerged as a discrete discipline in the United States, much of that enterprise too has been concerned with problems of reading. I do, then, address some of the central questions, methods, and consequences of both disciplines in the course of the book. But I make no attempt to provide a systematic account of the broad nature or achievements of experimental or educational psychology as such—or, for that matter, their blind spots and embarrassments. Experts may feel that in each case I have omitted too much. I can only apologize and point them to the histories that do exist.[9]

The Science of Reading and the Information Society

The Science of Reading traces the emergence, consolidation, and implications of a tradition of research from about 1870 to the present. This tradition arose out of what was known as *psychophysics*, a then-new experimental approach to psychological phenomena that was pioneered in Germany in the mid-nineteenth century. It was introduced into the United States by a handful of keen young researchers who had gone to Leipzig and Halle and returned to take up positions in American institutions, where they constructed instruments, developed laboratories, and taught generations of students. The first part of the story thus traces the creation of a scientific discipline. We then look at what the scientists of reading actually did—their laboratory techniques and practices—and show how their central concepts and tools circulated beyond academic institutions to reach almost every area of American life.

Both the questions that motivated these researchers and the answers they generated were of broad and fundamental importance. Schools and workplaces across the United States took notice, believing that solutions to their problems lay in the instruments that the scientists of reading invented and tinkered with. The science of reading began in the first era when corporations were becoming geographically distributed entities tied together by filing systems and other informational machines; industry and commerce required a skilled, literate workforce and demanded that the nation provide it; and advertising and other forms of market information governed their ability to make ever-larger profits. Mass newsprint was a powerful political agent, and all the more so in a country that from the start had been proud of its informed, well-read citizenry. Americans' lives were now structured around books, newspapers, magazines, posters, card indexes, folders, files, and all their associated paraphernalia—the complex, variegated, overwhelming, and fast-changing world of what the Belgian pacifist and universal bibliographer Paul Otlet christened *documentation*. The common denominator was that all those things had to be read. And increasingly Americans and their governors were anxious about the consequences, which meant that *what* could be read, *how*, and *by whom* were all matters to be managed. Hence mass education; but hence, also, the Comstock Act and other initiatives designed to uphold public morals amid a slew of unfettered information.[10]

The machines, theories, and practices of the science of reading therefore affected the lives of virtually every American in profound and inescapable ways. Citizens encountered them everywhere they went, in settings ranging from the nursery to the aircraft carrier, and from the bomber cockpit to the domestic kitchen. They learned to read from an early age by virtue of techniques that the science of reading underpinned and validated, and they sought to improve their reading practices in adulthood by employing that science all over again. Good reading could be identified scientifically; bad could be diagnosed, treated, and remedied. Writ large, this meant that the science of reading was also of critical salience for those, like Dewey and Lippmann, who worried about the culture of politics for the nation as a whole, because countless acts of reading collectively defined that culture. The science of reading therefore helped define the parameters in terms of which the great midcentury debate about democracy and publicity could take place.[11]

By the 1940s the science of reading had taken several forms, extending from a highly technical laboratory discipline to a sophisticated—and sometimes risky—field science. It was a major contributor to contemporary discussions on matters ranging from the segregation of southern schools and libraries to the management of the modern corporation and the politics of new media. Starting in the late 1950s, however, the science of reading underwent a radical shift. Having enjoyed broad respect for half a century, it found itself subjected to two sharp but distinct attacks. On the one hand, its experimental and instrumental traditions came in for severely increased criticism in light of the declining reputation of behaviorist views of human nature. A new discipline of cognitive science was in the offing, and it set itself against the allegedly authoritarian character of the older approaches, insisting that human learning processes were much more protean, exploratory, and constructive than any merely behavioral approach could grasp. One had instead to appreciate the complexity, autonomy, and freedom of the mind itself, and work to nurture those qualities in school settings. But at the same time, Rudolf Flesch's shocking *Why Johnny Can't Read* (1955), while likewise accusing the science of reading of being authoritarian and behaviorist, also assailed it for being insufficiently rigorous. Flesch claimed that it had spawned a nationwide education system that did not in fact teach children to read at all. His sensational book asserted that the science of reading was ill-conceived, out of touch, self-serving, and corrupt, and that only an approach that revived the earlier practice of "phonics" would actually teach reading as such. Setting parents against teachers and politicians against scientists, Flesch's diatribe launched a series of bitter "reading wars" that would continue for decades. They still flare up every so often to this day. Arising at much the same time, these two challenges created a severe crisis for the science of reading and its pedagogic applications.

The result was not only permanently damaging to the science of reading. It also fostered a deep, anxious uncertainty about the true nature, not only of reading, but of *learning* in general. To make things worse, in the years around 1960, champions of so-called programmed learning allied with inventors of a welter of automated teaching machines to proclaim a neobehaviorist revolution in the formation of American readers. Most of these machines stood athwart the emphasis on creativity, imagination, and exploration urged by advocates of a cognitive approach. But at the same time, a smaller array of automatic gizmos became available that attempted to encapsulate that more flexible and, in a term of the time, *autotelic* approach. Out of such efforts would come not only a revived science of reading, but also the first efforts to make machines that could themselves read. The fundamental conceptions that structured an emerging world of mechanized learning—or, as it soon became known, *artificial intelligence*—resulted from those efforts.

The story of the science of reading does not end there. In three major areas it has continued to the present. First, the traditional techniques of the science of reading found a new home, becoming central to the "science," not of reading per se, but of marketing. Second, in the new millennium, researchers aligned themselves with neuroscience and the technology of brain imaging. And, third, as information itself became digitized and networked, so the techniques of the science of reading were put to use to define key elements of "human-computer interaction." Every time we use a mouse to move a cursor between icons in a graphical user interface, we are operating with tools that originated in this transition of the science of reading to the new domain of digital information. And the ways in which we do so are routinely tracked—and even predicted in advance—by other tools derived from the same enterprise. In these ways and more, the science of reading continues to affect the everyday lives of us all.

Thinking about Reading and Reading about Thinking

Why did the science of reading arise when it did? What came before it? Curiosity about reading was hardly new in the mid-nineteenth century, after all. Attempts to represent, understand, and improve the practice of reading may be traced back far into antiquity. What had not existed, however, was a sustained research enterprise devoted to the subject. Still, earlier history did bequeath important themes and beliefs to the new enterprise. It will be helpful, then, to say something here about what those inheritances were, and how they emerged from the ways in which earlier generations had described reading.[12]

Many earlier reflections on reading did not overtly focus on its bodily or mental character at all. For medieval Christian clerics, for example, it was clearly important to come to some account of the nature of reading, but it need not relate much to the properties of the eye or the mind. The standard medieval division of scriptural reading into literal, anagogical, allegorical, and moral varieties displayed a quite different set of priorities. On the other hand, accounts of the embodied nature of perception were never far away if one knew to look for them. Classical, medieval, and Renaissance treatments of the art of memory, for example, did invoke notions of the mind and the senses, sometimes implicitly and sometimes explicitly. They were typically Aristotelian, in terms of the faculties of the soul, or Galenic, in terms of dietetics and the humors. Neoplatonist and alchemical writers in the Renaissance also went into detail, articulating how images in particular could produce powerful effects on the mind, such that the reader might be transported into ecstasy, insight, or madness. In such ways, a range of possibilities developed for articulating how characters on a page might affect a reader's mental, moral, or religious state.

In the Renaissance through seventeenth century, prevailing kinds of accounts drew on a broad-ranging literature devoted to what were called *the passions*. Passions in general were the responses that human beings experienced when confronted with some kind of experience. They ranged from curiosity and love to fear and hatred. The genre of works on the

passions was an extensive one, attracting authors from clerics to philosophers (including René Descartes, Thomas Hobbes, John Locke, and David
Hume), and the way of talking that the genre embodied would have been
familiar to all Europeans of the time. In essence, it emphasized the exercise of self-discipline to prevent the sensory imagination from imposing
too drastically on the mind's rational faculty. Although there was no one
way of classifying or explaining the passions, the consensus was that they
posed a serious risk of this kind. On the basis of such classifications, writers advanced extensive and detailed proposals for the management of
the self, and early-modern recommendations for reading practices were
generally conceived of as part of that repertoire.

It may seem odd that accounts of reading produced during the Scientific Revolution generally had little to say about the act of vision that
stood at the heart of the practice. One reason why the issue of seeing
per se was not a focus for this literature was that the optical theory concerned was largely uncontroversial, at least after Kepler and Descartes
had tackled the subject. Philosophers might well disagree about the fundamental nature of light and colors—and they did—but they more or
less concurred on the physics of the eye. They agreed that the organ was
a natural counterpart to the camera obscura, in which the lens of the
cornea cast an inverted image of outside objects onto the retinal surface at the rear of the eyeball. From there it was transmitted via what
were termed "spirits" (these were controversial entities) through the op-
tic nerve to the brain, where the mind apprehended it in a general forum
for sense-impressions called the "common sense," or *sensus communis*.
How exactly this happened, and how the mind perceived the impression,
were the unsettled questions. Compared to those, issues of optics were
relatively straightforward.

The model was further complicated, however, by the fact that Chris-
tians believed themselves to be living in a postlapsarian world. In that
world, the senses made mistakes. Not only that, but the body itself routinely generated false signals that appeared in the sensus communis and
might well be mistaken for true sensations. In perceiving impressions
there, the mind could not always tell whether those impressions were
accurate, or even whether they came from the outside world or from the
imagination. Every one of us, commentators warned darkly, was constantly fighting an internal civil war between Reason and Imagination,

and our everyday conduct reflected which of them had the upper hand from moment to moment. Luckily, people could be trained to discipline their passions and thereby give Reason a greater chance of winning. The way to do this was by developing "habits"—routinized responses to passions, which could accentuate the effects of the good ones and help reduce those of the bad. It was routine for early-modern people to believe that true freedom, and thence true virtue, depended on a quite rigorous self-disciplining or "moderation" (in the sense of moderating one's passions) of this kind.[13] Descartes and others said that it was just such a habit that allowed us to read characters on a page in a way that seemed automatic. Educational guides—above all, that written by John Locke, which proved immensely influential in the succeeding century—expanded upon this conviction. They argued that the essence of schooling was the inculcation of sound habits, on which a reasoning, civil adulthood might depend. Good reading was effectively one of those habits.

Perhaps the most important aspect of these early commentaries—it was certainly one of their purposes—was that they served to explain the strange and subjectively overpowering influence that reading sometimes possesses. In certain circumstances, worried moral commentators noted, reading could cause someone to lose track of time and space, to experience a liberation from conventional rationality, and even to undergo a personal transformation. Religious conversion was the key example of this in the post-Reformation era. Countless testimonies by religious "enthusiasts" spoke in terms of the Holy Spirit operating on a reader of Scripture to such transformative effect. But the context did not have to be religious: John Aubrey evoked a similar power in his famous anecdote about Thomas Hobbes encountering a copy of Euclid for the first time and falling "in love with geometry." A similar notion was later utilized to explain the kind of loss of self-control experienced by impressionistic German youths reading Goethe's *Werther*. These were moments when reading could hurl the reader into some new belief or state, seemingly bypassing the will. Later still, such transformative reading experiences exercised a powerful attraction in the Romantic era and beyond, when they were often explained in terms of mesmerism, or animal magnetism, which even became something like a science of influence.[14]

It seems that it was the need to account for—and hence control—such transformative experiences that called forth the first explicit, dedicated,

and extended treatments of reading as a physiological and mental phenomenon in the eighteenth century. To see where the matter stood at
that time, at the peak of the Enlightenment, one can look to a remarkably influential treatment by the Somerset physician and philosopher
David Hartley. Hartley elaborated Locke's account of epistemology into a
detailed philosophy known as *associationism*, in which he incorporated
an account of reading. He described how the sight of characters on a
page gave rise to specific kinds of vibrations in the body's sensory system, which produced quasi particles called *vibratiuncles*. The mind could
link these vibratiuncles together to form associative ideas, and over time
it was habituated to do so in advantageous ways. Hartley insisted that
moral thought and conduct depended on the educational processes by
which these habits were adopted. For him, the theory had practical implications, both in terms of how one acted on an everyday basis and in
terms of how one *learned* to act—for example, he proposed adopting a
particular kind of shorthand so as to incorporate the qualities of industry,
candor, ease, certainty, and thrift into communicative practices. Overall, though, his recommendations for readers remained largely the traditional ones, centering on the cultivation of moderation in the manner and
duration of one's reading. Intemperate reading, by contrast, would lead
even wholesome activities like science to become, in his words, "dangerous" and "Evil."[15] The associationist theory of reading thus neatly tied a
rather conventional canon of moral advice to a plausible-sounding par-
ticulate account of the mechanisms involved. It was successful enough to
be roundly satirized in Laurence Sterne's *Tristram Shandy*, and it would
substantially shape the notions of sensibility that pervaded the literature
of the late eighteenth and early nineteenth centuries.

The corollary of all this was that, in the new associationist world no
less than in the old Aristotelian one, reading was a matter of regimen:
of *mens sana in corpore sano*. On the one hand, physical exercise helped
make a good reader, and a good scholar in particular—an idea with ancient origins that would still remain very much current among theoretical
physics students in late nineteenth-century Cambridge.[16] On the other,
unsound reading could be hazardous to one's physical health as well as to
one's soul. Reading could be therapeutic, or it could make you sick. There
were physiological and psychological debilities associated with reading
too much or too hard, or with reading the wrong kinds of texts. Women,

inevitably, were thought to be especially subject to these effects, but it was by no means a condition restricted to female sufferers. For centuries, physicians and other medical practitioners were familiar with the symptoms associated with excessive or ill-disciplined reading. Patients came to them with upset stomachs, vertigo, fainting, headaches, breathlessness, insomnia, and more. In the Enlightenment, Rousseau's readers reported being driven to bed, contracting colds, and almost going mad; some feared that death itself might result from ill-judged reading experiences. In England, readers responded with almost equal outbursts to the epistolary novels issued by the printer Samuel Richardson. And writers were not slow to draw conclusions about what this meant when reading scaled up across populations. Writ large, in collective terms reading could bring about enlightenment or it could trigger a descent into sectarianism and fanaticism.[17]

By around 1800, then, generations of ethical advice, philosophical description, and educational practice had insisted that reading was a matter of the passions, variously related to human physiology, and that because of this it had profound moral, mental, and medical consequences. Good citizens were citizens who had not only learned to read but had done so by means of the adoption of good habits of self-discipline. In the nineteenth century, as the various disciplines making up natural history, natural philosophy, and the mathematical sciences coalesced into the enterprise of modern science, this congeries of views would be constituted into a formal practice known as *ethology*, or "self-cultivation." Ethology was the science underwriting educational practice at the time when mass education first arose, and it was touted by its proponents as the essential component in the making of a civil society. The prime mover in creating ethology is generally held to have been John Stuart Mill—a philosopher who credited his own recovery from a mental breakdown to a transformative experience of reading Wordsworth's poetry, which he interpreted in entirely Hartleian terms. Mill advocated a rigorous analysis of what he called "the internal culture of the individual," and in that analysis he included a series of reflections on the character of reading. He retained an associationist psychology and the traditional emphasis on habituation—which, in the wake of Romanticism, was by this point a somewhat dated commitment. Remarkably, however, he then inferred from these premises and his own experiences the conviction that a good reader did not consciously experience individual characters on a printed page at all. More precisely, Mill acknowledged that those characters must be "perceived and recognized" in some sense, but he argued that the reader was not aware of this, and certainly did not *remember* perceiving them while concocting interpretations. This unconsciousness was a product of good habituation. As a child learning to read for the first time, Mill reflected, one enunciated the written syllables aloud, and each time one did so one underwent "a kind of electric shock." Through repetition, this process of shocking became so rapid, so habitual, that it grew too fast for conscious apprehension. Adult reading thus rested on a "chain of association" that was subjectively independent of basic sensations. And this fundamental process, he reckoned, must be accorded a central place in any viable attempt to understand "the course of human thought."[18]

Mill issued these remarkable reflections—perhaps most notable, as we shall see, for their counterintuitive claim that readers did not consciously proceed character by character—just before the advent of the science of reading to be addressed in this book. Not by coincidence, it was also the moment when the advent of industrial communications transformed the public culture of the nation. Steam presses used the paper churned out by mechanical mills to produce print in quantities never possible before, and stereotyping allowed for new degrees of uniformity in texts themselves. All this occurred, moreover, within a particular political economy, hailed by Charles Babbage and Andrew Ure as the "philosophy of manufactures." So at the same time as advancing his ethology, Mill also repeatedly trumpeted the importance of print, and books in particular, for the fate of civilization. Books, he announced, were "the medium by which the ideas, the mental habits, and the feelings, of the most exalted and enlarged minds are propagated among the inert, unobserving, unmeditative masses." That sounds like an assertion of simple diffusion. But although sound commonwealth depended on print, he averred, it must also rest on good understandings of creativity and reception—in effect, of writing and reading. These sentiments echoed through his writings of the 1830s and beyond. Mill's celebrated argument for the primacy of freedom of individual expression in *On Liberty*, published in 1859, has to be read in the context of his lasting anxiety about the political economy of communication within which such expression must take effect. "This is a reading age," Mill remarked, "and precisely because it is so reading an age, any book which is the result of profound meditation is, perhaps, less likely to be duly and profitably read than at any former period." As a direct consequence of steam printing, the public "gorges itself with intellectual food, and in order to swallow the more, *bolts* it." A book might therefore produce no more "durable impression" in a reader's mind than an apparently ephemeral newspaper article. Mill announced himself in favor of state subsidies for authors of quality, partly because they would otherwise lie at the mercy of an industrial publishing industry that subordinated everything to the consuming power of the masses.[19]

The echo of traditional regimen advice was clear in Mill's warning. But so was the insistence that in such a radically new context that traditional advice took on new meaning and importance. For his friends Mill devised an elaborate regimen that would produce a proper "habit," which involved reading a prescribed number of pages—twenty, fifty, or a hundred—as early in the morning as possible, and supplementing this with further exercises that would not "fatigu[e] the mind." He constructed a detailed list of works, to be read with different levels of intensity and at different times of day.[20] At the public level, he advocated the establishment of state-financed libraries as an essential component in "national education," dedicated to imbuing not only the *ability* to read, but an interest in actively doing so. Education for the proletariat must be "mental cultivation," he declared. Ethology, as "the science which corresponds to the art of education," was to be the *philosophia prima* corresponding to this art of self-formation. It was Mill's candidate for an "Exact Science of Human Nature." From it, he hoped, a comprehensive social science might also develop. In fact, his renowned *System of Logic*, seen today as a classic in formal philosophy, was in his own eyes a long argument for this kind of "self-culture," on which a future democratic politics might be based.[21]

I have drawn attention to Mill's worries here because he represents something of a connecting link between older notions of mind, body, and reading, on the one hand, and, on the other, the enterprise to be traced in this book. But Mill was far from alone in voicing these kinds of fears, of course, and in succeeding decades commentators on public reading habits revisited them repeatedly. Appalled Victorians denounced the "vice" of "overfeeding" at the trough of the newsagent W. H. Smith's as "a vulgar, detrimental habit, like dram-drinking." The practice, it was said, would "end by enfeebling the minds of men and women, making flabby the fibre of their bodies, and undermining the vigour of nations." It remained common to lament the "dreaming or reverie" into which ill-disciplined reading could pitch the debauched. These were not, it should be explicitly stated, merely metaphorical notions. Improper reading *really did* deter sound physical exercise in the eyes of these writers, and it *really could* throw off the balance of one's mind and body, leading to the danger of thoroughgoing atavism. The language of addiction, demoralization, and degeneration was loud, explicit, ubiquitous, and insistent.[22]

The Psychophysics of a Reading Age

Why could poor or excessive reading have such catastrophic effects? What causes linked an apparently inoffensive and leisurely activity to personal and collective doom? Not only were the mechanisms of reading still unclear—the doctrine of association notwithstanding—but the same habit was apparently both so powerful that it could drive people to madness and, mutatis mutandis, so malleable that two different readers could respond quite differently to the same text. What *was* this practice, anyway—how did one acquire it, practice it skillfully, and master its effects? The experimenters who devoted themselves to the topic of reading from the 1880s to the 1930s believed that a plausible answer to such questions must be sought, ultimately, in the objective measurement of what might be thought the most trivial of experiences. The first third of this book describes what those experiences were, and how researchers tried to detect, analyze, and, at length, reform them.

At the heart of the science of reading that those researchers developed was an apparently tiny and unimportant observation. When we read a line of text, in general we feel that our eyes pass smoothly along the line. The experience is all but universal, at least in Western nations that use alphabetic scripts. As we shall see, even quite informed writers (like Marshall McLuhan) have invoked it. But in fact this smooth movement almost never happens. A reader's eyes typically proceed by a series of abrupt jumps, usually numbering around five to seven across a line of a normal-sized book. Between those jumps they remain more or less stationary and focused on one point in the line. Subjectively, readers have no experience of this staccato set of motions, and so they long remained unremarked. It was only in the 1870s that they were brought to the attention of ophthalmologists, and a couple of decades after that that the public in general heard about them. By then the champions of a rising new science had made them their focus. They showed that a science of reading was necessary because subjective experience was not to be trusted.

It is conventional to attribute the discovery of this curious phenomenon to a Parisian ophthalmologist named Louis Émile Javal, who published a note about it in one of a series of articles he issued in 1878–79 on "the Physiology of Reading."[23] Javal christened the jumps *saccades*— literally, *jerks* (a word that was sometimes used in English-language work)—while the moments of stasis soon came to be called *fixations*. The science of reading is conventionally said to have originated with this discovery. And that this conventional view is plausible is shown, paradoxically, by how uninterested Javal himself was in the subject. He relegated his remarks to a minor note at the end of the last publication in the series. He certainly did not see saccades as the key to a new scientific tradition. In fact, it is not even clear that he himself believed that the phenomenon was real.

Javal credited the discovery of saccades to one Lamare, an otherwise mysterious figure—it seems that we do not even know his forename— who had apparently experimented for a while in his laboratory. This Lamare had found that an assistant could simply *see* the eye's jerks as he read, or feel them by placing a finger over a closed eyelid. Later, at the laboratory of physiologist Charles-Émile François-Frank, Lamare built an apparatus that connected his own eyelid to a drum so that its tiny movements produced audible sounds. This generated a discernible series of clicks as he read across a line of print on a page, the number of which could be recorded by a technician. He estimated the frequency and length of saccades on the assumption that they were equal, and concluded that these motions might well cause "fatigue" in the eyes—an effect that he thought might be worsened if the movements were not regular.[24] The hesitant Javal later proposed using a kymograph (a device in which a stylus inscribed a line onto paper mounted on a rotating drum in order to trace a varying quantity such as pressure over time) to show the effects. So did his fellow Parisian Edmund Landolt, who invented an instrument specifically for this purpose that he called a Kinophthalmoscope.

Interest in saccades soon grew beyond Paris. In Prague, for example, Ewald Hering—a longtime rival of Hermann von Helmholtz in matters

visual—invented a very similar tool to Lamare's, with which he too noticed the telltale sounds. British researchers made similar observations
at much the same time. Above all, in Germany practitioners of the fast-
growing field of psychophysics latched onto the topic. Psychophysics
practitioners were dedicated to creating a precision science of human
responses to stimuli, hoping to build from their data a new science of
human nature. Here was a small-scale bodily phenomenon of undoubted
psychological importance, which seemed tailor-made for the kind of precise measurement that they liked to pursue. If an objective science of the
mind must be based in quantified experimental facts, as was their credo,
then here, it seemed, was a very promising area in which to get them.
They also appreciated them in light of what they already recognized as a
tradition of attempts to observe eye movements in general, often in situations where the body was destabilized and put into motion. A "rotative
couch"—a kind of centrifuge designed to address vertigo—that Erasmus
Darwin had proposed and James Watt designed in around 1800 was a
key instance, which was later repurposed by Raymond Dodge to test the
capabilities of early aircraft pilots. Helmholtz, Franciscus Donders, and
Johann Benedict Listing all developed technologies that might enable the
movements to be reproducible in a laboratory. So the psychophysics pioneers could draw on a number of existing experimental techniques. By
the 1890s the psychophysics of reading was something of a hot topic in
the making, and the possibility of measuring saccades to arrive at an objective quantity of human perception was a major reason why.[25]

In short, before the late nineteenth century there had been a long history of attempts to understand what happens when a human being reads,
and this history had been hugely consequential for the history of culture
because of its applications in education, religion, and politics. The language of passions and the practice of personal regimen were common
to all of them prior to the 1870s. What is notable is that if we take this
long sequence of representations and, as it were, use our imagination
to "turn the world upside down," what results is a history of concepts of
the body and its relations with the mind and the world. It extends from
ancient arguments over vision (intromission and extramission theories),
through camera-obscura theories of the eye, to accounts of the passions
in the Renaissance and of mesmerism and similar "powers of mind" in the
eighteenth and nineteenth century. These should be taken not simply as

matters of curiosity but as foundational elements of the experiences of readers that it would be absolutely essential to grasp in order to understand *anything* about what it was for them to read. In the 1870s and 1880s, this became not only true but explicit.

Steam, Stereotyping, and Science

At this point the convergence of three widely noted cultural trends made the creation of a sustained research enterprise devoted to reading at once more likely, more desirable, and more achievable. The first of these was the reconfiguration of the intellectual disciplines that occurred in the decades after the French Revolution and the embrace of research in the German university tradition following the creation of the University of Berlin in 1810. By the 1820s, the rise of specialist and technically demanding enterprises, coupled with the engagement of research with industries, and the Romantic repudiation of eighteenth-century encyclopedism, had given rise to a widely shared concern at the disunity of the natural sciences. It was in the German lands that philosophers took the lead in addressing this sense of chaos, forging new alliances across what were otherwise discrete technical disciplines. English visitors such as Samuel Taylor Coleridge and Charles Babbage returned full of admiration for the burgeoning communities of naturalists there, and aspired to reproduce something like the research culture of *Naturwissenschaft* in Britain. Commentators sought some physical principle of "connection," "correlation," or "conservation" of "power" or "force" (or, in German, *Kraft*) around which a newly coherent constellation of disciplines might coalesce. It was from this set of concerns that William Whewell, the Cambridge polymath, coined the term *scientist* as a general moniker for the new, specialist researcher—no longer a philosopher, let alone a "priest of nature" (like Robert Boyle and Isaac Newton a century and a half earlier), but a technically adept expert devoted to one subdiscipline and committed to advancing it.[26]

The second trend was the transformation in print that occurred in the same years. The steam press and stereotyping were both introduced in the early nineteenth century; they were first adopted at scale by the newspaper industry in the 1820s, and by the 1850s they were having a major impact across the publishing industry in the United Kingdom and United

States. For all that printing had long been regarded as the most important single force behind the Protestant Reformation, secularization, and the emergence of modern societies, workers using hand presses could print only perhaps two hundred impressions an hour, and edition sizes were typically of the order of a thousand copies. (A few genres, almanacs in particular, were made in much larger numbers.) Steam printing—with its accompanying papermaking and other technologies—was a revolutionary change. By the 1850s, the *Times* in London was being printed at a rate of ten thousand copies per hour. In the same generation, the adoption of stereotyping freed printers from the prospect of having to keep type locked up in forms if they wanted to produce more copies of a given work, and thus drastically reduced costs for successive impressions.[27] With the ability of railways to convey the products of the press rapidly from town to town, cheap print saturated society in a way that had never before been possible. What the cultural consequences of this would be nobody knew. But it was clear that they would be massive, and to understand them some account of the practice that lay at the heart of the new printing revolution was sorely needed.

This was also, third, the era of great political reforms. In Britain, the Chartist movement eventually gave rise to more than one Great Reform Bill, and by the 1870s a general election was happening with, for the first time, effectively universal male suffrage. On the Continent, the revolutions of 1848 first promised radical reform and then brought it down to violent defeat, but by the 1870s, again, the end of the Second Empire in France and the unification of Germany were raising fresh hopes of a more representative governmental system. The question of enfranchisement immediately, necessarily, and obviously implied a question of public knowledge. In every major Western country, the late nineteenth century saw major expansions of schooling for the general population, both to answer the need for a responsible electorate and to sustain the workforce that industrial capitalism required. At the core of the new mass education systems was the ambition to foster mass literacy. Reading, in other words, was what mass public schooling was for.[28] Whatever further or higher aims school systems might have, teaching children to read was the fundamental sine qua non to which they were devoted. The campaigns for mass literacy would be decades long. In light of them, contentions about the nature and power of reading took on a broad social significance that they had not really had before. One set of questions concerned how to measure objectively the extent and quality of an individual's reading—and, at scale, those of populations. Another concerned how best to teach reading to those—adults as well as children—who could not read at a desired level, and then to maintain their identities as "readers" for life. Both demanded newly robust and reticulated accounts of the skill itself and its acquirement.

And it was in the United States that these trends converged most consequentially. The nation itself had been founded in a revolutionary moment for which the idea of an "informed citizenry" had been inspirational. The very identity of the country was bound up with this idea, and the establishment of a vibrant publishing industry since the 1820s had reinforced its centrality. The publishers of New York and Philadelphia had spent the central decades of the century vying with each other to produce literature and news in ever-greater volumes and at ever-cheaper prices, establishing commercial and credit networks that extended across the hinterland. It was a matter of pride that the American industry thrived in the absence of any international copyright agreement, which allowed this kind of demotic information age to take shape. (The publishers drew a sharp national contrast with their London counterparts, who, they charged, produced elegant books in small editions and at high prices.) In this democratic nation, citizens must be readers too. But by the third and fourth quarters of the nineteenth century nobody could ignore severe fault-lines that this attractive vision of democratic dynamism had tended to obscure. Industry was fast-growing, and it needed a population of adaptable workers quite different from the old "hands" of Dickensian factory lore. The capitalism of the era was producing large-scale, spatially extended corporations that relied upon massive information management systems in the forms of files, microfiches, telegraphy networks, and more. These too required extensive and skilled literacy if they were to be sustained. The tabloid press was widely denounced by upper-class society as venal and riddled with falsehoods. And, above all, the fact could no longer be evaded that a large part of the population had, until Emancipation, toiled in slavery within a society that had treated the very act of teaching them to read as a serious crime. There was simply no way to uphold America's prized conception of itself as the world's one modern, industrial, and free democracy without finally confronting these disturbing paradoxes. The nation needed to make reading as close as possible to a universal practice among its citizenry. And it desperately needed new ways to think about the place, nature, and consequences of reading itself.[29]

The puzzle of reading was thus made an urgent question by a convergence of potentially revolutionary processes in the peculiar context of America in the wake of its Civil War. The explosion of printed materials of all kinds (not just books, newspapers, and magazines, but everything from advertising billboards to business cards), the aspiration to sustain a mass democratic polity, and the demand for a trainable industrial workforce were all forces that excited both great hopes and fearful apprehensions. They could not be ignored. But traditional arguments about reading and culture seemed utterly inadequate to their scope, scale, and importance. In the mid-nineteenth century, however, not only were major new challenges like these at hand; new ways to take them on were also being created. Because they occurred in the same generation of intellectual upheaval that saw the invention of that new and promising figure, the scientist, American society naturally turned to that figure for answers. What was needed, in this age of science, was a science of reading.

1

A New Science

On September 21, 1882, an intense, ambitious, entitled, and rather melancholy young man arrived in Baltimore, ready to take up a fellowship in philosophy at what was then the most exciting university in America. James McKeen Cattell was twenty-two. He had spent the last two years studying, as many ambitious young Americans then did, in Germany, at Göttingen and Leipzig. Now, thanks to the finagling of his influential father, he had been appointed to a highly desirable position at the institution that was effectively recreating the German-style research university in the United States. It was, as he wrote in his journal, all that he could have asked for. But Cattell was not excited by the prospect. Morose and dispirited, his strongest feeling was one of aimlessness. "I sit here restless and unhappy," he noted, "and see youth and health slipping from me, and alas I do not know what I want."

Cattell's doubts were nothing new. He had long confided similar feelings to his journal. The son of a respected Presbyterian minister who served as president of Lafayette College in his home town of Easton, Pennsylvania, he had had a privileged but rather isolated upbringing. His family were relatively wealthy, the descendants of Quakers on his father's side and Ulster protestants on his mother's. He had been educated at home and had accompanied his parents to Europe when they traveled to survey technical schools that they thought Lafayette might emulate in building its own programs. James naturally matriculated at Lafayette himself. He started early, flourished intellectually, and graduated in 1880 with a precocious interest in Comtean positivism and a taste for empiricism in the style that the nineteenth century liked to associate with Francis Bacon. The graduation ceremony had barely concluded when he left for Europe. His journey, financed by family wealth, was meant to polish him for a business career, but he soon had other ideas. Arriving in Liverpool, Cattell wandered through England and the Netherlands before arriving in Germany, where he ended up in Göttingen. The sleepy medieval town was often recommended to young Americans keen to learn the language, because the local accent was accounted unusually refined. A robust little community of US students consequently existed there. But Cattell, who had been something of a loner at Lafayette, remained one in Göttingen: he would walk the old city walls alone before dawn, and his membership of the student dueling society, despite his strong predilection for sports in general, did little to cheer him up. The only bright spot was his encounter with Rudolf Hermann Lotze, a philosopher and psychologist (and historian of aesthetics) who was one of the university's celebrity scholars. Lotze had been a leader of the generation of scholars in midcentury who had sought to relinquish the old quarrels over vitalism, materialism, and idealism that had characterized Romantic philosophy in the German lands. He championed a physiological approach that promised to supervene such categories. His *Medicinische Psychologie* is sometimes reckoned the first modern book on psychology. For Cattell, the prospect of a scientific psychology—one that could be slotted into the Comtean hierarchy of sciences that had won him over at Lafayette—was enormously attractive. So when his father urged him to send a paper to Johns Hopkins so that he might earn a fellowship that Cattell senior was needling the university to offer him, he agreed, resolving to write about Lotze's philosophy.

Cattell started the paper on Lotze, but he soon fell back into ennui. The work languished. And then disaster struck. Lotze, who had been lured to a position at Berlin, died suddenly, and Cattell found himself cast adrift. "It seems probable," he wrote, that he was destined for "crime or madness or suicide; perhaps all three."[1] His mood was not helped when an insurance company refused to provide him with a life policy on the grounds, it politely explained, that he was likely to die young. Cattell moved for a while to Leipzig, home of the greatest of Germany's psychophysicists, Wilhelm Wundt, but he did not stay, and before long was passing through

Geneva and on to Paris. By New Year's Eve he was actively contemplating shooting himself.

But Cattell did not commit suicide, and he did finish his essay. He finally mailed it to Baltimore from Italy, before making unwonted haste to Britain to embark on the voyage back to the United States. The fellowship turned out to be waiting for him when he made landfall. He accepted it, and by September he was in Baltimore. At once Cattell found that the discipline of philosophy was in disarray there. The university's professorship in the field was vacant, and it was proving hard to find someone to occupy it who could pass religious muster. Three temporary lecturers were meanwhile serving as stopgaps, one of whom was Charles Sanders Peirce. Cattell seems to have paid little attention to Peirce's teaching, but he did become secretary of his Metaphysical Club, which met regularly to discuss matters of intellectual culture. In any event, Cattell found himself much more interested in the physiology laboratory, where he might be able to put into action some vague plans for a science of psychology that he had envisaged in Göttingen. Cattell launched himself into a series of quantitative experiments, seeking, as he would later often put it, to provide psychology with the kind of rigorous, precise, factual foundations that the physical sciences enjoyed. In particular, he seized on the subject of perception. And he focused above all on that most everyday perceptual practice of the modern citizen: reading.

By the 1880s American reading was big news and big business. The nation's publishing industry had become mechanized, financialized, and, to an extent, centralized. The major New York publishers were often no longer manufacturers of books—they hired printing firms for that—but cultural capitalists, speculating in stock and copyrights. They operated for the first time on a truly continental scale. Rail and telegraph systems gave them national reach. Stereotyped and steam-printed books—and, even more, newspapers and ephemera—had become ubiquitous and affordable. No citizen could escape their reach or their influence. For the first time in history, it was really true that print—and therefore reading—had become a defining component of a nation's everyday life. Indeed, it was perhaps only in the two generations between 1880 and the outbreak of World War II that United States culture was truly characterizable as a culture of print, because after that electronic media would come into their own. In the meantime, printed materials tracked both the nation's emergent identity and its disarticulations. Newspapers thrived and carried authority as never before or since, and the advertising industry grew into an autonomous cultural force. The middlebrow press and an emergent literature of business vied for intellectual authority with the new research universities that came into being after the founding of Johns Hopkins in 1876, and with the nascent ecosystem of academic publishing that they fostered by means a new generation of "university presses." Much of the information infrastructure of scientific and humanistic knowledge that we still use today dates from the ferment of print and publishing in this period.[2]

If that ferment made it vital to understand what reading was and how it could be managed, what made a *science* of reading conceivable as a way to meet that need was a radical recasting of disciplines that had occurred in the middle third of the nineteenth century. The Romanticism of the first part of the century had retreated by around 1830, but the encyclopedic convictions of the Enlightenment that had preceded it no longer offered an attractive alternative. Technical specialization and increasingly strong ties to industrial, state, and military interests meant that the old ideal of the naturalist as a "priest of nature" had little appeal either. The enterprise of natural philosophy had broken apart once and for all. As it did so, over the course of the century a panoply of new would-be sciences found champions. Some would survive (crystallography, conchology, biology, ecology); others would not realize their founders' ambitions (electro-biology, animal magnetism, and others). Responding to the sense of dislocation that accompanied these changes, in the mid-1830s the British polymath William Whewell coined the term *scientist* as a general term for the practitioners of such fields. At the same time, investigators in several of these areas sought to articulate some principle of "connection" or "correlation" of natural powers, as a basis on which the various technical sciences might be unified. Whewell's first published announcement of his new term came in a review of one of these efforts, Mary Sommerville's *On the Connexion of the Physical Sciences* (1834). By the 1870s, the strongest contender for this unifying role was the principle of the conservation of energy. And an essential concomitant of the embrace of energy as the new axiom of scientific unity was a commitment to the precise measurement of key physical constants. Such measurement was pursued in a new generation of physics laboratories—the best-known being the Cavendish at

Cambridge, run by James Clerk Maxwell—and it demanded unprecedent levels of technological and personal discipline. Here, it seemed, could be found the essence of science itself. By the 1880s, an investigation of reading would therefore surely aspire to be accounted scientific, and that meant that it would have to look like this.[3]

By the time Cattell arrived in Baltimore, then, there were reasons why it was extremely important to understand the phenomenon of reading, reasons why any attempt to investigate that phenomenon should be a science, and reasons why that science should be experimental, instrumental, and quantitative. But it had not yet come into being. By this point, however, a possible model for such an enterprise was already being pursued. It had arisen, unsurprisingly, in Germany, where the new ideals of institutional scientific research were most prized. It was called *psychophysics*, and Cattell already knew a lot about it.

If you were an ambitious young American in the late nineteenth century, you were well advised to travel to Europe and learn at one of the best universities of Britain, France, or Germany. Sure enough, students did go, in large numbers, to pursue all kinds of disciplines. But for those interested in the mind in relation to the body and the wider world, psychophysics stood out. Nowadays remembered as the precursor to experimental psychology, psychophysics was a laboratory enterprise developed in the mid-nineteenth century to circumvent old debates about materialism and idealism by scientifically investigating human responses to stimuli. It was named by Gustav Fechner, one-time professor of physics at Leipzig, who claimed to have come up with its founding principle of a connection linking the physical world to mental responses while lying in bed one morning. Among its most renowned protagonists was Germany's leading scientist, Hermann von Helmholtz, a famously tireless exponent of a vast variety of sciences (the only one he had not mastered, Carl Ludwig said, was comparative anatomy) and a fierce advocate of the notion that it was a scientific researcher's "duty" to seek rigorous perfection in his work.[4] Its foremost practitioner of all was not Helmholtz, however, but Wilhelm Wundt.

Trained as a physiologist at Heidelberg and Berlin, Wundt had worked for seven years as Helmholtz's assistant before moving to Leipzig in 1875. It was there that four years later he founded the world's first psychophysics laboratory. He was not only a prodigiously productive scientist in his own right, but also a dedicated teacher, who taught thousands of students and advised more than 180 doctoral dissertations. These graduate students worked on their own projects, of course, but they also acted collectively to build up an armamentarium of laboratory instruments, techniques, approaches, and questions, and it was these that defined the scope and character of psychophysics in the Wundtian mode. The influence of this enterprise would be wide, deep, and consequential, in the United States as well as in Europe. At its heart was the measurement of very short time intervals, particularly as they related to human sensory responses.

Empiricist philosophers had long asserted that the origins of all knowledge lay in sensations; the most famous statement to this effect was John Locke's *Essay Concerning Human Understanding.* They had devoted a great deal of care to attempts at classifying the sensations and explaining the processes of association by which the mind forged ideas from them. One common assumption, however, was that the experience of an exterior phenomenon occurred simultaneously with its manifestation in the sensus communis. Psychophysicists took that assumption as a challenge. Wundt and his peers were finding that there was a measurable delay between stimulus and response, and they resolved to investigate it by means of precision laboratory instruments.[5] The stock in trade of their new science was therefore the quantified measurement of reaction times, in an enterprise called *mental chronometry.* Seeking to drive the resolution of their measurements down to the region of a thousandth of a second, they developed what has been called a "cult of precision," comparable to that which existed in the most advanced physics laboratories of the time, such as Maxwell's Cavendish. Their hope, indeed, was to provide a physics-like underpinning of precisely defined facts on which a rigorous science of the mind could be built. By the 1880s, they had developed a repertoire of laboratory instruments for this end.[6] In fact, by this time the best-funded and most lavishly equipped laboratories in any scientific discipline were to be found, not in physics, but in experimental physiology.

Beyond the quantitative measurement of reaction times, psychophysicists also began to produce automated *traces* of phenomena. They designed self-registering instruments that could convert physiological signs—pulse rates, respiration, and the like—directly, without the intervention of a researcher, into lines marked on paper or film. These lines both represented experimental values precisely and, when examined for deficiencies in shape and smoothness, made any peculiarities in an individual case visible. The origin of such graphic machines may be traced back through the history of the "method of curves," as it was called, to around 1800, the best-known early example being the "indicator" diagrams that the pioneering engineer James Watt had his steam engines produce to show their work rates. The subsequent mechanization of curves owed much to the machines of the second industrial revolution— the kind of thing Charles Babbage noted when he hailed the ability of machinery to produce identical objects in huge quantities by marginalizing human craft. These mechanical diagrams provided visual registrations of the value of roads, canals, and railways in terms of a unit of human labor extended over time. In Berlin, Helmholtz had adapted such instruments to measure the mechanical work done by frogs' legs, and from there they had become central to psychophysiological work in many settings. As Robert Brain contends, the point was not only that such automated traces were putatively objective. For the psychophysicists, the human body itself was a kind of machine, so the development of sensing and recording technologies (including gramophones, telephones, and telegraphs) partook of general evolutionary processes. Moreover, Watt's indicator diagrams had not only told steam engine operators about the work their machines were doing, but also revealed any faults a particular machine might have. The traces displayed problems as clearly visible deviations from replicable lines. The traces produced in psychophysics laboratories, similarly, could reveal significant idiosyncrasies and pathologies in the people being experimented upon. By about 1900, the sophistication of such self-recording graphic devices had become impressive enough that researchers fantasized about their replacing a good deal of writing altogether, not only in laboratories but also in factories, hospitals, and other institutional settings. As the renowned physiologist Étienne-Jules Marey put it, automatic machines manifested a kind of "natural writing," as if the phenomena themselves were communicating to researchers. And Cattell's new teacher at Johns Hopkins, the pioneering American researcher G. Stanley Hall, agreed. The "graphic method," Hall told readers as early as 1879, was "fast becoming the international language of science."[7]

Intoxicating Reading

Johns Hopkins University was at the vanguard of the new discipline in America, and G. Stanley Hall was at the vanguard of Hopkins's scientists. Any young American wanting to address the leading questions of the time would still be well advised to take ship and seek out Wundt, but failing that, the best place in America to start was in Baltimore. This was the place to create a scientific analysis of reading.

When he started at Johns Hopkins, Cattell knew that the German movement of psychophysics was developing a repertory of instruments that could measure very precisely the times that human subjects took to do certain basic things—react to a stimulus, say, or perceive a color. They could resolve durations of time into units as small as milliseconds. He quickly made his way to a house near the campus where he could find such devices. There, Hall—holder of the first doctorate in psychology awarded in the United States, and a man who had studied with the great German pioneers, including Wundt himself—had set up a "physiologic-psychological laboratory." Cattell became one of his first students, taking his place in a select handful that included Joseph Jastrow and John Dewey. He launched himself into the labor of experimentation. But he soon found himself feeling uneasy about the enterprise. Cattell's first serious experiments seem to have arisen from the doubts he began to feel about the central claims of laboratory psychophysics itself. A problem with psychophysics, he now conjectured, was precisely that it used such distinctive apparatus. Its equipment was complex, labor-intensive, and intrusive; a normal person in a laboratory was thus inevitably in an "abnormal state." The combination meant that any findings might not reflect "the conditions of real life." So Cattell instead tried to think of experiments that would use only simple and unintrusive devices, which he could use to measure times that someone "ordinarily" took to do everyday tasks. Reading was just such a task. Cattell decided to measure how long it took for him and his friends to recognize letters of the alphabet printed on paper. It was here, then, in April 1883, that four rather unusual laboratory subjects—Cattell, Dewey, Hall, and Jastrow—carried out America's first dedicated experiments in the science of reading.

What Cattell did was to repurpose a standard curve-tracing device from psychophysics called a kymograph. In a standard kymograph, a pen traced a line onto a chart mounted on a steadily rotating drum, producing as it did so a record of the variation of some measured quantity over time. Cattell effectively reversed the device's information flow. He took a kymograph with a fifty-centimeter circumference and set it to rotate at a constant rate (which he established by using a tuning fork to trace waveforms on a smoked sheet atop the drum of the device). He then affixed white paper onto the drum, on which he had printed roman-typeface letters spaced at one-centimeter intervals. Between the kymograph and the reader, he placed a black screen, in the middle of which was a one-centimeter-high horizontal slot of variable width. With the slot's width set at one centimeter, the reader could see one character at a time through the slot; with it set at two centimeters, the reader could see two; and so on. He then corralled his three volunteers, and by varying the speed of the kymograph he ascertained the shortest time in which each one could read thirty to forty characters correctly. From there it was easy to calculate the average time per letter. Then, in a second set of experiments, his readers moved away from the kymograph and read printed passages aloud in six different languages: English, German, French, Italian, Latin, and Greek. Each reader had to be fluent in at least one of these languages for his experiment to proceed, but then Cattell's volunteers were an elite bunch.

As he proceeded with this attempt to measure the reading of letters, Cattell also found his interest piqued by another kind of influence. This one was far more risqué. With what he later described as a youthful "spirit of adventure," he did what a number of venturesome researchers were doing in these years and started to experiment with "stimulant and intoxicating drugs." The adventure began when a friend introduced him to hashish. Cattell found the experience transformative—it more or less cured him of his melancholy—and decided to make it the starting point for a program of experimental exploration in parallel to his psychophysics of reading. The result was a series of experiences that combined high scientific ideals with adolescent wonder and that soon merged with his reading trials. For example, he tried drinking a bottle of wine and then writing out passages of Coleridge by hand, to see how inebriation affected handwriting. What about chocolate, concentrated caffeine, cannabis, hashish, ether, morphine? For some of these drugs, he later recalled, he imbibed "perhaps the largest doses ever taken without suicidal intent." Under the influence, Cattell felt, he became "two persons—one of which could observe and even experiment on the other." That was a remarkable version of objectivity, voiced only three years before Robert Louis Stevenson's Dr. Jekyll transformed himself into Mr. Hyde in a process that looked very much like drug addiction. His aim was to exploit this psychological disintegration to notice and record facts about the effects such substances had, even as they were having those effects. The intoxicated Cattell thus wrote poetry and music; he drew pictures; he even drew a diagram of the brain. He read Shelley's *Adonais* while under the influence of hashish and enjoyed it, a fact that he attributed to his having virtually replicated the state Shelley himself had been in when writing the work in the first place. His devotion to psychophysics notwithstanding, much of his experimentation involved reading experiences like that.

As his disciple Walter Dearborn later remarked, the psychophysical experiments that Cattell did with Dewey, Jastrow, and Hall were "one of the first steps taken in the analysis of the mental processes concerned in reading." As he proceeded in these trials, moreover, in his notebooks Cattell began to extrapolate from them a path of future progress in science itself. Positivist to its core, it envisaged the objective observation of "facts" leading to an ascent toward full scientific understanding of the world, as individuals reiterated the development that humanity itself had traced out over its long history. The diligent and precise recording of quantitative data on the time it took to read characters—a prime example of what he would later call *psychometry*—would thereby become the basic foundation on which a truly scientific modern world could be constructed. Thrilled by this insight, he resolved to devote himself even more wholeheartedly to "physiological psychology," so as to gain "physiological, psychological, ethical and metaphysical" perspectives on his experiences. The ideal of positivistic progress in science, teaching, and philosophy had, he said, finally provided him with "something to believe." It gave Cattell "something to work for."

And he made a major discovery in the process—one that would indeed prove foundational to the science of reading that ensued. What Cattell found was a simple but striking fact. A reader, whether reading aloud or silently, perceived groups of letters—words—faster than individual characters. In later experiments in Germany he would go further, and declare that whole sentences were taken in faster still. In reading a sentence, he argued, a single "will-act" sufficed to perceive and interpret. And this unanticipated finding mattered all the more because he could be confident that it was not an artifact of the kind of highly artificial setting that psychophysics tended to involve. The distinguishing feature of Cattell's experimental design was that it used simple methods designed to approximate as closely as possible "the condition of actual life." As Dearborn related, he had taken care to ensure that his conclusion was no mere product of artifice. What had happened in Hall's house would happen anywhere reading took place.

It was clear to Cattell then that his trials could and should have consequences far beyond the laboratory. That he focused on the issue of speed was not only a consequence of his knowledge of psychophysics. He undertook these experiments at a time when the question of how to make it faster and easier for readers to perceive characters was regarded as a pressing one. The steam press and automated typesetting were churning out printed words with unprecedented speed and intensity, so the question of how fast one could read seemed central to the fate of the citizenry in an incipient modern age. Every American, at least in the eastern-seaboard cities, was confronted by a constant array of posters, newspapers, cheap novels, pamphlets, magazines, flyers, and tickets—an explosion of printed characters that had no historical precedent, and which was both hailed and decried as the phenomenon of its age. "One must learn to read in self defense," Cattell himself would write in 1909. "If ninety per cent. of the population carry pistols, it will not do for the remaining tenth to go unarmed."[8] In this context, America was already generating a flood of proposals to rationalize orthography and typography so as to facilitate fast reading—Theodore Roosevelt himself endorsed them. Cattell's ambition extended to the same realm.

In this light, it was significant too that Cattell took the trouble to publish the reading times registered by his subjects as individuals, rather than simply presenting calculated means. That was unusual, if not unprecedented. Cattell's psychophysical predecessors generally assumed for their experiments that they were seeking one true quantitative value, which would be a property of all competent humans. Differences were in that case mere errors. Cattell, by contrast, seems to have seen value in these differences from the very early days, even though he did not yet make his rationale explicit. It is possible, as he himself later claimed, that he noted the variations because in his role as a trainer of athletes—he remained avid about sports—he was used to logging the performances of individuals. "It seemed as natural to measure the rate at which people read as the rate at which they ran," he recalled; "to place them in order of merit for psychological traits as for eligibility to the college football team on which I played or the baseball team of which I was manager." He did nothing with the individual numbers at the time, however, beyond reporting them. Yet later he and his followers would come to see them as constituting the very point of the exercise.[9] With rapid reading becoming an essential acquirement for modern life, it would not be long before a capability to measure the differences between individual readers' speeds would promise radical consequences for the education of future children.

But that was in the future. After his initial experiments on reading were complete, Cattell wrote up a brief account of this "study of the conditions of reading and time of certain mental processes" and submitted it to the administration at the university, confident that it would be grounds for his fellowship to be renewed. The position was a prized one, and it promised a glittering philosophical career. But now disaster struck again. His application for renewal was denied. It is not entirely clear why. Cattell's experimentation with drugs may have made him somewhat notorious, and it seems that Hall was less supportive behind the scenes than in Cattell's own presence. It certainly appears that Cattell himself could be abrasively ambitious at times. At any rate, the decision was final, and it provoked a sudden and ferocious sense of betrayal on Cattell's part, which fomented a deep and lasting breach with the university. He retreated to his family's home in Easton, and then set off with them for Europe on another voyage. Meanwhile, the Hopkins fellowship descended to the next person in line. That happened to be Cattell's friend and collaborator in reading trials, John Dewey.

The Psychophysics of Reading

Cattell headed for Leipzig, with his brother, a trainee chemist, in tow. There he entered the orbit of Wilhelm Wundt for the second time. It would mean far more to both men than the first. Wundt encouraged him to resume his Hopkins work and continue trying to measure the times taken by simple mental processes and associations. He should relate them, Wundt proposed, to practice, attention, and (an important concept in the context of energy science) fatigue. Setting up the experiments in his own rooms, the reinvigorated Cattell launched into the work with gusto.

Cattell began where he had left off in Baltimore, by investigating the legibility of "letters, words and phrases." He first sought to distinguish between apperception and enunciation times by reducing the size of the kymograph slot to less than one centimeter. After that, he had collaborators read sequences of letters or words in six languages, each passage being drawn from a major work in that language: Swift's *Gulliver's Travels*, Rousseau's *Émile*, Boccaccio's *Decameron*, Tacitus's *Vita Agricolae*, and Plato's *Apologia*. He timed them by stopwatch as they read one hundred letters or words, and then as they read five hundred.[10] He also had his volunteers read the texts backward, and then read sheets of individual, disconnected words and characters that he had had made. The point of doing so, he recorded, was to distinguish the times needed to recognize letters and words, and to indicate how those times varied if they were parts of coherent words and sentences. He found, among other things, that women read faster than men, and that reading a foreign language was always slower, even if the reader did not subjectively experience it as such. (Such a measurement, he suggested, might provide an objective way for schools to measure students' abilities in languages.) But the most striking and consequential finding was this: it was always faster to read a letter if it formed part of a word, and always faster to read a word if it formed part of a sentence. And the difference was not inconsiderable. The times involved were, he stated, "very strikingly shortened." As to why this should be, he reasoned along the lines he had first pursued in Baltimore. It was because "the words and letters are not singly apperceived, one after another, but a whole group of them is grasped in one mental process." Here, he concluded, was exactly what psychophysics needed. It was a result of such solidity that it could serve as the counterpart of laboratory facts in contemporary physics. Moreover, it concerned a practice that occupied precisely the line between body and mind, reflex and thought.[11]

This work became the basis of a continuing series of researches, for which Cattell would earn his PhD and a valuable place in Wundt's inner circle of assistants. An important element, despite his earlier reservations about intrusive devices, was his invention of new precision instruments.

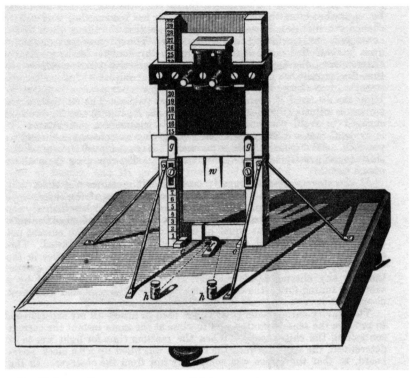

1.1 Cattell's Gravity Chronometer. J. M. Cattell, "The Time Taken Up By Cerebral Operations," *Mind* 11, no. 42 (April 1886): 223.

Most notably, Cattell devised what he called a "gravity chronometer" (fig. 1.1). This was a machine that revealed a piece of text to a viewer for a brief, precisely controlled period of time. It was perhaps the forerunner of the tachistoscopes that were later to be commonly used in the science of reading. He was very proud of it, ranking it as the most important achievement of his career to that point, and writing excitedly to his parents when he took delivery of a refined version that he felt "like an author with his first book."[12] The device soon became a standard piece of equipment in Wundt's lab, and the pioneer of medical informatics John Shaw Billings acquired an early example for the Army Medical Museum in Washington, DC. Cattell aimed to use it to gauge a reader's response when asked to name a character or word seen fleetingly in its window. To that end, he added two more devices—the "lip-key" and "sound-key"—

that were held between the lips of the reader. Linked electrically to a chronoscope, these devices could log the exact time of a spoken utterance.[13]

Instruments, to be sure, were always a mainstay of Wundt's enterprise. One that Cattell introduced the great psychophysicist to, incidentally, was the typewriter, which soon became a vital enabler of Wundt's legendary productivity.[14] But Cattell saw his as significantly different. His instruments, he would insist, stood out from the conventional repertoire of the laboratory. Not only were they formidably precise, but they also—and more importantly—relatively robust, simple, small-scale, and unobtrusive. As always, he emphasized that their scale and simplicity allowed the practices of everyday life to be measured without impediment or distortion—a vital consideration for an action like reading. Viewing the experimental scene in figure 1.2, the modern reader may be forgiven a certain skepticism about this claim. But it needs to be judged in the context of other experimental equipment that could be far more ungainly, and we shall see several examples later of far more extravagant machinery. Still, in inventing the gravity chronometer and its accompanying

1.2 Cattell's apparatus for experiments on reading. Clark University Archives and Special Collections, Robert Hutchings Goddard Library.

paraphernalia, Cattell inaugurated what would become a central element of the science of reading. As that science developed over the next century, the importance to it of a tradition of instruments and their management would be hard to overstate. Indeed, Cattell himself brought the first such devices to the United States. In 1887, long after leaving his graduate work behind, he returned to Leipzig and ordered a number of new devices from the same instrument-maker, Carl Krille, who had made his original gravity chronometer, initially for use in the laboratory he was then intending to establish in Britain. But that Cambridge venture soon folded, and they accompanied him when he left for Philadelphia. They became the central instruments of Cattell's laboratory there.[15]

Cattell briefly sought to extend his experiments beyond reading per se, looking at the time it took to recognize colors and images too—an ambition that would recur with later researchers, as we shall see. But reading remained his chief focus. In particular, Cattell used the gravity chronometer to measure for how long a letter had to be exposed before it would be registered by the eye. He took 15,000 discrete timings, using twenty-six English and German words of four to five letters each and another twenty-six words of eight letters or more. (Such heroic tolerance for tedious repetition, incidentally, would become a hallmark of the science of reading, which in this regard reflected its origin; William James once remarked that Wundt's program could only have arisen in "a land where they did not know what it means to be bored.") The key result was once again that it consistently took less time to read words than individual letters, but now Cattell added that the consciousness seemed to "grasp" more letters at a time when they formed a word than when they did not. It would also grasp more words when they formed a phrase. "The sentence is taken up as a whole," he reported: "if it is not grasped, scarcely any of the words are read; if it is grasped, the words appear very distinct." He had found something similar in the Baltimore kymograph trials, but the chronometer now proved to him that it was a real, consequential, and measurable phenomenon. And it had major implications for the practice of reading itself. Cattell recognized immediately that his claim should be taken as powerful support for the then fashionable practice in schools—already taking root in America, but until now lacking scientific warrant—of teaching children to read by having them identify whole words. "Children are now generally taught to read words as words, and are not required to spell out the letters," he remarked: so "it is well to *prove* that we read a word as a whole."[16]

Beyond this, Cattell also used the chronometer to see how different characters and fonts affected the speed of reading. German Gothic characters were the worst, of course, but within English too he found that different fonts, and even different letters, had different legibilities. This, he pointed out, was "of the greatest practical importance." "Reading is one of the largest factors in our modern life," he observed; but it was "a thoroughly artificial act," and if done poorly it could cause real damage to the reader. Children in modern cities were already suffering from myopia, headaches, and "weak" eyes, and no wonder, given the "immense tension" reading put on the eye and brain in a culture saturated with print. Society needed to be on its guard "lest these diseases become hereditary." So it was vitally important to relieve children of such tension—and thereby potentially save the race itself—by using only "the printed symbols which can be read with the least effort and strain." Schoolbooks in particular should be typeset on principles derived from experiments of this kind. It was "more of a hurt than help to the eye and brain" to use both upper-case and lower-case letters, for example, so one set should be jettisoned. The letters themselves should be simple geometrical forms without a combination of thin and thick lines. Punctuation marks were hard to see and "quite useless"—they should be replaced by spacings corresponding to intended pauses. Removing the dot from *i* and using λ instead of *l* would help too. And the letters *C*, *Q*, and *X* should be eliminated altogether. "Our entire alphabet and orthography needs recasting," Cattell concluded. The total time wasted every single day because of the obscurity of the letter *E* alone was, he said, "incredible," and the eye- and brain-stress it caused were "still more appalling." A revolution in character-sets would not only enable everyone to read faster, but also "teach us to think more clearly."[17]

These were radical claims. And Wundt was certainly impressed by this intense young American, capable of such a combination of precision and ambition. Cattell grew in his estimation until he became his assistant, and he went on to obtain a PhD in his laboratory. He was not the first American to work in Leipzig—that had been Hall—nor the first to gain a PhD there, but he was the first whose doctoral work centered squarely on experimental psychology, and Wundt seems to have held him in higher regard than his other American collaborators. He accepted Cattell's minor criticisms of his own work with good grace and encouraged him to publish his findings in Wundt's own house journal, *Philosophische Studien*. Cattell's first published work duly appeared there in 1885. It described the Baltimore and Leipzig experiments on reading in detail, and added new material on color perception. Cattell sent a copy to his parents, proud that he was finally the author of a scientific paper as well as an instrument. But he added a characteristically catty note about Johns Hopkins, where the research it described had begun: it contained, he told them, "the work they did not give me my fellowship on, and which Dr. Hall wanted to appropriate."[18]

Doctorate in hand, Cattell moved on from Leipzig to Great Britain and to Cambridge University. He had a vague plan to study medicine, mainly because it guaranteed a career. But his heart was not in it. Instead, he took up an affiliation with St. John's College and began maneuvering to establish an anthropometry laboratory, hoping to set it up under the aegis of the Cavendish Laboratory, then the world's most important site for experimental physics, which was now headed by J. J. Thomson. Captivated by the prestige and intellectual community of Cambridge, he almost certainly harbored ambitions of becoming the university's first faculty member in experimental psychology. Along the way, he became a member of the Aristotelian Society, a prestigious intellectual group. In early 1887 he read his first paper for the society, choosing the topic of "The Way We Read"; we do not know what it contained, but it almost certainly drew on his Baltimore and Leipzig work.[19]

Cattell's relatively brief time in Britain gave him the opportunity to cultivate connections with the champion of anthropometry, inventor of the idea of eugenics, and cousin of Charles Darwin, Francis Galton. They had met even before Cattell arrived at St. John's, and Galton had already given an extensive account of Cattell's gravity chronometer to his Anthropological Institute in London. The gentleman scientist had not hesitated when he heard that Cattell would be moving to Britain; he sent him a social invitation immediately, and they soon became friends and allies. At this time Galton was running an ambitious anthropometry laboratory in the Science Museum in South Kensington. The lab had initially been established as part of the International Health Exhibition in 1884, and it was devoted to the gathering of data by measuring the citizens who visited it.

One purpose was to detect individual problems such as color blindness. But the greater vision for Galton was statistical and eugenic. That is, it had to do with the cultivation of improvements to the "race" itself, and the reduction of what were seen as flaws, through the management of heredity. One consequence was that Galton, unlike Wundt, did not assume that differences recorded between individuals' results in anthropometric experiments were unimportant variations around a common value. On the contrary, he focused on them. He proposed that "mental tests," as he called them, could be devised to discern the different capacities of particular classes, races, and even individuals, and that practical policies could be designed on the basis of the numbers they generated. That was why his laboratory measured some 13,000 people before it closed in 1891. Although the numbers never meant very much in the end—largely because the principle of selection was ill-defined—the ambition behind them was inspirational to Cattell, who would dub Galton "the greatest man whom I have ever known."[20] He became a eugenicist himself, convinced that "lines of descent of superior intellect and character may be bred"—so much so that he would later offer his own children money to marry the progeny of professors. Later still, Cattell would be found insisting that every true psychologist had to imbibe Darwinism, and that only if they did would theirs be the discipline to demonstrate the unity of science itself.[21]

Eventually, Cattell did manage to set up an enterprise at the Cavendish—in just one small room, which he was told he had to equip himself. There he proceeded from his earlier work on individual readers to a series of experiments at scale on "mental association." These he pursued in partnership with a widowed teacher of mathematics at a London girls' school, Sophie Bryant. Bryant's involvement was substantial; she wrote a number of books on the practice and philosophy of education. Together the two recruited hundreds of subjects, ranging from students at Cambridge (men) and Bryn Mawr (women) to pupils at Bryant's school and elsewhere, in hopes of doing a kind of Galtonian anthropometry of readers' minds. Before very much could be done, however, Cattell's time in Cambridge came to an abrupt end. By late 1888 he had heard rumors that a lectureship was finally being considered for him; but he was freshly married and felt that he could not afford to live on the meager salary being envisaged. Besides, he had been offered a far more lucrative and auspicious position in Philadelphia. The choice seemed obvious. Cattell was en route back to the United States before it became clear that the rumored job in Cambridge had never existed anyway.

The Science of Reading Comes to America

Cattell moved back to the United States in 1889, becoming Professor of Psychology at the University of Pennsylvania—not the first such position in America, but certainly one of the first handful. He started work at establishing a laboratory of experimental psychology, situated inside the biology lab, with Krille's instruments as the foundation. But it was again to prove a brief sojourn. These were expansionary years in America's universities, and they were racing each other to set up psychology programs. Two years later Cattell was lured onward to Columbia, with the promise of not only a laboratory—including its own in-house instrument-maker, no less—but a whole new department of experimental psychology. And this time he stayed.

During the ensuing decade of intense work, Cattell built Columbia's laboratory up into what he boasted to be the only true rival to Leipzig in the United States. As he did so, he used his position to champion the use of quantitative anthropometric tests, akin (and in some respects identical) to Galton's, to appraise the abilities of students and measure the effects on them of education. An original proposal for large-scale testing of this kind at Penn had never got off the ground, but at Columbia he advanced an even more ambitious one that did. For several years, then, starting in late 1894, every incoming student at the university was subjected to Cattell's battery of "tests of the senses and faculties," and every departing student required to repeat them. What would emerge, he hoped, would be an objective measure of the impact of an undergraduate's college education. All that reading must surely make a difference, after all, and for the first time the difference would be demonstrated.[22]

Word of this initiative spread quickly, and for a brief time the kind of Galtonian program Cattell created at Columbia was à la mode across the United States.[23] Joseph Jastrow, Cattell's old Hopkins friend, championed it at Wisconsin and then at Chicago. It is not hard to imagine why university administrations would see value in a scheme to quantify the intellectual gains that their colleges offered to students. Furthermore, the tests acquired a public presence. At the 1893 Columbian Exposition in Chicago, Jastrow and Cattell's Columbia colleague Franz Boas installed a set of anthropometric instruments to test both visitors and, opportunistically, the "primitive" peoples who were taking part in the exposition itself. They ended up gathering data from some 100,000 people. Meanwhile, elsewhere in the exposition, G. Stanley Hall organized a congress on "Experimental Psychology and Education," which triggered the emergence of what would become known as the "child study movement." Hall's movement was dedicated to using psychophysical procedures to measure childhood progress. A little later, a laboratory specifically devoted to detecting "defective" children was set up at the University of Pennsylvania. And in the later 1890s, several initiatives were launched to use anthropometric tests in attempts to identify differences between the races, although the most ambitious of them, a proposal at the American Association for the Advancement of Science in 1896 for a systematic survey of the "white race" to be conducted under Cattell, Boas, and others, never got off the ground. The effort overall was nevertheless impressively wide-ranging and large-scale, and it attracted much popular interest.

For all that they were touted as "mental tests," the procedures that these programs employed did not include assays of "intelligence" as it would later come to be understood. In one sense, this was not for want of trying. Cattell himself originally wanted to include a full array of reading tests at Columbia, in the manner of his Johns Hopkins experiments. But as it turned out, only one such test was included. The rest were jettisoned, partly for reasons of time and partly because he felt that the experimental protocols for examining mental associations were still unsettled. So Cattell's students were given a sheet of five hundred printed characters and asked to mark as many instances of the letter A as possible, but that was as intellectual as their ordeal got. Otherwise the tests they endured either sought to measure powers of vision, hearing, and memory, or they aimed to record putatively basic facts like reaction times. Cattell was adamant that the process was still worthwhile. As he said so often, this was the way to generate the kind of objective, factual foundations that psychology needed in order to become "an exact science" and merit a high rank among the positive disciplines. Besides, it would allow for assessments of two questions of urgency and importance: how a person's various traits of senses and mind depended on each other, and the relative importance of heredity and environment in forming individuals.[24] He now saw his old experiments on readers as prototypes of these tests.

By 1895–96 Cattell was president of the American Psychological Association, and in that role he created a Committee on Physical and Mental Tests in order to standardize testing for college students at large. He served on the committee himself, arguing for maintaining the emphasis on lower-order phenomena like reaction times and character-perceptions rather than what he called "more complex mental processes." But it should have alarmed Cattell that even Galton was by now questioning his single-minded focus on anthropometry, and a reaction was soon brewing against his positivism. The group was not unanimous, as it turned out, and a minority report appeared, arguing for more tests of a specifically psychological character. Jastrow concurred in the committee's majority report, but with some reluctance. As it transpired, the outcome proved to be something of a harbinger. Harvard professor Hugo Munsterberg, himself an ex-Wundtian, now emerged as the psychological herald of a rival movement to Cattell's. Munsterberg even publicly warned the educational profession against what he called "the danger from experimental psychology," first in a speech to the Schoolmasters' Club of Boston, then in an article in the *Atlantic*, and then—even more pointedly—in a paper published in the *Educational Review*. He took as a foil a newly published and extravagant manifesto for psychophysics, Edward Wheeler Scripture's *The New Psychology*. Munsterberg declared roundly that he had "never measured a psychical fact" himself, and did not believe that such a thing would ever be done. In his view, the whole enterprise that Cattell personified was based on misunderstood premises. The laboratory work of the anthropometrists, Munsterberg announced, simply could not tell teachers "anything which is of direct use to you." Cattell might have wanted to insist that his science of reading, at least, had direct consequences in the classroom. But he was at a disadvantage, because Munsterberg actually taught classes on psychology for teachers themselves, so he could appeal directly to the very audience that stood to be affected.[25]

Cattell was exasperated to see Munsterberg's attack, and all the more because it was published in a middlebrow magazine whose readers, he believed, could not possibly understand the issues at stake. This, he may have felt, was the very problem of contemporary American reading culture exemplified.[26] He drafted an angry response, which would surely have inflamed the debate had Munsterberg himself not persuaded him to withhold it. A few months later, when tempers had calmed down, Cattell finally published a toned-down version. But the challenge had now been issued, and others were soon raising questions too. Edward B. Titchener—yet another Wundt protégé—and his students published pointed claims that it was not obvious what Cattell's tests were really tests of. For example, the young Guy Whipple—from whom we shall hear again—complained that all those reaction-time experiments actually measured nothing beyond reaction times themselves. The bitterest blow of all then came from one of Cattell's own assistants, Clark Wissler. After the Columbia testing program had been running for several years, Cattell gave Wissler the task of analyzing its results. To his surprise and dismay, his analysis demonstrated statistically that there was no significant correlation among the test results at all. So, for example, the test that involved reading the letter A on a printed sheet did not correlate in any way with the results of any of the other tests. Moreover, none of the tests correlated with students' academic performances either. By its own standards, Cattell's program had succeeded in producing definitive, objective, precise, and quantitative evidence. The trouble was that it had now been proved to be of no significance whatsoever.[27]

Although Cattell's anthropometric program came to little—as had Galton's, in fact—testing itself persisted. Titchener was instrumental in fostering a new discipline of "educational psychology," and testing became a central part of that discipline at the hands of Titchener protégés like Whipple. The new system of testing, however, was based not on Cattell and Galton's anthropometry but on the far more word-centric protocols of the French psychologist Alfred Binet. It tested "intelligence," not some hoped-for bodily proxy for mental strength. Yet one of the central problems of intelligence testing, as it turned out, was that it was reliant on subjects being able, skilled, and willing readers, so it was never entirely clear what a reported "Intelligence Quotient" (IQ) measured either. Educational psychology and testing will both recur several times in future chapters of this book, because practices of testing—for intelligence, reading abilities, and other properties and skills—became a constant element in the science of reading and its alternatives.

In the end the soul of Cattell's psychophysics of reading lay not in a set of concepts, theories, or facts, but in an armamentarium of instruments, housed in specific institutions and minded by skilled practitioners sharing a common practical tradition. The instruments and techniques that Cattell brought to America—and the community of experts equipped to use them—were now in place. So was a strong motivating interest in the fundamental questions that Cattell had asked, sometimes implicitly: what happens when a reader confronts a page of text? What is so powerful about that moment? And how can modern society monitor, shape, and manage it, while still maintaining a commitment to freedom and individual autonomy? There were few more important problems than these to face the first age of mass media and participatory democracy. Although attention has long centered on the analyses offered by prominent public thinkers like Cattell's friend John Dewey, the scientists of reading would from now on offer their own distinct, and in practice more consequential, answers—not necessarily in prominently published newspapers and magazines, but in the form of machines and practices. Their science owed its origins to the convictions of Cattell and his students, and in some ways it would continue them. But it would also develop in very unpredictable directions, and it would come to exercise enormous influence over every aspect of modern society, economy, and culture.

Science, Education, and Democracy

It is not always easy for twenty-first-century readers to grasp the excitement generated at the time by work like Cattell's at Columbia. Even in the 1890s, it was increasingly seen by rivals as narrow, pedantic, and trivial. But this work was exciting for those who championed it, and at first the many ordinary citizens in Britain and America who flocked to its centers believed that it offered something new and auspicious. They saw in it a scientific future for themselves and the nation.

The enthusiasm derived partly from the credence that some of the basic tenets of positivism still enjoyed in the 1880s and 1890s—a credence that was much more widespread and deep-seated than is generally recognized now. In that light, to provide objective, quantitative, experimental data of the kind Cattell had in mind was to create the only proper condition of possibility for a new science. And this would be a science, moreover, that according to positivist schemes of the hierarchy of the disciplines should sit almost at the pinnacle of the pyramid, atop physics, chemistry, and biology and beneath only sociology.[28] As Cattell himself put it in an 1887 article entitled "The Time it Takes to Think" that was initially published in Britain and then reprinted in *Popular Science Monthly*, experimental facts were essential to bring psychology, "the last of the sciences," up to par. "Just as the astronomer measures the distance to the stars and the chemist finds atomic weights," Cattell remarked, "so the psychologist can determine the time taken up by our mental processes." He implied it would be relatively easy from there to develop insights into the laws governing the association of ideas, which Cattell believed to be "no less fixed than the laws prevailing in the physical world." And then one might answer the age-old problem of the mind and body by finally eliminating the possibility of a radical separation between the two on the basis of the conservation of energy. That done, such work even offered the possibility of an effective extension of life itself. "It must be borne in mind," Cattell announced, "that the faster we think, the more we live in the same number of years." In his own case, he reported, it took one-eighth of a second to read a letter and one-seventh of a second to read a word. Cattell believed that such intervals were likely to shorten "in the course of evolution" and that therefore in the future "'we will live so much the longer in the same number of years.'" His experiments might accelerate this process, allowing modern people to cram more reading—more thought—into their lives. That was why the science of reading should be made the foundation of all sound "methods of education and modes of life."[29]

Cattell's interests in wider "modes of life" was not only experimental. Nor was it confined to the laboratory. From at least the 1890s, he was convinced that scientists had to take an active part in cultivating public interest and support. To him that meant managing the business of publishing itself, and he dedicated himself increasingly to the jobs of editing and publishing scientific journals, both for specialists and for broader publics. As he put it at another time of stress in the scientific public sphere, in 1925, the United States might enjoy enormous popular interest in scientific matters at any given moment, but sustaining a culture of research required that that support be continuous. That meant that scientific research, even of the most esoteric and laboratory-based kind, depended on the press, and the scientific community must be alert to all that this implied.[30] This was the kind of perception that lay behind his dedication to science publishing, which started with his foundation of *Psychological Review* in 1894. The move was partly undertaken to provide a venue independent of his old Baltimore foil, G. Stanley Hall, who edited the *American Journal of Psychology*, which was then the country's only journal in the field. More representative of a broader credo was Cattell's rescue and revival of *Science* in the same year. Originally founded with the support of Thomas Edison, and then relaunched with that of Alexander Graham Bell, *Science* had always had the ambition of becoming a counterpart to Britain's *Nature*. But it had been one of many such journals launched in a short period, and like the others it had languished. As its new head, Cattell soon proved himself a canny impresario of intellectual commerce. He installed a prestigious editorial board of scientists, bolstering the journal's authority, and arranged to have the printing done by a plant he discovered in Lancaster, Pennsylvania, that had been devoted to local newspapers and could offer cheap rates. Under his leadership it would become America's leading venue for public science. As it did so, the printing operation, now named the New Era Printing Company, became a leading specialist in scientific periodicals. In 1901 Cattell incorporated the Science Press to publish the journal, which by then was affiliated with the AAAS. He would continue to be instrumental to both *Science* and Science Press for decades, becoming America's most important science and education publisher. He eventually moved to Lancaster himself to spend his last years living in an apartment above the printing shop.

Cattell conducted his last substantive research at around the time his testing program at Columbia fell victim to its own success. In retrospect, it has always been assumed that the failure of the anthropometric program was so devastating that it ended his research career. But at the time this was far from evident. In the late 1890s, far from seeming demoralized, he was set to launch another innovative research endeavor. This project would be dedicated to exploring perceptual thresholds and the relations between energy, fatigue, and human vision. However, it did not get very far, in large part because work in the very science of reading that he had inaugurated undercut its premises. But both the abortive program and his new role as a mediator between science and its publics underlined the seriousness of the questions that he and the scientists of reading he trained wanted to pose. For that reason it is valuable for us to follow Cattell as he negotiated these shifts in fortune.

As his anthropometric venture came under pressure in the late 1890s, Cattell envisaged pursuing another major research enterprise. It involved reviving his earlier work on reading—and, more, its extension into perceptions of colors and shapes—in a different context. His inquiry had to do with a puzzling fact of everyday experience related to the eminently psychophysical question of perceptual limits. Experimental trials had shown that the energy of a light ray sufficient to produce the smallest perceived sensation in the eye corresponded to 0.0000000000000000075 horsepower (plus or minus a factor of ten depending on the observer, he added diligently).[31] Measuring such values was hard. But even above that level there were questions of perception which psychologists had failed to reckon with. In particular, our bodies were constantly in motion, and so were our eyes. At a time when box-camera photography was the popular fad of the day, that suggested a problem. The images cast on the retina must be similar, Cattell thought, to "those of the photographic plate when the camera is constantly moved hither and thither." Should they not be blurred, then? And indeed, when black and white patches on a color-wheel rotated rapidly in front of an observer, the perception was that they blurred into a uniform gray. With colors, however, the experience was much more complex, and it varied by individual. Moreover, if the movement were only in the observer and not in the outside object, then this kind of "fusion" seemed to cease altogether. Suppose, he suggested, that you looked rapidly across a shelf full of books, sweeping your eye past fifty spines in a tenth of a second: Cattell claimed that you would be able to perceive the text on every spine distinctly. (His assertion that the eye really did sweep continuously is intriguing, by the way, because it made the case quite different from that of reading a line of text.) If the same row of books were to move at an equivalent speed past your stationary eye then they would blur together. Why the difference?

To Cattell this question of perception and relative motion revealed something fundamental. He reckoned that it proved that "all the phenomena of color vision" were not explicable in terms of retinal images at all. Instead they were "cerebral phenomena" through and through— that is, they were not "organically fixed" but the results of "individual experience and adjustment." That we sometimes perceived a sequence of discrete signs as a continuum showed that "our perceptions are not 'copies' of a physical world, but are mental phenomena dependent on utility and the whole content of present and past experience." He even denounced the idea of a perception being a copy of the world as "absurd." Perception, rather, was something active, a *skill*. It had to be learned, and different individuals perceived differently (and so did an individual over time) because their learning experiences had varied.[32] Moreover, Cattell suggested that the reason why perception must be a learned skill was that it was a product of evolution. A "savage" seeing a panther running toward him would benefit from perceiving the object in motion, he pointed out, rather than from isolating particular limbs. So the modern human organism had *evolved* to perceive differently depending on whether the eye or its object were moving. Perceptions became "the result of experience and utility" not only over individuals' lifespans, then, but over the entire history of the species. In other words, "our perceptions in no wise correspond to the physiological processes in the eye, but are what they should be for our safety and convenience."[33] Returning again to the relative legibility of the various letters, he reiterated that the commonest of them, *E*, was the least perceptible of all (fig. 1.3). What did this mean if perception was evolutionary? Not only did it imply a continuing need to change the standard letter shapes, but such a fact might have major implications for our scientific knowledge of human nature itself.

In the end, Cattell did not explore that question any further, because his ambitious project did not get off the ground. One reason for that, however, was that its founding claim turned out to contradict flatly the science of reading. Even before his plan was published, the leading experimenter in the new field, Raymond Dodge, had realized that Cattell's notion of perceptual fusion was at odds with his own signature discovery. As we shall see in the next chapter, Dodge believed that readers did not actually perceive letters on a page at all while their eyes were in motion; they only "read" at those points when their eyes stopped. So much, then, for Cattell's experiment of looking quickly over a row of books. His grand interpretation of the evolutionary, learned nature of perception—one with such great implications for education and society—had emerged, in his own telling, from reflecting on a phenomenon that on Dodge's account did not exist at all. For all that Cattell insisted that his broader contentions

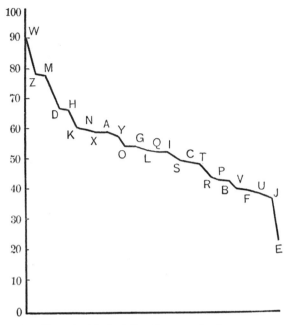

1.3 Cattell's graph of the legibility of letters under the gravity chronometer. J. M. Cattell, "The Perception of Light," in *System of Diseases of the Eye*, ed. W. F. Norris and C. A. Oliver (Philadelphia: J. B. Lippincott, 1896), 1:525.

remained valid, he was forced to concede the point, in an admission that was no less disappointing to him for being honorably made.[34]

The Impresario of Science and the Nature of Scientific Publics

In all likelihood, Cattell could have found a way to rescue his psychophysical program. Dodge himself was not intent on destroying it, and in succeeding decades Cattell would occasionally venture into print with advice and proposals that retained its core commitments. But the setback, along with the failure of his Columbia testing program, occurred as he was increasingly committing himself to a much more directly focused assault on the existing institutions of learning and knowledge, and this turned out to dominate his next few years. At Columbia in particular, Cattell became a major thorn in the side of the president and administration, arguing—often in the pages of his own journals, or in tracts published by his own press—that the institution had lost its way in becoming too emulative of corporate culture. For example, in 1906 Cattell used *Science* to publish an acerbic review of the rise of autocratic university governance and a call for universities to adopt a form of representative democracy; he then reprinted it separately in 1911 and circulated it widely to scientists with an invitation for comments; and in 1913 he published it a third time along with their responses, under the title *University Control*. Through strategic actions like this he pressed the case that the all-powerful presidents and boards of trustees that characterized American universities at the time were a betrayal of democracy, both within the institutions and in the nation more broadly. Their rule was typically justified in the name of "efficiency," he noted, but he maintained that in practice it mainly produced mediocrity and conformity, coupled with a crudely pseudo-Darwinian mode of exploitation. This absolutist style of governance arose, he thought, because elite universities had grown excessively large and complex, and because they had become dependent for support on "the modern knights of industry." They emulated the methods of commerce, when commerce itself was characterized by industrial titanism, so it was no wonder that they became gargantuan and opaque. "Our methods of communication, transport and trade, of manufacture, mining and farming, have led to the doing of things on an immense scale," and the individual had been subjected to "crude commercial standards." He argued for something of a return to the form of the original medieval universities, which had been "extraordinarily unhierarchical, democratic, anarchic," and had had no administration whatsoever. The optimum size of any academic unit was ten to twenty faculty, he suggested, this size being "prescribed by a psychological constant." The great foundations of the Gilded Age needed to be curtailed and academic freedom and tenure asserted. The scientist of reading had become an activist of information politics.[35]

Cattell was capable of being politic as well as principled in his campaigns, but he could also be harshly undiplomatic. It was hardly constructive to liken the typical university president to a rabid dog that had to be shot, for example, or to remark, as Cattell did, that he had suggested to his young daughter that she name her doll "Mr. President" on the reasoning that he would lie in any position in which he was placed.[36] Over time, such behavior steadily lost him allies in the higher echelons of Columbia. Eventually he was ejected from the university altogether, the occasion, if not the real cause, being his opposition to American entry into World War I; the university administration even officially accused him of treason. (Edward Thorndike's wife noted flatly in her diary that he had been dismissed for this offense.)[37] A few colleagues—Dewey first among them—stood up for him, but he was never given another chance. A few years later he did win a legal settlement in the wake of his dismissal, in a case that became celebrated in the annals of academic freedom. But rather than use the opportunity to try to reenter academia, Cattell joined with two of his ex-students and fellow scientists of reading, Thorndike and Robert Sessions Woodworth, to invest his award in a new kind of commercial venture. Called the Psychological Corporation, its intent was to market psychological science—testing in particular—as a service to public, commercial, and educational institutions. The company struggled in its early years, but it survived as a pioneering effort in early market and industrial research. We shall encounter it again.

In 1913 Cattell articulated the point of all this activity in a speech to the American Association for the Advancement of Science on the theme of "Science, Education and Democracy." He lauded the AAAS for recognizing the need to develop a science of education—the most important of all applied sciences, he declared, and yet at that time the most "backward and empiric." What made such a science urgently necessary was the advent of modern society. He placed this in an evolutionary perspective extending back beyond humanity itself. In prehuman and prehistoric orders, he claimed, there had been "a kind of democracy," in which every individual was prima facie equal; but in historical times "privilege" had become the dominant system. In order to sustain favored social classes possessing the abilities and resources to advance science and the arts, the masses had had to work incessantly. But science had now transformed that situation. It had "recreated the world," outlawing "the exploitation of children, sex slavery, industrial servitude, [and] kleptocratic classes." Modernity thus made universal education possible; but it also made it requisite for "a true social democracy." Such education had to be structured according to the topics and, especially, the "methods" of modern science. So the provision of education and scientific research had become "the chief concern of a democratic society." The United States now led the way in elementary, high school, and college education; over 96 percent of children ten to fourteen years old were in school, and "reading and writing have become, like air and water, the common heritage of all." This was the foundation of democracy's future.

Although the fate of modern America apparently rested on knowledge about reading, Cattell was notably caustic about conventional assumptions on such lines. He thought schools were actively harmful, for example, because they tended to produce "the conventional and the commonplace." He disdained claims that reading could transform the self—that was an antiquated Romantic conceit. On a collective level, too, he was skeptical about complacent claims that mass literacy produced social harmony. Sensationalist news coverage was much more likely to excite "national hysteria" than "solid homogeneity." He even suggested that for most people the principal benefit of reading was as a substitute for drink. "The effect of reading the newspaper or current novel," he explained, was "similar to that from a small dose of alcohol or opium." We know that he delayed teaching his own children to read for several years, much to their later chagrin. Writing too was no panacea in his eyes. The best he could say about it was that "its acquisition, while likely to be harmful to the immature nervous system, is less destructive than learning to read." What was needed, then, was a thoroughgoing, top-to-bottom revolution in learning.

Cattell's proposed alternative to prevailing educational methods was a combination of his friend Dewey's philosophy, his own psychophysics, and some hard-edged eugenic Darwinism. For Cattell, Darwin had proved that evolution rested on differences that individuals possessed at birth. The greatest service of education, then, was not to expand children's store of knowledge—their capacity for that was preordained—but to match talents to roles. The extent of people's differences, and the degree to which education could fit them to their appropriate situations, were "scientific problems on the solution of which depends the future of social and political institutions." Indeed, Cattell, like Thorndike, was given to saying that discovering "what may be done to fit the individual to his environment"—rather than learning as such—was the *point* of education.

As was often the case with early eugenicists, there was a social edge to Cattell's view. For all its undoubted racism (Cattell could remark almost casually that "there is not a single mulatto who has done creditable scientific work"), it was not indulgent of traditional hierarchies. One could never tell whether the members of an elite merited their positions unless the social effects of privilege could be excluded, and doing that was in his view the true purpose of schools. The people on whom society depended would "do equally well whether they learn to read at twelve or at six," he remarked, but schools could *reveal* their natural talents. Yet here was precisely where the current system was lacking. The nation's schools were wedded to outdated, sexist, and class-preserving ways. The universities, too, polished the sons of the rich, while trade colleges channeled the offspring of the poor into becoming "cheap skilled labor for exploitation." And yet, Cattell noted, "the road to democracy may be paved with bad intentions." Quoting Woodrow Wilson (without naming him), he observed that the universities themselves—and the new wave of state universities in particular, with their openness to men and women—could be seedbeds of change. Cattell wanted a massive funding boost for them, urging that scientific and medical institutions be given the stratospheric sum of four billion dollars annually from a new federal income tax. And a great national university should be created in Washington, DC, to embody the ideals of a great democracy—to be followed by an international university to extend those ideals beyond the US. At the same time, two hundred dollars per child should be invested to educate the next generation, with salary increases for teachers to raise their status and attract "the wisest men and women of the country" into the profession. The fact that Americans spent twice as much on alcohol and prostitution as on schools must end. Furthermore, within the schools themselves the existing methods had to be transformed. "Book learning" was based on "a bad psychology." A new "science of education" must arise, based on the new scientific psychology. Only these policies would permit "a stable and complete political democracy" to come into being.[38]

By the 1920s, then, the old eugenicist and champion of a psychophysics of reading had become America's strongest and most influential voice for a vision of science and education in a democratic order. In the interwar years, Cattell (fig. 1.4) remained the country's most important scientific gatekeeper, constantly agitating to sustain a dynamic public realm in which the claims and arguments of scientists and educators would be aired, circulated, and defended. He ran *Science* for some fifty years, all told, substantially creating the norms of editorship in the field. Only once was his regime really threatened. In the mid-1920s the Lancaster printing

1.4 James McKeen Cattell. University Archives, Rare Book and Manuscript Library, Columbia University Libraries.

company changed hands and abruptly increased prices, throwing *Science* into crisis. Efforts to persuade the Carnegie Foundation to step in and create an operation around it on the model of Oxford University Press fell through—perhaps because of the foundation's commitment to the broader campaign for public reading to be described later. In the end the American Association for the Advancement of Science rescued the magazine, while Cattell took over the Lancaster printing business himself. He reestablished it as the Science Press Printing Company and continued to manage its ventures well into the 1940s. Science Press branched into many other science publishing ventures too, including the venerable *Popular Science Monthly*. Particularly notable for the science of reading was *School and Society*, a weekly to which Cattell was devoted and that bristled with articles about his own valued subjects of America's problematic schools, academic freedom, and the perils of "absolutist" systems of administration. It became a key vehicle for many of the debates that shaped knowledge about reading for the next decades.

By the time World War II broke out, Cattell had long been the doyen of American science communication. He had not only sustained his stable of publications but also served for a decade as president of Science Service, a scientific news agency founded by the newsman Edward Scripps in 1921. Throughout these decades, many of the key arguments of the science of reading that we shall encounter in the following chapters appeared under his editorship, or at least were debated in the pages of his journals. His last public foray, appropriately enough, would be a correspondence campaign against the censorship of scientific research during the war. By then he had retreated from New York and was living in a lavish apartment complex atop the Lancaster plant. When he died there in 1944, he would be remembered not for the science of reading, but for the reading of science.[39]

2

The Work of the Eye

Americans in the years around 1900 frequently looked at their own society with awe, admiration, and pride, as well as with a certain horror at its excesses. Just a century or so after it had become an independent nation, the United States had become decisively, cruelly modern, with a complex infrastructure of transportation and communication, and an energy industry equipped to power a new century of progress. Its democratic system promised to inspire the world. True, just a generation after the Civil War, Reconstruction had stalled amid blood and terror, and the behemoth monopolies that dominated the economy were widely decried. Promise and reality were starkly different. Progress was the faith of the day, but degeneration its perennial fear. The question of reading lurked behind all this like an unsolved equation in an engineering problem. How could this immense social machine—reliant as it was on sophisticated systems of engineering, management, finance, commerce, and politics— survive and flourish, if its people could not read the instructions?

In the 1880s and 1890s James McKeen Cattell and a handful of others had posed the question of reading in a new way. If anything, the testing regimes that emerged from that tumult to flourish in the new century accentuated the need for an answer. The proportion of seventeen-year-olds who were high-school graduates might have risen sharply—from 2.5 percent in 1880 to almost 50 percent by 1940—but tests furnished hard, quantitative evidence that vast numbers of them still found reading difficult and laborious.[1] Corporations, the military, and the state all grew increasingly alarmed at what the tests revealed to be a stubbornly persistent problem. Just a generation earlier, the quest for literacy had been central to the modernization narrative that Americans had told themselves: create a literate citizenry, they believed, and a free, prosperous, and stable democratic order would result. The twentieth century realized that that belief—the belief that drove the rise of mass education—was much too simplistic. *That* the citizenry be literate was not enough; *how* they practiced their literacy mattered profoundly.

At the onset of that generation, the instruments, methods, and sensibilities of German psychophysics came to the United States, and they did so in the form of a science of reading. As is always the case when sciences move across continents, however, there was more to it than a straightforward transfer. In effect, the science of reading emerged from the conjunction of experimental science with the anxieties and desires of Gilded Age America. Americans worried about industrial efficiency, fatigue, and the social disorders of a democratic republic; and they increasingly did so in terms of social Darwinism and anxiety about racial destiny. The science of reading grew to reflect and respond to those concerns. The author of the field's first major book, Edmund Burke Huey, was particularly frank about this. Huey ascribed his own initial motivation for entering the field to two causes. One was a question raised in the laboratory in 1898 by his colleague Guy Whipple about whether silent reading must involve "inner pronunciation." That question implicitly invoked the contemporary preoccupation with making work of all kinds as rapid and efficient as possible—nonverbalized (or "visual") reading might well be far faster and less wasteful. It also prompted the thought that the reading process itself might "mirror the processes of thinking," and thus provide a gateway for an ambitious experimenter to address the very nature of cognition in the lab. So far, in a sense, so Cattellian. But the second cause was larger in scope and more evidently consequential. It concerned what Huey called the "peculiar fatigue" caused by reading. Doctors had diagnosed enervation as one of the products of excessive reading for centuries, he knew, and had produced medical remedies for it. But his own reasoning went along quite different lines, running from the distant evolutionary past to the dangerous present and future.

This habit to which we subject ourselves and our children for so considerable a part of the time is an unnatural one, intensely artificial in many respects. The human eye and the human mind, the most delicate products of evolution, were evolved in adaptation to conditions quite other than those of reading. Such functionings as reading requires not having been foreseen in the construction of these organs, we need not be surprised that our continued and careless exercise of these unusual functions causes fatigue and, in very many cases, certain dangerous forms of degeneration. The very evident inheritance of some of the more disastrous effects, such as myopia and nerve exhaustion, warns us of the danger of race degeneration from this source.

We should worry about bad reading habits, then, not merely because they were inefficient or personally wearisome but because they might have catastrophic consequences on an epochal scale. Reading was a matter of eugenics. And that meant it fell under the rubric of what was then called "mental hygiene," a field of work that ranged widely over matters of school sanitation, ethics, administrative rationality, curriculum design, and furniture engineering. Indeed, the last of the four sections of the classic work that Huey wrote on the subject was entitled "The Hygiene of Reading." It was this association that not only set the science of reading apart from prior accounts of reading itself, but also aligned it with other emergent disciplines of the time—in particular, educational psychology and psycholinguistics.[2]

In this fraught context, scientists of reading rested their controversial hopes and claims mainly on two instrumental technologies. Both of them emerged originally from the psychophysics laboratories that gave rise to Cattell's initiatives, but both underwent significant modifications in the new field, after which they would remain in constant use for decades. The first machine was the *eye-movement camera*.[3] This was designed to trace in fine detail how someone's eyes moved while reading a line of text on a page. The second was the *tachistoscope*.[4] It showed a series of letters or numbers to a reader, either through some kind of window or by projection onto a screen, for a very short but precisely defined duration. Used in conjunction, these devices supposedly captured the two main elements of the activity of reading. The eye-movement camera recorded the jagged motion of the eye as it traversed a line of print, and the tachistoscope revealed how well a reader perceived letters, words, or phrases during the pauses in that motion.

The reliance on these machines, however, was double-edged, especially in the early years. They were ingenious and, at their best, astoundingly precise. But they were also persnickety and hard to standardize. In practice, during this time they were generally bespoke, Rube Goldberg-esque devices, often built for particular experimental programs. They had shared features, but no two were identical with each other, and when crucial distinctions rested on differences of a hundredth of a second or a fraction of a millimeter, small variations mattered. To give one example, a protégé of Wundt's named Julius Zeitler argued around 1900 that the eye latched onto certain dominant letters rather than whole words as such, in stark contrast to Cattell's claims, and he maintained that he could prove this by reducing exposure times to an absolute minimum. That made the tachistoscopic duration potentially critical, and being sure of this was no simple matter when the device might in practice take a finite duration of time to reveal a character or word.[5] The machines were replete with glitches, too, like the distractingly loud noise made by a circuit-breaking mechanism in early gravity chronometers—something almost never remarked upon in formal papers, but evidently a real issue in practice. They were also simply large and ungainly. So these were troublesome devices. And yet, around these machines were built experimental, educational, and therapeutic programs that at their peak affected hundreds of thousands of Americans directly, and in all likelihood hundreds of millions indirectly. Arguably, by the 1930s they had redefined what it was to read in the United States.

The new science took root through the circulation of a fairly small number of people, partly because of their rare expertise in handling these precision instruments. These were the experts (in the first generation they were all men) who could be relied upon to design the machines, look after them, coax them into working reliably, and get from them consistent and meaningful traces. As they moved between the new American research universities, taking their tools and their expertise with them, these people left behind not only instruments but also bands of trained researchers. And these disciples were sustained by a shared sense of the vast social and psychological importance of the work that they were doing. So it was that the movements of the first experimenters gave rise to a distinctive set of instruments, practices, questions, methods, and convictions. Many of them relied on a fairly small and select group of publication venues to communicate their work, too, central to which were the printing and publishing operations of Cattell and his sometime friend and later foil, G. Stanley Hall. It is not an exaggeration to say that virtually every single paper of consequence in the emerging field appeared in a venue associated with one of these two. In this sense, the science of reading was a publishing enterprise as well as a laboratory discipline. As such, it had a certain reflexivity to it: a science devised in response to the rise of mass print, it flourished through key figures' exploitation of print itself. Each paper that appeared in Cattell's publishing realm reinforced the sense that this was indeed a science, which could be made the foundation of a radically reformed and improved culture of reading.

A significant contrast—a gulf, in fact—consequently arose between the everyday work of the scientists of reading and the stakes that they believed to be at issue in that work. On the one hand, they spent their time generating minutely precise measurements of quantities that were all but imperceptible, such as a momentary pause in the motion of a reader's eye as it passed over a line of prose on a particular page. But on the other, in their grander moments they were convinced that civilization itself depended on those measurements. If democracy were to survive—something that came increasingly into question as the new century advanced—then it would require a populace with the ability and propensity not only to read, but to read quickly, fluently, accurately, responsibly, and at length—and to repeat that practice, day in, day out. Traditional schooling might have generated mass literacy, but it had failed to produce a reading populace that could be trusted to exercise its skills in this manner. There was only one good way forward. America's future depended on getting the science of reading right. And that was why the minutiae of laboratory measurements mattered.

The Instruments of a New Science

Among the thousands of young Americans who went to Germany in the late nineteenth century to attend universities, a few followed Cattell and immersed themselves in the research culture of psychophysics. Of this number, a handful committed themselves to analyzing the act of reading. It was this small group who on their return would establish and consolidate the new science in the United States.

The science of reading coalesced in America around five itinerant Americans in particular: Walter Fenno Dearborn, Edmund Delabarre, Charles Hubbard Judd, Raymond Dodge, and Edmund Burke Huey. Their paths were notably similar, and they came to know each other well. Briefly, Dearborn studied under Cattell at Columbia, before proceeding to Göttingen, where he earned a medical degree with a dissertation translated in 1906 as *The Psychology of Reading*. He then continued to work on what was increasingly being called "educational psychology," moving from the University of Wisconsin–Madison to the University of Chicago, and then on to Harvard.[6] Delabarre attended Brown and Amherst before studying in Berlin, Freiburg, and Paris, as well as with William James at Harvard; he then returned to become Brown's first professor of psychology. He would later train the renowned Edward Thorndike. Judd went to Wesleyan before following in Cattell's footsteps as the second US citizen to get a PhD in experimental psychology under Wundt in Leipzig. He spent the early 1900s at Yale, but in 1909 moved to the University of Chicago, where he would become a figurehead at the Department of Education for almost three decades. As we shall see, Judd created a hugely influential school of research at Chicago, which would be continued into the 1930s by protégés William Gray and Guy Buswell. Dodge graduated from Williams College before heading to the University of Halle, where he worked with the Helmholtz disciple and Kantian philosopher Benno Erdmann; he ended up spending most of his subsequent career at Wesleyan, where he took a position that had briefly been Judd's. And finally, Huey, like Cattell, began at Lafayette College, before he proceeded to Paris. There, he paid homage to Javal, by now aged and blind. He collaborated with Erdmann in Berlin before eventually taking up positions at Clark and then at the Western University of Pennsylvania (today's University of Pittsburgh). Huey's 1908 book *The Psychology and Pedagogy of Reading* would constitute the first definitive statement of the new science's scope, claims, and purpose.[7]

At the laboratories of Wundt, Helmholtz, Erdmann, and others, these men acquired through hard work and acculturation the personal discipline, the mechanical skills, and the intellectual commitments that leaders of a new laboratory science would need. And as they returned one by one to the United States and took up positions there in the country's fast-expanding university system, so they deployed these lessons to build new laboratories and forge new research networks. In particular, like Cattell they brought back instruments. Very soon, these devices—eye-movement recorders and tachistoscopes in particular—became fundamental to their endeavor, such that its claims to credibility and its viability as an enterprise depended on their everyday management.

Initially, the more compelling of the two instruments was the eye-movement recorder. A precisely calibrated device of this kind could reveal where in a line of text the reader's eye paused, and how fast it moved between those pauses. And it would generate an automatic record of the pauses and movements, typically in the form of a crooked line on a film strip. The resulting trace provided a compelling visual record of a reader's abilities and propensities. Although experimenters' papers were full of tables of ostentatiously precise numerical measurements of things like saccade lengths, the traces were valuable tools in their own right—graphic representations of human virtues, frailties, and diseases. Experimenters—and, after them, teachers and clinicians—could quite literally *point* to a trace of this kind, in the confident belief that it displayed the essentials of a reading habit. They could then discern how the habit might be improved, perhaps by reducing the number of retrogressive motions, or by training the eye to make fewer stops in a line. Interestingly, these early machines were almost entirely restricted to tracing horizontal movements across the page, on the assumptions that the reading eye would not move vertically very far and that any vertical motions it did make were unimportant. Not all agreed with these assumptions, as we shall see, but the linear record was a notable characteristic of the early eye-movement devices.

While the eye-movement recorder addressed issues of motion, the tachistoscope addressed perception when the eye was momentarily stationary. It revealed how much someone could read in a brief duration (less than half a second), implicitly when the eye was, relatively speaking, at rest in one of its stationary "fixations." Providing a fleeting glimpse of a letter, word, or phrase, it could perhaps isolate the act of readerly perception itself, separate from any influence of memory, reflection, and the like. As with eye-trackers, in due course tachistoscopes would be deployed in attempts not only to measure a reader's performance but to improve it. If a reader could be trained to take in more letters in a tachistoscopic exposure, the argument ran, then they would perceive more in each fixation during everyday reading, and their general reading performance would increase. But at first the aim was simply to understand the ability to perceive what was on the page by measuring the field and accuracy of perception in a fixation.

The most important early deviser of both instruments was Raymond Dodge. Dodge came from a quite different background from Cattell. Born just over a decade later, in 1871, he was the son of an apothecary and small-town preacher; he later recalled spending much of his time as a child in the workshop behind the family store, making devices of various kinds. Lacking Cattell's family fortune, he funded his own education at Williams partly by working in the library. There he came across the works of Immanuel Kant, and by the time he graduated he was devoted to Kantian philosophy. After a precocious paper on that subject failed to win him a scholarship at Harvard or Columbia, he resolved to travel to Europe instead in pursuit of a doctorate. Dodge chose not to go to Leipzig or Berlin, but instead headed straight for the University of Halle and the scholar who had edited the volume of Kant that he had read as a student, Benno Erdmann. It so happened that that semester Erdmann was holding a seminar on the psychology of reading—we do not know exactly what it covered—and Dodge decided to sit in on it, even though he feared that his still-inadequate German would preclude his making any real contribution. This proved a momentous decision. During one of the sessions, Erdmann expressed exasperation that there was no instrument in existence that could expose a complete word simultaneously and for a precisely defined moment to both eyes at the same time. Such a device would be a huge boon to any experimental research on reading, but the best psychophysicists had not produced one, and he feared it was impossible. Dodge took the lament as a challenge. As he later put it, it became his "initiation into experimental psychology."

The issue was that in existing machines, Cattell's gravity chronometer being a good example, the letters were revealed by a moving screen. The time it took for the screen to fall was certainly short, but it was still a real duration. It was possibly sufficient for the eye to register a letter during its emergence, and if so then this would render measurements imprecise.

Dodge hit upon the idea of using a rotary shutter instead of a screen, because the aperture could be increased from zero (and reduced back to zero) without the letter being legible during that tiny duration. He soon demonstrated a crude version to Erdmann using the letter H, and found to his delight that the professor was impressed enough to fear that this was "some sort of American hocus-pocus." After a further year of work, Dodge had a purpose-built tachistoscope incorporating a chronometer ready to go. It became the central element in the next two years of collaborative research that he and Erdmann undertook on reading, and it had pride of place in the book that the two then cowrote. After that it served as the model for the tachistoscopes that researchers of reading would employ for decades.[8]

In succeeding generations, Dodge was primarily associated with his tachistoscope. But in fact he played a key role in devising eye-movement instruments too. The problem he faced in this context was that although methods like Lamare's audio device did exist for registering the fact that the eye moved irregularly, no method achieved the psychophysical aim of producing a legible trace. For that, one needed some instrument that could amplify and record the tiny motions of the eyeball itself. At the very end of the nineteenth century, two techniques arose to meet that need. The first was devised by the German researchers Eduard Rählmann and August Ahrens and then refined by Huey and Delabarre. It involved connecting the surface of the eye itself mechanically to a tracing device. In Huey's version, the reader's head had to be held in a stabilizing frame, its precise position being maintained thanks to a mold that preserved an imprint of the teeth and thereby fixed the mouth in position. The cornea was made insensitive by holocaine or cocaine, and Huey then placed a small plaster-of-Paris cup on the surface of the eyeball, the cup having been molded to fit and sandpapered down to be light and thin, with a hole drilled in the center. An aluminum pointer that was connected to this cup then responded to every movement of the eye, tracing a record onto smoked paper mounted on the surface of a rotating cylinder. To measure speed, an electric current running to the cylinder was interrupted by the vibrations of a powered tuning fork, such that a spark from the pointer's tip displaced a dot of soot at each regular interruption. It all sounds distinctly disturbing, but Huey expressed confidence that the reader was generally unfazed by this apparatus and that "reading proceeded as glibly and easily as could be desired." And it worked, at least to the extent that he managed to generate what was the first published record of an eye moving during reading (fig. 2.1).

Once he had the contraption functional, Huey tried it out with a variety of line-lengths, font sizes, and rates of reading. The traces showed fixations and saccades very clearly, he found. They also hinted that the eye stayed roughly on the horizontal line of the text being read. The eye made about four stops per line, and it generally did not stop anywhere on its reverse sweep to the next row. But occasionally it would pause near the end of that motion, presumably in order to get its bearings, and this happened more with longer lines. In general, smaller type, or reading at a distance, tended to increase the number of pauses. Huey also found that a typical reader's eye did not actually pass across a whole line of text from beginning to end. In fact, it generally started at a position slightly indented from the start of the line, and it made its last fixation at a place indented rather more from the end. Further, a good reader tended to fall into a regular pattern of pauses when reading a block of text, so that the

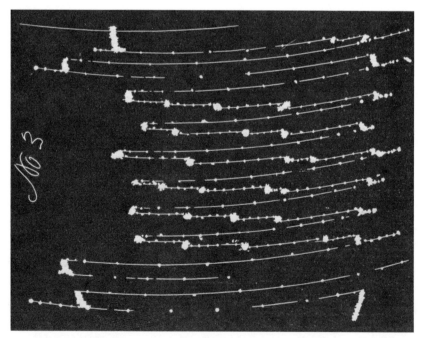

2.1 The first recorded "'spark' record" of eye movements. Edmund Burke Huey, "On the Psychology and Physiology of Reading, I," *American Journal of Psychology* 11, no. 3 (April 1900): 290.

fixations would more or less line up vertically down the page. The pauses themselves might last varying amounts of time, but the movements themselves would be fairly consistent, as if forming a natural rhythm. He also found that when trying to read as fast as possible, readers made fewer and faster pauses, and larger jumps—but the durations of the jumps did not change, and most of the time spent in reading was devoted to the pauses. Fast readers therefore performed less "eye-work," as he put it. Their movements were presumably less fatiguing, both because they were rhythmical and because they covered larger angles more smoothly.

This mechanical kind of tracking was problematic, as Huey realized. The mechanism itself was distracting, its weight was likely to cause fatigue, and the momentum of the rod was sufficient to make it overshoot. Other researchers therefore experimented with a second technique that involved using a light ray instead of the physical arm to create the trace. It was Dodge, again, who succeeded. He did away with the eye-cup altogether and had a ray of daylight reflect off the cornea of the eye itself, tracing a line onto a descending photographic film. Like Huey's, Dodge's machine was rather a gimcrack machine at first—"homely" was his own word for it—and was built on-site out of everyday objects. The cylinder that cushioned the fall of the film was a repurposed bicycle pump, for example. Later, he and others would refine the instrument. But for the time being, it worked well enough, and it obviated all the problems of mechanical levers (Fig. 2.2).[9]

By the time Dodge got his machine working reliably, he was back in the United States. He returned in 1897, bringing his instrumental expertise with him, and took up a teaching position at Ursinus College in Pennsylvania. But the job proved grueling, and as soon as he could he escaped to Wesleyan University, where he took over a philosophy program that was already strongly identified with the approaches of Wundt and James. That was thanks to Charles Hubbard Judd, who had graduated from Wesleyan in 1894, gone on to Leipzig, and returned to inaugurate a new laboratory and a research program devoted to vision and optical illusions. But Judd had just moved on to a professorship in experimental psychology at NYU, so Dodge was hired to replace him. He immediately launched himself into a sequence of laboratory courses that would continue for decades. These classes would in effect map out a program for the science of reading in America. The first was on the psychology of language; that was followed

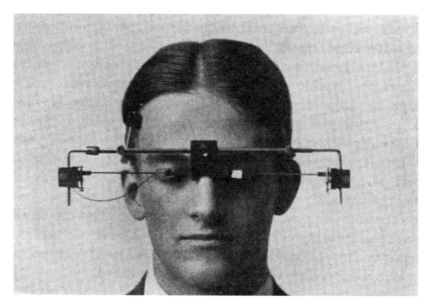

2.2 Raymond Dodge's device for revealing eye movements. R. Dodge, "A Mirror-Recorder for Photographing the Compensatory Movements of Closed Eyes," *Journal of Experimental Psychology* 4, no. 3 (June 1921): 172.

by courses devoted to eye movements, the visual perception of motion, the psychology of reading itself, and fatigue. As they proceeded, Dodge published a stream of papers based on the courses' laboratory experiences. The most prominent of them was an account of eye movements, in which Dodge revealed the details of his light-based eye-movement machine (figs. 2.3 and 2.4).[10]

As researchers like Dodge moved up the promotional ladder and relocated to more prestigious institutions, so tachistoscopes and eye-movement recorders moved fairly quickly through American academia. After Dodge introduced his photographic technique at Wesleyan, for example, he followed up at Yale in 1903 with the first binocular device; he then introduced it to a state nutrition lab in Boston. Meanwhile, Dearborn built his own instrument to Dodge's design at Columbia. He then took it with him to Madison, and on to Harvard. And so on: by a kind of hopscotch, instruments and makers moved from institution to institution across the country. By the mid-1930s, there would be eye-movement recording machines in use at (at least) Wesleyan, Yale, Columbia, the University of Wisconsin–Madison, Harvard, the University of Chicago, the

University of Texas at Austin, Cornell, the University of Kansas, Western Reserve (Cleveland), the University of Nebraska, Bryn Mawr, Oberlin, the University of Minnesota, the University of Iowa, the University of Southern California, and Lehigh, as well as at Cambridge University in Great Britain. And what was true of eye-tracking machines was true too of tachistoscopes. It is harder to trace their spread, because they were generally simpler and more multipurpose devices, and they attracted less fanfare. It is likely that all major experimental psychology laboratories had them. But it is telling that by the mid-1930s companies were beginning to compete to sell their own proprietary versions for analyzing readers. The American Optical Company's Metron-O-Scope, in particular—a tachistoscope designed to display parcels of text in sequence before an audience of students—was distributed to schools in the thousands before World War II. And a genre of subsidiary machines and publications arose alongside these devices, including research papers, to be sure, but also popular treatments, policy recommendations, teacher-training guides, and textbooks.

As these technologies stabilized, so their practical significance became manifest. For the first time in history, practitioners could claim to

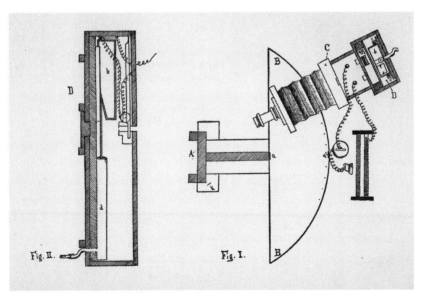

2.3 Raymond Dodge's early eye-movement camera apparatus. R. Dodge and T. S. Cline, "The Angle Velocity of Eye Movements," *Psychological Review* 8, no. 2 (March 1901): 150.

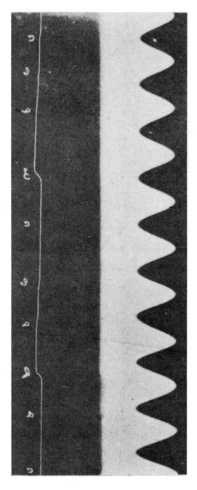

2.4 An early trace of eye movements produced by Dodge. R. Dodge and T. S. Cline, "The Angle Velocity of Eye Movements," *Psychological Review* 8, no. 2 (March 1901): 145–57, plate 1.

apply scientific instruments to a human in the very act of reading and arrive at a precise, accurate, meaningful, and consequential account of how well they were doing it. The vertically oriented film strips on which eye-movement recorders traced their linear records of the saccades and fixations of readers could be viewed to arrive at accounts of the fluency, efficiency, fatigue, and disabilities of readers. For example, a reader might make more pauses than average, backtrack during a line, or pause every time he or she moved to a new line because of losing track. Or the two eyes might not move in sync with each other. In cases of dyslexia, too, these traces could provide a firm diagnosis—or so Dearborn argued, but not everyone agreed, and as we shall see in chapter 3 dyslexia was such a complex condition to deal with that not all scientists of reading were sure that it even existed. At any rate, each trace was full of information to those who knew how to interpret it, providing evidence of talents unrealized or problems undiagnosed. And, importantly, to its champions the machine implied that the conditions it revealed were *correctable*. If poor reading was a matter of undisciplined eye movement, then retraining the eye to move "better" would improve a reader's skill. Eye-movement recording thus led directly to interventions, be they pedagogical or ophthalmic.

Those interventions were acknowledged to be sorely needed, thanks again to the distinctive demands of modern life. This was a point that Dodge himself made explicitly and repeatedly, and which he reiterated in public forums. A December 1901 article in *Harper's* on "the act of vision," for example, offered an excellent opportunity. Dodge insisted here that the magazine's middlebrow readers had to come to terms with this vital new experimental science, "in these days of rapid transit and weakened eye muscles," because evolution alone had not equipped human beings for the twentieth-century sensory environment. Trying to view an external object while one was riding in a streetcar or train, for example, was unnatural and extremely fatiguing, to the extent that doing so continuously would be "ruinous to the delicate muscles of the eyes." A modern researcher *knew* this, he said, with scientific confidence. Eye-tracking machines revealed it clearly. Anyone valuing their "general vitality" should therefore keep their vision focused strictly within the carriage. Reading a book might be acceptable if the ride were smooth and the type clear; a newspaper with heavy type was preferable. But one must take care to close one's eyes at every jolt. On the other hand, looking out of the windows at the passing billboards and shop advertisements was at best reckless. Seats that provided a view out of the side of a streetcar were "menaces to the public health." They were surely destined to be outlawed, Dodge advised, and "the sooner the better!"[11]

Physiological Aesthetics, the Mind, and the Nature of Experiment

The eye-movement instruments invented at the end of the nineteenth century by Dodge and his contemporaries generally shared one characteristic that reflected their purpose as reading-science machines: they only recorded horizontal motions. Except within very narrow limits, these devices were not designed to record any motions of the eye above or below a line of type. But for one champion of the new science this was a serious deficiency. Charles Judd heard about the new instruments in the late 1890s and was immediately intrigued by them. However, Judd's major interest at that time was not reading, but the viewing of images. In repurposing the instrument to serve that interest, he created a device that inspired profound reflections on the recursive nature of experimental knowledge and human subjectivity.

Judd (fig. 2.5) was a struggling junior academic in his late twenties when he took up the science of reading.[12] He had been born in 1873 in British India, the son of two Methodist missionaries who had been living near Lucknow. Both parents were sickly, and they returned to America in 1879. He attended high school in Binghamton, New York, where he found himself fascinated by books about evolution, anthropology, and biology, which ran counter to his parents' strict biblical literalism. When his father and mother both died, he abandoned their hope that he would become a minister. Instead he matriculated at Wesleyan to pursue the sciences. There he encountered the charismatic psychology professor Andrew Campbell Armstrong—a decisive influence on Dearborn and Thorndike too—and fell under his spell. Wesleyan had no psychology laboratory, but Armstrong took Judd to the physics building so that they could try out experiments. He also brought his young protégé with him to the American Psychological Association's annual meeting at Columbia University in 1893, where Judd got to see Cattell's psychology laboratory and set eyes on a "galaxy" of intellectual stars including William James, Dewey, Munsterberg, and Cattell himself. It was also Armstrong who first proposed to him that he investigate the topic of visualization, and they collaborated on a paper on the subject—Judd's first piece of scientific writing—in Cattell's *Psychological Review* in 1894.[13] By the end of his college time, Judd was regularly going to the laboratory at Clark University some eighty miles away to brush up his German and gain experience (partly as a subject) in the psychology laboratory there. He soon needed both, because on graduation he promptly set sail for Germany, heading for Leipzig and the laboratory of Wilhelm Wundt.

The reserved Judd did not take to German student life, with its rituals, hierarchies, and formal conviviality. He found himself fraternizing

2.5 Charles Hubbard Judd. University of Chicago Photographic Archive, apf1–03073, Hanna Holborn Gray Special Collections Research Center, University of Chicago Library.

mainly with a handful of Americans, and especially with a Californian named George Stratton, who was at roughly the same stage in his own career. The two agreed to serve as experimental subjects for each other's research. Still, there were advantages to being antisocial, and Judd made the most of his time. Work in the Leipzig laboratory was, he discovered, hard, rigorously time-disciplined, and ascetic. That reflected the persona of Wundt. As Judd later recalled, Wundt was "a prodigious worker," dedicated to a daunting daily routine of writing "ponderous tomes" (with the typewriter Cattell had given him), examining theses, and—often the same thing—editing authors' contributions to his journal. His community idolized him, and by the same token they cultivated a strong sense of allies and foes in the discipline at large. Workers still respected Cattell, for example, for his dedication, but they disdained William James, partly for having the temerity not to be a Leipzig graduate. The focus of the group was on space- and time-perception, so Judd followed suit. He raced through his doctoral work, completing a thesis on the perception of distances by touch in early 1896.[14] He then spent the rest of that spring meeting regularly with Wundt in person—something that otherwise could be a rare event. Judd had resolved on translating Wundt's work into English, and Wundt insisted that this could only be done if the text were kept under his close control and printed in Leipzig. He wanted to monitor every word through the press.[15] Judd found the experience draining, but it left him more impressed than ever by Wundt's robust empiricism, by the "historical method" of his later years, and by his conviction that higher mental processes could not be explained simply by "piling up great collections of lower mental processes." Unusually—and surprisingly, for someone who would become so identified with the grind of laboratory work—he came to see Wundt's work on speech and language as his most important achievement, because it made psychology a science of the social.

That summer, Judd sailed back to the United States. He returned to Wesleyan and for two years taught a variety of courses there, including one in experimental psychology. At the same time, he labored to write papers for publication, and made sure to attend American Psychological Association meetings. The efforts bore fruit in 1898 when he was appointed as a professor in the School of Pedagogy at New York University. It seemed like he had arrived. And once in Greenwich Village, the young scientist tried to devote himself to his calling. He soon made the humbling discovery (repeated by many since) that doctoral training had not equipped him to tell experienced teachers how to teach better, so he devoted months of mornings to visiting schools in a quest to find out what actually happened in them. At the same time, he spent afternoons teaching and experimenting, using phonographic and other devices to record traces of the processes involved in speech and handwriting. And he drafted a book on the subject, only for publishers to reject it as too arcane.[16] But before long everything soured. Judd had had the misfortune of arriving as a newcomer in an academic environment that turned out to be riven with discord. Cattell's Columbia was far from the only university to be seething with calls for internal revolt at this time; it was a tumultuous moment at NYU too, and Judd found himself caught up in a raucous campaign for reform. As he later remembered, he gave "intense and unremitting attention to university politics." It proved professionally disastrous. Lacking Cattell's reputation and clout, he had overreached. In mid-1900, only two years after arriving, he was forced to resign.

Initially at a loss, Judd was saved by friends who hastily arranged a position for him as a professor of psychology and pedagogy at the University of Cincinnati. The prospect necessitated more months of cramming, only to land him in another institution in chaos. He had to devote all his time there to what he later called "economic rescue work." But it was at least a base, and before long he got a chance to get away when an opportunity arose to move to Yale to teach introductory psychology. It meant a reduction in salary and a demotion, but he accepted forthwith so that he could escape the drudgery and institutional politics of Ohio. Not that Yale had escaped the furor afflicting other universities—its campus too was in a state of institutional civil war—but at least it had unparalleled resources for research. In New Haven, Judd would have not only a laboratory and a cohort of graduate students but also a dedicated workshop with a mechanic able to construct machines. And he had learned his lesson. He would keep his head below the parapet this time.

It was at some point during this time of anxiety and transience, punctuated by cramming, insecurity, and febrile politicking, that the science of reading piqued Judd's interest. That it should do so was surely unsurprising, given his laboratory expertise and his desperate need to find in that expertise something useful to tell teachers. But in fact the connection arose through an unexpected channel: his erstwhile collaborator at Leipzig, George Stratton. While Judd was flailing on the East Coast, Stratton had returned to California and founded an experimental psychology laboratory of his own at Berkeley. He was fast becoming famous, mainly for his idea of using inverting glasses to show that the mind was the dominant agent in visual perception and could adapt to inputs that were flipped upside down. But at the same time as undertaking these self-trials, Stratton also launched himself into what seemed at first a quite different subject. That subject was what his contemporaries called "physiological aesthetics"—a discipline that is now rather obscure, but that many at the time saw as portending what one British critic called an "invasion" of the realm of artistic experience by the dispiriting forces of science.[17] To repel that invasion, Stratton took the science of reading and applied it to visual interpretation in general. Judd would take it back again.

Physiological aesthetics was associated with philosophers like Herbert Spencer and George Santayana, as well as with a crowd of ambitious European aesthetes and self-proclaimed champions of a scientific avant-garde. Among these, Stratton's major foil was the Canadian evolutionary writer Grant Allen, whose *Physiological Aesthetics* (1877) had advanced particularly aggressive claims for the approach. Among other things, Allen, Santayana, and their allies claimed that the reason people experienced aesthetic pleasure when they perceived certain smooth shapes and lines was that their eye movements in tracing those lines were smooth and efficient. It was a matter, they said, of muscular economy: one used less energy and felt less fatigued. That certainly seemed to match subjective experience, Stratton conceded. It *did* feel like his eyes moved more freely over such lines. But were the eye's movements really like this? After all, fluent readers felt the same thing in perusing a line of Shakespeare, but people like Dodge and Huey were showing that their eyes actually moved in a jerky, staccato fashion. As soon as he heard of Dodge's use of photographic film to record such movements, in fact, Stratton realized he could make an instrument to prove his point. Dodge's device only tracked horizontal motions, of course; but Stratton had an ideal helper close at hand to remedy that. This helper was a colleague in engineering, the photographic pioneer—he produced some of the earliest X-ray images—and intrepid explorer Joseph "Little Joe" LeConte. Working together, Stratton and LeConte soon succeeded in modifying an eye-movement machine to record in two dimensions. Stratton then persuaded friends to sit in front of the machine and try to trace with their eyes some simple shapes that he had drawn on paper—a circle, a rectangle, and a long, elegant S-shaped curve reminiscent of William Hogarth's "line of beauty." He hoped to show that artistic appreciation and reading had one critically important thing in common: each was not what people felt it to be.

The images that Stratton ended up publishing achieved exactly that. They showed very clearly that when directed to follow a smooth line, the eye's path was not smooth at all. True, his pictures did not reproduce the eye movements themselves, as he was careful to note. The original photographs were too small and faint to be published, and issues of distortion had to be reckoned with too. Besides, they were composite still images showing where the eye had moved over a given time span, not records of motion as such. But they were good enough to make his point. They did show "certain grosser features of the eye's action," he insisted, and those features clearly included halts and abrupt changes in direction. That was all he needed. The eye clearly stopped at points that might or might not be on the target line and traced between those points an at best crude and jagged approximation to the line. The eye's path was angular, uncertain, "tremulous." Even where it did by chance trace a "graceful" arc for a moment, that arc bore no relation to the one on the paper. In short, eye motions were "a libel on the figure we perceive."

Viewing an aesthetic design was indeed like reading a line of fine prose, then, but in its very indeterminacy. A design on paper was "a help and guide" to the eye rather than anything more decisive, and *attention* (which led to interpretation) must be distinct from *fixation*. The experience of smoothness and ease that viewers felt was consequently an "introspective illusion." In that light, it was a prime instance of a point that Cattell and his students constantly hammered home against their rivals, to the effect that introspection should not be accounted evidence. Moreover, Stratton's pictures further showed that the eye did a better job tracing straight lines than curves. Judging by "mere *ease* of ocular movement," then, people should prefer straight lines and rectangles to curves and circles; but no art critic believed that straight lines were more beautiful. In short, Stratton concluded, the muscular activity of eye motions when encountering a beautiful image was no more connected to aesthetic experience than the leg-muscles were to the feeling of pleasure one experienced when walking through the main art gallery in Dresden— not a chance reference, incidentally, as that gallery was the home of a particularly fine Raphael that American tourists were exhorted to visit more than any other painting in Europe (and Cattell had in fact made the trip himself). The eye muscles merely placed the eye in various positions from which it took what Stratton called, in a very early use of the term, "snapshots." The brain then constructed the aesthetic object from those snapshots. In that light, the experience of aesthetics confirmed what his inversion lenses had already told Stratton. Just as the mind governed perception itself (as Cattell was contending at just this time), so it could and would countermand even the strongest muscular sensations of the eyes.

"The aesthetic object is not furnished ready-made by the sense organs," he concluded. "It is a spiritual creation."

Stratton published this first foray in 1902—it appeared in the same two-volume Festschrift for Wundt in which Cattell published some of his last research. A couple of years later he returned to the topic to add that optical illusions, too, could not be explained in terms of eye movements. And the same was even true of symmetry, which was often reckoned a key element in aesthetic experience. This last case mattered because a follower of physiological aesthetics would have claimed that symmetry produced a calm, restful moment of "equipoise," with the eye balanced motionless between the symmetrical elements. Shown an image of a fine Greek vase, however—almost the epitome of aesthetic symmetry— Stratton found that an observer's eyes were not stationary at all. They wandered all over the picture. They tended to alight on symmetrical elements only when the viewer was in doubt about whether they were in fact truly symmetrical after all—a moment, then, not of equipoise, but of "dissatisfaction and unrest." Indeed, Stratton remarked, "one is struck by the almost grotesque unlikeness between the outline observed and the action of the eye in observing it." This hardly meant that one had to abandon all idea of relating aesthetics to physiology, and Stratton's own instinct was to agree with the otherwise unfavored theory of British couple Violet Paget ("Vernon Lee") and Kit Anstruther-Thomson that the solution lay in the larger organs being stimulated into sympathetic motions based on remembered experiences.[18] But his conclusion was nonetheless damning, and to a degree rarely encountered in the rarefied rhetoric of scientific prose. "The mythology of the eye-muscles," he pronounced, "will some day make an interesting paragraph in the history of delusions in psychology."

It was Stratton's work that led Judd to adopt eye-movement machines himself in earnest for the first time. He had already been thinking about optical illusions—they were a standard issue in psychology at the time— and he immediately recognized this work as a major advance, with profound implications for "the moral and religious purposes of life."[19] Stratton and Dodge together had "opened up the field," he declared. But Judd immediately realized that the two programs had complementary problems. Dodge concentrated on horizontal movements, which he recorded on a plate falling vertically to produce a photographic record of motions over time. Stratton exposed an entire photographic plate at once in an otherwise dark room, with the aim of getting a single trace of movements in two dimensions. There was no way to use Dodge's device to record in two dimensions, and no way to use Stratton's to track changes in position over time. Time and space: what Judd needed was a single machine that would combine both.

He found the solution in a radical new media technology of the time that existed precisely to meld time and space. Moving-picture devices were all the rage in the years around 1900. They existed in various forms, some of which would soon evolve into early versions of cinema. Judd gained access to one of these machines in the form of the Edison Company's "Kinetoscope." This was a wooden box with a lens eyepiece on top, inside of which a fifty-foot-long reel of 35 mm film moved across a sequence of rollers. As it did, the film displayed moving images to a single viewer peering in through the eyepiece. Judd took two Kinetoscopic camera machines to the Yale laboratory workshop and had a new kind of eye-movement instrument made. Rather than a single plate, each camera produced a series of photographs taken in rapid succession on a continuous filmstrip. Judd added a kymograph to record a timeline on the edge of the strip, and timed the two cameras so that their frame exposures would overlap. Then, instead of trying to track a reflection on the eye's surface—which Stratton had found a major source of distortion—he hit upon the technique of applying a dot of paraffin-based "Chinese white" (a popular pigment) to the cornea. This produced a much more accurate trace, and it did not require any of Huey's cocaine. Readers simply sat in a specially designed chair, rested their chins on a bolster, and fixed their heads in position by clamping their jaws down on a mold in the manner pioneered by Dodge. Judd then added a pair of spectacles with polished steel bearings on the rims to provide points of reference on the photographs, so that any minor head movements could be eliminated. (Fig. 2.6 shows Judd himself within the contraption.)

Once a volunteer had read a text in Judd's machine and a film had been exposed, Judd and his researchers would take the film and place it in a projector. They would cast the successive movie frames one by one onto a drawing-board hanging from two ropes attached via ceiling pulleys to counterweights. A sheet of paper was tacked to the board, and they marked the images of the spectacle beads on it to form points of

2.6 Charles Judd tries out his apparatus for experimenting with readers in 1904. C. H. Judd, C. N. McAllister, and W. M. Steele, "Introduction to a Series of Studies of Eye Movements by Means of Kinetoscopic Photographs," *Psychological Review Monograph Supplements* 7 (1905): 1–16, plate 1.

reference. Then they would chart and number the Chinese-white dots, advancing from frame to frame and adjusting the board slightly each time to match up the reference points. What emerged after this laborious process was momentous. Judd had finally realized the long-standing ambition of the scientists of reading: he had obtained eye-movement photographs, as he put it, "without any retouching at any stage of the process." The entire record was thus freed from human distortions. Here, at last, was a truly objective record of that fleeting and capricious phenomenon, the reading eye in action. When reduced and reproduced for publication, his procedure had the disheartening result of making the labels so small that one needed a magnifying glass to read them. But it was worth it.[20]

Judd considered following Stratton into the debate on aesthetics. But he decided against that, instead returning to his previous preoccupation with optical illusions. In a series of trials, he subjected students and colleagues—and, in the manner of experimentalists at the time, himself—to his new instrument while they puzzled over designs like the famous Müller-Lyon figure. Judd asked viewers to move their eyes across the image while experiencing the illusion. He found that the relationship between eye movements and an illusion varied: for some images it seemed that the movements were important, for others not. That was puzzling. But here the prior history of his instrument suggested a solution. It was possible, he suggested, that the inconsistency might be due to the ingrained habit of reading. Maybe people were so habituated to reading from left to right that this tendency overrode the guidance of the image. In general, it seemed that illusions, like Stratton's aesthetic experiences, should be explained more in terms of the mind than in terms of eye movements. Yet Judd also found that over repeated viewings, as the subjective experience of an illusion gradually disappeared, so eye movements *did* become predictable.[21] He found that he could even draw a "curve of practice" representing the decreasing efficacy of an illusion. So a "habit" of looking at an illusion did come to be formed. And as the illusory effect vanished, Judd concluded, "one really sees the figure in a new way."[22]

Judd extrapolated from this radical conclusion a further point—one that related reading to learning, and hence to the very nature of knowledge. John Dewey had criticized the simplistic distinctions that early psychologists tended to draw between stimulus, idea, and act: in reality, he had argued, all three were coordinated in an "equilibrium of consciousness." Judd's instrument implied that a very similar conclusion should be drawn from eye movements. As much as those movements reflected whatever was being looked at, they were also "in no small sense" responses to the experimenter's instructions. Every figure drawn on paper served as a mere "guide" for a set of motions that were equally initiated, and quite possibly defined, *by the experimenter*. Indeed, Judd found that some of his volunteers did not move their eyes at all until instructed explicitly to do so. The insight threatened not only his own experiments but the entire science of reading as it was then conducted. To an extent, it reinforced the point that the key to both overcoming illusions and reading fluently was the formation of habits; that was why training was so important. But it also suggested another question. Did it make a difference if learner-readers *knew* what habits they should be forming? Perhaps training itself was recursive. There might well be a distinction to be drawn between "recognized" and "unrecognized" habits, such that explicit, critical instruction might work better than training-by-doing. "Put in its pedagogical form," therefore, Judd's experiments on illusions became experiments on the nature of learning itself. By that token, the science of reading could be regarded as a program of experimentation on the very foundations of knowledge. He finally had something he could tell a schoolteacher.[23]

So it was that Judd spent much of his time at Yale doing experimental research on reading and learning. His first book, *Genetic Psychology for Teachers*, used his experiences at NYU and Cincinnati in an attempt to show practicing teachers that learning to read was "a process of social inheritance," and that arithmetic and handwriting worked the same way. For years he worked closely with school authorities in New Haven and across Connecticut to evaluate and amend classroom practices. At the same time, regular technical reports of experimentation under his aegis appeared in the *Psychological Review* monographs series. One of their most notable achievements was the creation of a standard mode of graphical representation to record the locations of the eye's fixations on a line of text being read (fig. 2.7). Others including Dodge and Dearborn had earlier attempted to establish such a standard, but it was Judd's convention of using vertical lines to mark the fixation points that became the disciplinary default, presumably because so many of the twentieth century's leading researchers passed through his laboratory and learned it there.

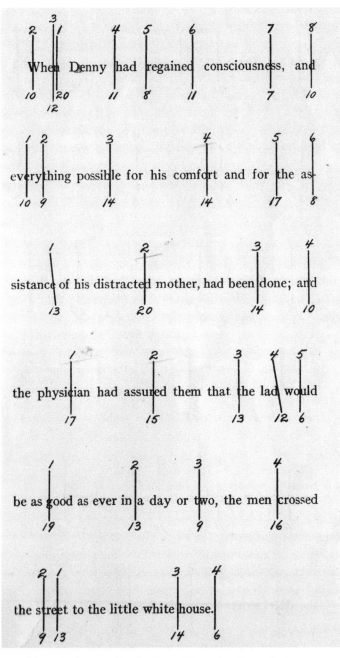

2.7 Judd's mode of representing fixations while reading. The reader here was a graduate student in Judd's unit. Each vertical line represents a fixation; the numbers above the text indicate the sequence of fixations, those below the text their durations. C. H. Judd, *Reading: Its Nature and Development*, Supplementary Educational Monographs II:4 (Chicago: University of Chicago Press, 1918), 17.

By 1907, Judd had achieved promotion to full professor and, it seemed, safety as an admired Yale authority. Yet about a year later he abruptly abandoned New Haven. He relinquished his position and moved halfway across the country to take up a position in charge of the School of Education at the upstart University of Chicago. He was ambitious to achieve more than was possible at Yale. Fine though its facilities were, at that time Yale remained largely an undergraduate institution—a finishing school for eastern elites, and a place from which students were sent to Germany, Oxbridge, and elsewhere to train as research scientists. For someone keen to establish a new scientific discipline—and a discipline, at that, with aspirations to reach every home, school, and college in the country and transform the lives of millions—that was not enough. Judd needed both research collaborators and an institutional pipeline of doctoral students whom he could train and disperse about the nation. It would help, too, to operate at a university linked to a school for children that had been created by John Dewey and was devoted explicitly to pedagogical experimentation. Chicago had all three, as well as a situation within what was then America's most dynamic and expansionary metropolis. Judd would have much less freedom to do his own research, but that was far less important in 1909 than it had been in 1900. What he wanted to do now was to scale up.

But what was this enterprise that he wished to establish? It so happened that Judd had a chance to answer that question. In 1909 he served as president of the American Psychological Association, and that December, six months after his move to Chicago, he stood up to address the gathered eminences of the discipline at the association's annual meeting in Boston. Judd took the opportunity to declare what he thought the field of psychology was and was not, and what it should become. The speech, entitled "Evolution and Consciousness," was notable for two reasons, given the nature of the science of reading as what might seem a pettifogging laboratory exercise. First, Judd insisted that *consciousness* not only mattered but must be the central topic of psychology. He mounted a frank and unsparing attack on the central principles of behaviorism, remarkable for being voiced almost before behaviorism itself existed. And, second, he based this insistence on a thoroughgoing evolutionism, but one that offered no succor to social Darwinists. Consciousness, he maintained, had evolved out of protoplasmic irritability to become an organizing principle in the intelligent organism, rather like homeostasis. It had given rise to a second world of ideas and representations expressed in language and writing, by virtue of which the organism no longer had to adapt to an unforgiving environment but could master and tame that environment for its own ends. Thanks to reading and writing, humanity could undertake "a reorganization of the environment in conformity to its own imaginings." The real environment was now an environment of words, and it was in this "world of words," not in some presocial nature, that humans lived, acted, and died. There had arisen "a new realm of fact." The entire history of technology, science, and art amounted to the emergence of this reality, such that "writing and coins and bills of exchange" had become the shaping principles of human nature itself. Judd even confessed that he sometimes sympathized with anti-Darwinists who separated humanity from the rest of Creation. There was perhaps something of the old Methodist scion in that, no less than in Judd's conclusion that civilization was now a matter of "moral and intellectual propriety" rather than animal instincts. Psychology must therefore be at the core of modernity, as the central social science. And its essential task, Judd declared, must be that of explaining how consciousness related to this new realm of words via the motions of the body. That work would now be done in Chicago.[24]

Edmund Huey and the Evolution of Reading

For the moment, we shall leave Judd there, en route between New Haven and Chicago, full of plans to shape his chosen science and use it to revolutionize the education of America's masses. This is a good moment to pause and take stock, considering the state of the science of reading at the moment when it was about to take on that grand mission.

As we have seen, the experience of modernity was what made the science of reading so critically important. To live as an American citizen in the years around 1900–1920—especially if one resided in a city, as more and more did—meant being exposed to a relentless deluge of visual and other stimuli, and in particular a torrent of printed words of all types and sizes. Information came at you fast, thanks to the news and advertising industries, which in turn relied on the telegraph and telephone. Early cinema added to the sensorial battering. Citizens themselves moved fast, too, and Dodge's warnings about reading on streetcars reflected fears about the long-term effects of such speed. Just as Frederick Winslow Taylor created "scientific management" techniques to try to update the practices of work for this technological world, and psychologists designed testing systems to measure people's intellectual fitness for it, so the scientists of reading wanted to upgrade the most essential modern skill of all. We can see what this meant—we can get an idea of the deep convictions underlying the endeavor, and grasp the extent of the practices involved—by looking at Edmund Burke Huey's *The Psychology and Pedagogy of Reading*. Published by Macmillan in 1908, the book reflected researches inspired by both Cattell and G. Stanley Hall. It originated partly in a trio of papers issued in Hall's *American Journal of Psychology* in 1898–1901 and partly in an article on "hygienic requirements in the printing of books and papers" from Cattell's *Popular Science Monthly*. The two came together to become a powerful manifesto for the new science.[25]

Edmund Burke Huey (1870 or 1871–1913) came from a small western Pennsylvania township, where his father was a physician and part-time schoolmaster; after his mother died and his father moved to Illinois, he was raised by his grandparents. Like Cattell, he attended Lafayette College, and the two of them were acquaintances there. He then taught Latin for a while in Wilkes-Barre—an experience that may have influenced his later disdain for the educational role of classical languages—before deciding to begin doctoral research. His initial hope of working with Cattell at Columbia fell through when Cattell was unable to obtain a stipend for him, and Huey went to Clark University instead.[26] Hall was president of Clark at the time, and under him the college had become a hub of the new experimental psychology; one classmate of Huey's was Lewis Terman, later the champion of intelligence testing and eugenics. Hall had fostered an unusual intellectual environment for such students, intense but untrammeled by disciplinary formalities. They were encouraged to pursue projects that might not be conventional, just as long as they had the potential for real consequences. He also insisted on what he called "elbow learning"—students and teachers working closely alongside each other—and had his group air arguments at a regular Monday salon at his home.[27] But rather than Hall it was the leader of the psychology laboratory, Edmund Sanford, who became Huey's immediate adviser. Sanford had already published on the perception of small typefaces, and he encouraged Huey to pursue the newly exciting field of reading. Sanford also volunteered to become one of the readers subjected to Huey's eye-movement instrument. Huey received a PhD for that work in 1899, and it was soon published in Hall's *American Journal of Psychology* as the material on which his book would be based a decade later. In the meantime, for a couple of years he taught psychology at a "normal school" (the equivalent of a teacher-training college today) in Minnesota, before heading for Europe, where he visited both the aged Javal and Erdmann. After a spell at Miami University and a turn back at Clark as assistant to Sanford, in 1904 Huey finally got the chance to inaugurate an experimental laboratory of his own at the Western University of Pennsylvania, the predecessor to today's University of Pittsburgh. He was expected to start a school of education too, though, and the workload proved debilitating. Huey found it draining enough that he had to take time off to recuperate.[28] But it was also during his period there that he published *The Psychology and Pedagogy of Reading*.

Huey's book exemplified how the science of reading twinned the most minute and pedantic of laboratory practices with the grandest cultural ambitions. To provide a full account of "what we do when we read," he declared, would be "the acme of a psychologist's achievements." It would mean that he had succeeded not only in describing the workings of the mind, but in telling "the tangled story of the most remarkable specific performance that civilization has learned in all its history." His book therefore presented not only an account of the current science of reading in close detail, but a sweeping history of the practice of reading itself extending back to the origins of humanity. The point was to advance a theory that explained the one in terms of the other, with consequences that would extend to the formation of every modern American.

The key to that theory was the principle of evolution. That every aspect of nature, society, and culture was in constant flux, evolving through complex processes of competition and adaptation, was the premise of Huey's book. The problem that the science of reading existed to solve arose from the fact that defining features of modern life were out of kilter with the evolutionary history of humanity. Reading itself was an ancient invention, to be sure, but only recently had laws for compulsory education led to its becoming "the most striking and important artificial activity to which the human race has ever been moulded." A large part of the waking time of every citizen was taken up with it. "One is seldom out of sight of some sort of matter to be read," Huey pointed out, "and having formed the reading habit it has become second nature to read all that appears." Printed materials were ubiquitous and inescapable stimuli, each of which triggered the reflex of reading. In effect, print defined the environment of modern life. And yet, he warned, this increasingly inescapable habit was unnatural and "intensely artificial in many respects." The eye and the mind—"the most delicate products of evolution"—had evolved over long periods in conditions where nothing like reading could possibly exist. As a result, the central problem of modernity was that it constantly demanded that the human frame contort itself to a habit for which it was not prepared. That was why reading led to "fatigue," and even, all too often, to disability. Effects such as myopia and "nerve exhaustion" warned of "the danger of race degeneration." So it was vital to the well-being of both individual and race to understand exactly what happened to the brain, nerves, mind, and eye when one read. "What may be done to avoid or minimize the dangers that come with this most universal and most artificial of habits?"

Huey began with the phenomena of saccades and fixations. Until he began his work, he recalled, they had attracted little attention. But they were the gateway into understanding the act of reading. For example, repeated small movements of the eye were particularly fatiguing, and that might explain why children tended to bring a page up to the eye to read it. Doing so increased the angular sweep in any given saccade, and hence reduced fatigue. But it did so at the expense of increasing the child's risk of developing myopia. His own eye-movement experiments suggested that fast, fluent readers covered longer distances in each saccade, and therefore performed less "eye-work."[29] Furthermore, good readers fell into a "motor habit" of pausing at similar points in each line as they read down the page, which gave rise to a "regular, rhythmical movement" that again reduced fatigue. The speed with which one read depended largely on "the ease with which a regular, rhythmical movement can be established and maintained." Good, fast reading was, then, almost a matter of dancing with the eyes.

Huey recounted his own 1897–98 experiments on "the work of the eye" in detail. One vexed question was whether a reader perceived anything during saccades. Dodge's work implied not. It was well-established that impressions produced by a rotating drum would fuse together into a blur at a rate of roughly thirty to sixty per second, and even slow reading involved saccades that exceeded this rate. Huey suggested that one could confirm this by moving a piece of printed paper rapidly from side to side and seeing how the letters blurred together, although he noted Cattell's claim that a reader did not experience "gray bands" in normal reading. Huey himself was inclined to reserve judgment, particularly because he felt that in practice reading involved bringing to bear a large repertory of prior experiences, which allowed textual sequences to be recognized even if not precisely "read" letter by letter. But in any case, it might not matter. The perceptions made during pauses would be much stronger, and they would convey meaning—and everyone agreed that in general, the mind neglected "raw sensations" in favor of those that carried meaning. So as the eye moved across a line, a "strong memory image" of a preceding pause probably persisted into the next one, such that the two would combine to drown out any perception that might occur during the intervening motion. Sensations during motion were at most confirmatory, if they happened at all. It was a special case of the evolutionary habit of ignoring stimulations and sensations that had no meaning for the organism. The mind simply ignored fleeting sensations, just as it typically ignored, say, the pressure of clothing on the skin. So Dodge was right, but Cattell too might have had a point.[30]

But if the main perceptions occurred during the eye's pauses, how much could a reader take in during each "peep" of this kind? Less than people tended to suppose, Huey argued. Using tachistoscopes, he and others could show that a reader generally perceived a roughly circular range of two to three lines and two to three words around the fixation point, but beyond this only an occasional outstanding shape (a capital letter, say, or a word with an unusual form) could be made out. That was what Dodge and Erdmann had found, and it matched what Helmholtz himself had seen a generation earlier when he had asked a subject to identify letters lit instantaneously by an electric spark.[31] Huey himself employed a tachistoscope set at an exposure of a sixty-sixth of a second to match the duration of a fixation. Cattell and others had experimented with even shorter times, however, and had found, oddly but intriguingly, that they produced better perceptions. At a hundredth of a second, "more could be read, or the same amount could be read more easily, than when the exposure was longer." Perhaps afterimages explained this curious fact.

What all this implied to Huey was that an "apperceptive filling in" was occurring all the time as one's eyes moved across the page. The reader was constantly making inferences from clues they picked up, and this inferential process was at least as important as the straightforward sensations produced by optical vision. Huey suggested that a habit of *predicting* forms was critically important here: at each fixation, the words to the right of the area of perception, although only dimly seen, were "helped into consciousness and preserved in memory by associative links from those that are clearly seen." A reader constantly "filled in mentally" what was coming next from what could be clearly perceived. That meant that practical reading was a matter of memory as well as of perception. It was like recognizing a friend from a glimpse of his hat, Huey said. In principle, he added, readers' eyes could take in as much as half a line of text in one go. In practice they paused more often than that, with the number of fixations per line varying greatly with the kind of text being read. A lot therefore depended on familiarity and on how much "rhythmic swing" a text had. And this kind of inference comported well with the findings of other scientists of reading. Recall, for example, Cattell's claim that we read in units of words, phrases, or sentences, not individual letters; he had found that his was true even to the extent that a reader would concoct a plausible sentence from traces received in a flash and be convinced that this sentence had been read. Dodge, too, had found that a reader would recognize a word even when the individual letters were too small to be individually identified (Huey was a little skeptical of this).[32] Although the science of reading seemed to focus so much on physiological measurements, then, what it showed was that reading was *not* a matter of mere vision. It was a matter of remembered, predicted, and inferred *meanings*.

That point applied at a deeper level than the mere recognition of characters. As by now should be familiar, Huey and his peers felt that in normal circumstances we did not really read characters at all—that "reading may and must go on by other means than the recognition of letters." So everything depended on the associative connections that a reader would bring to the page by virtue of previous experiences.[33] Over time, practiced readers developed a "stock" of ideas that served to cue recognitions when they encountered certain signs, and this process involved all kinds of short-cuts that could not be rendered scientifically visible. So, for example, one might immediately recognize the word *reading* if printed horizontally, but not if printed vertically—something that was rather baffling if one assumed that reading proceeded on a character-by-character basis, but not if it were a matter of recognizing patterns. The implications extended to "a very important section of general psychology," Huey declared, which "must here be worked out" afresh. An account of learning itself must be "constructed in the new."

The central question that the science of reading posed was therefore this: how did we perceive—how did we come to know—*anything*? That "stupendous problem," Huey affirmed, "rises at every turn."[34] What he insisted upon—the central, fundamental argument of the whole laboratory science of reading, at least on his account—was that perception was never merely reception. "Perceiving is an *act*, a thing we *do*, always and everywhere." It must never, under any circumstances, be accounted "a mere passive sensing of a group of passing sensations or impressions." It was an act, moreover, of both body (muscles, nerves) and mind. That meant, again, that an act of reading was not only dependent on experience, but also always partly a *projection* from within. As one's eyes moved across a line of text, any given printed word was always liable to be "preperceived" on the basis of an existent "'set' or 'predisposition,'" and real reading was more a matter of preperception than of conscious perception. This was why practice was so important. The more experienced a reader became, the greater his or her repertory, and the faster and more efficient the preperceptions brought to bear. Moreover, with practice one could concatenate these discrete acts into automatic habits, making them into one automated apperception, which would occur much more rapidly. To read a new word for the first time was a slow and fallible process, to be sure; it was like learning a new way of serving in tennis. But repetition freed the mind from having to attend to all the details.[35] As with learning a tennis stroke, by practice the mind could "unitize" discrete steps into larger wholes, and then it could attend to these wholes rather than to the individual traces. And this seemed to be what readers were doing when texts were flashed up in front of them via tachistoscopes. They recognized words, phrases, and even multiple sentences from their earlier experiences. In short, when we read, our mind "projected" words onto the page, "somewhat as a lantern might throw them outward upon a screen." Indeed, Huey added, "it might be said that the mind furnishes the screen as well."

So it was that Huey came to what he considered an especially important aspect of the laboratory science of reading, "of the greatest importance practically and pedagogically." This was the use of its instruments to measure the *rate* of reading. It was well known that readers read at very different speeds, and that faster ones tended to understand what they were reading at least as well as slower ones. There was no real correlation between intellectual ability as such and reading speed—Huey could quote experiments on "several highly distinguished men in science and literature" to that effect, and he had a mathematician friend who could read a 320-page novel in just over two hours. He insisted that reading rates could indeed be substantially improved in general. But the rhythms that readers adopted tended to be established when they were first learning to read, so improvement would require a program of measuring the reading rates of schoolchildren systematically, across different grades and for different kinds of texts. If this were done—and acted upon—then it could have "the greatest value in lessening waste and in increasing effectiveness both in reading and thinking."[36] In other words, society must now rise to the challenge of accelerating reading or risk disaster, and the science of reading offered the key.

But why had the world found itself in this predicament, needing to accelerate a population's reading as such a matter of urgency? To answer that question, Huey left the laboratory behind and ventured onto the terrain of history. He provided a long narrative of the development of reading since the paleolithic era—perhaps the first such comprehensive history ever written, and one not matched for ambition until the emergence of a community of historians devoted to this subject in the late twentieth century. His point was that both the distant origins of the practice and the progress that it had seen over centuries had immediate implications for the present. This was so in two senses, both of them essentially grounded in evolutionary explanations. In the first place, Huey maintained that the development of both reading techniques and the forms and layouts of books had been shaped by an endless evolutionary process, such that their current forms included vestigial elements that could and should be jettisoned. In the second place, however, and complicating that notion, Huey was a convinced advocate of the principle then known as *recapitulation*, according to which the evolutionary development of a species was reiterated in the growth of an individual in the present. For example, embryologists might argue that the human embryo recapitulated the evolution of humanity during its growth in the womb. The idea was associated primarily with German zoologist Ernst Haeckel, but it was more loosely applied in fields such as anthropology, and it was also prominent in the literary works of authors like Émile Zola. Although it was often identified as Darwinian, in fact there were important distinctions between recapitulation theories of this kind and Darwinism sensu stricto, not least in the sense that recapitulation tended to suppose a more teleological, progressive course, and one in which acquired characteristics could be inherited. And Huey's own immediate inspiration for adopting recapitulation theory was not Haeckel, but almost certainly G. Stanley Hall.

In the 1890s, Hall's own career in academia had rather stalled, and he had seen the core faculty of his prized department at Clark University poached by the University of Chicago. He had turned instead to promoting the "child study movement," a program for gathering massive amounts of questionnaire data on children around the country. It was explicitly a recapitulationist project. Hall himself described it as "partly psychology, partly anthropology, partly medico-hygiene," and it embraced accounts of "primitive peoples" and animals alongside those of children. A central tenet was the belief that an educational regime should be designed to encourage the expression of such recapitulation in the growing child. Huey's own first work in the science of reading had taken place at the peak of this movement in 1898–99, and it retained that Romantic flavor even though Huey's book appeared only years later, when Hall's project was on its last legs.[37] That may be why he did not choose to emphasize the connection, instead invoking the work of the first professor of anthropology at Oxford, Edward Tylor. Like Tylor, Huey believed that he could discern a uniform line of development in human culture, traces of which could be seen in the particular histories of, say, the Maya, the Egyptians, or the Ojibwe. Reading and writing in early prehistory must therefore show "much the same course of development that has been observed among later peoples and that is going on to-day among savage races." This same sequence was then reiterated once more in every child who learned to read in modern America. "Mankind began his reading with picture-books and his writing with picture-making," Huey claimed—"just as the child likes to begin." Child and race alike began with a language of bodily gestures, which then gave rise to drawings, and these were much the same the world over. And here, it is worth noting, was an excellent reason for the scientist of reading to seek out children to study: the reason was precisely *not* to see what a child's reading was in itself, as a distinct practice worth addressing in its own right, but to use it as a key to understanding the evolution of civilization. Children, he asserted, tended to "live over again the use that the race has made of gestures." Child study was therefore a form of historiography. From a combination of child research and archaeology, one could come to understand the gradual progress of human communications practices, from pictures, through pictographs and ideographs, to alphabets proper. Puzzles found in today's children's books thus made manifest the ancient transition between picture-writing and words, "the highest intellectual effort of one period in our history coming down, as so often happens, to be the child's play of a later time." The historical process, like the personal, must involve the development of skills for good reading: good memories for symbolic meaning, for example, and a "clear grasp of association."[38]

Huey then turned his evolutionary eye to the page that a reader looked at. The book itself—the codex—had arrived early in the Common Era, as had the division of the page into columns. Paper followed. For the standardization of letters, he delved back into antiquity again to provide examples of Roman capitals, cursive, and uncial scripts, but he argued that it had in fact been the first printers who had had the greatest impact on letter forms. In Germany, printers after Gutenberg chose to adopt a Gothic script, "and German readers still suffer the consequences." But the English took a "beautiful running script" from Italy, giving rise to a set of symbols easier to read than any other. And so, Huey said, "by the slow processes of evolution, through variation and selection, the characters used in reading have developed through the ages."

Huey found it remarkable how little role "conscious purpose" had played in this historical process. Modern letters had been an unplanned "growth," he felt, and like an evolved organism they had retained "from remote antiquity" various "useless but interesting marks of their origin." And the same held for the printed page too, the layout of which had started as a haphazard matter of "protoplasmic free arrangement" akin to children's cartoons but become more formal and organized over time. This language of protoplasm was not adopted casually; it reflected contemporary understandings of the substrate of evolutionary processes that had arisen in the very early history of the earth and continued to fill living cells in all organisms. "The protoplasm of total meanings expressed in sentence-wholes," Huey continued, had similarly evolved into a complex array of discrete characters and sound elements.[39] The division into words and paragraphs, too, had originated in the Middle Ages, with punctuation marks arising in early modernity. Varieties proliferated. "And so, step by step," he concluded, "was evolved the modern book."

The history Huey recounted, then, was one of at least three evolutionary strands—of reading, of typography, and of the physical book—all proceeding inextricably from each other. And all were recapitulated in modern children and "savages," as could be made manifest through appropriate scientific investigations. Thanks to this triple helix, in modernity "reading and the reading habit" had become "practically universal, in all civilized countries." But at the same time the "tremendous modern development of reading" came with "disquieting" consequences. That point would loom increasingly large as he proceeded.[40] Because, of course, reading was still an unnatural skill. It had to be acquired, and the crux of the evolutionary processes he portrayed came at the moment of its acquisition. The point of application for Huey's science of reading, therefore, and for his history of reading too, was the school, and the elementary school in particular.

If reading itself had always been mysterious—"almost as good as a miracle"—then the processes by which one learned to do it were even more so. No wonder, then, that premodern societies had accorded the skill such reverence that it virtually became "a holy office" performed by "individuals who possessed divine powers." The legacy of this sacralization was visible to Huey in every school in the country: as Dewey had said, reading was "the Fetich of the Primary Grades." The problem was that pedagogy too inherited ancient prejudices. What Huey called the "A B C method" had originated with the Greeks and Romans. It relied on getting the child to learn first the names of individual letters and then the ways of combining them into sounds and syllables. There were even "mechanical devices" to help with it.[41] But alternatives had long existed, based on sounds or on pictures, and in the nineteenth century different methods had proliferated with the push to create a literate society. Samuel Worcester's *Primer* in 1828 had put the learning of words before that of individual letters for the first time. Others followed, until by 1870 the "word-method" was being adopted by "progressive teachers" across the country. But at the same time phonic and phonetic approaches had multiplied too, and various new kinds of symbol—such as a much-ballyhooed "Scientific Alphabet" along the sort of lines that Cattell had favored—had attempted to capture the sounds that conventional spelling ignored. Yet another approach, the "sentence method," had begun to be adopted around 1870 and become fairly widespread by 1890. What was at issue between these various "methods" was, in part, the very familiar one to researchers like him of the proper unit of reading: was it to be the letter, the word, or the sentence? After examining more than a hundred different manuals and systems, Huey concluded that a word method was most widely employed, but that no one approach really predominated. The situation across the country's schools and households was one of lamentable chaos.

So Huey undertook an investigation. Books advocating one or other method were being marketed in the tens of millions, he found. The *New England Primer* alone had sold more than three million copies, and had had "a profound influence on the moral and religious thought of the whole country."[42] Even that was surpassed by Webster's *Spelling Book*, which had first appeared as early as 1783 and had sold some forty-seven million copies by 1847. Since then, cheap volumes of excerpted texts, of which the most famous were the McGuffey Readers launched in 1850, had multiplied incessantly. Primers, spelling books, and readers—an entire industry now existed, continental in scale and replete with monopolizing aspirations. The books it generated were often pretty, but they were designed to appeal to the adults who purchased them, not the children who had to put them to use. And for those children, Huey warned, all too many of the books were miserable failures. "An old curiosity shop of absurd practices" now obtained, each clamoring for attention in what amounted to a dismally uninformed marketplace.

Huey's appraisal of the contents of these books was likewise almost unqualifiedly negative. He found them moralistic, tedious, and ill-suited to learners. Not only were these books bad; they were so bad that no child should ever be permitted to see them. They might have been designed specifically to spoil a child's chances of becoming a good, fast, and creative reader. "The most striking thing about at least three-fourths of them," he wrote acidly, "is the inanity and disjointedness of their reading content." They made no attempt at all to reflect what anyone would naturally say about anything, instead adopting a patronizing tone apparently designed to talk down to children. No wonder the young scorned them. The source of the problem, Huey thought, was that they were based on a false concept of what they were trying to teach. The common priority of these books was to impart, not reading as such—that is, not the creative, subtle art that his science revealed reading to be—but a quite distinct skill: the pronunciation of isolated words. They presupposed a culture of spoken enunciation, not silent reading. In consequence, they got the natural procedure of reading exactly wrong. They made the letter precede the word and the word precede the idea. They taught a crude mechanical skill, "at the expense of the child's formation of natural habits of reading, of using language generally, and of thinking." Instead of these stodgy primers, Huey proposed publishing a line of books designed to inspire children and, as he put it, "impregnate the soul with the race's highest ideals and tastes." He had in mind medieval sagas of a distinctly Wagnerian bent. For want of such a project, Huey maintained that even if one of the existing books ever did achieve its aim perfectly, it would not have "taught the child to *read*" at all. In fact, it would have been "most likely to permanently unfit him for intelligent, natural reading."[43]

The science of reading thus found itself in direct confrontation with one of Gilded Age America's most notorious commercial behemoths: the textbook industry. The chaos of approaches that Huey described was the product of a jungle of competing, and quite frequently combining, publisher conglomerates. It put these companies into the position of judges, deciding what should be taught, how, and at what cost. It was time, Huey declared, for science to have a say. Just what *did* a child do normally during reading? An objective answer to that question would be essential "in order to plan a natural and economical method of learning to read." Nobody had thought to try teaching "purely visual reading," for example, which would be faster and more accurate than a practice based on speech. Current schoolbooks' very formats militated against students developing that all-important rhythmical "motor habit" of the eyes, forcing them instead into a cautious and halting style that was hard to overcome in later life.[44] It should be possible to reduce fatigue and increase speed by redesigning the page itself, so as to create a format "that can be read with one-fourth of the eye-work required by the page of the present." The industry needed to be prodded into uniting around these goals. And above all, society had utterly failed to train people in the effective and creative use of books in the plural—Americans lacked the skill of using indexes, catalogs, and libraries as tools for thinking. Only with that skill could they exercise "selective reading" in a modern, autonomous, and efficient way. Huey was effectively calling for printers, publishers, and bibliographers all to embrace revolutionary change at once, and to do so on the basis of his science.

He did at least have two examples of success to point to. Huey contrasted his depressing account of the norm to the practices in place at two notably progressive institutions: the school attached to Teachers College at Columbia University, and the Institute adjacent to the University of Chicago (the institution later named the University of Chicago Laboratory Schools). Both were heavily influenced by John Dewey. In New York, young pupils were being encouraged to make their own books by connecting together stories from a popular work, and their attention was directed first and foremost to the thoughts being conveyed, at the level of the sentence rather than the word or letter. Phonetics were kept clearly distinct and used as a subsidiary tool, principally to help with enunciation. And an explicit aim was to educate children to make choices about *what* to read, rather than just teaching them *how* to read in the literal sense. In Chicago, meanwhile, children were learning to read by following their own desire to discover and tell something. The motor for learning was to be their own interest. They would do experiments, or work outside in a garden, and then report on what had happened, with the teacher writing sentences on a blackboard. These sentences would then be printed by older children, again generating a self-made book. (Huey did not say how they were printed, incidentally, and there seems to be no record nowadays of what the practice involved, but it is eminently possible that the school had a hand press.) The collection of these books then formed a library that evolved over time and could be consulted by future generations. "Thus reading and writing and drawing are learned in the service of what the children are doing as a social community," Huey enthused. "Reading is not made an end in itself, and does not gather the mannerisms and the débris of technique that accompany reading done for its own sake and by 'Reading's' own special methods." The result was that these third graders would eagerly read anything that was offered to them.[45]

For Huey, what these two Chicago and New York institutions did right was that they taught reading as a form of thinking. Instead of drilling children into a discipline of dronish anti-imagination, they fostered generations skilled in making meaning from pages. Schools in general did the opposite, he believed, with the aid of all those soulless primers. The results of the prevailing approaches were (at best) a superficial reverence for books as containers of culture and an epidemic of myopia and other "degenerative tendencies" associated with poor posture and "the overworking of the associative mechanism." And that was even before one reckoned with the sheer, debilitating *inefficiency* of it all. As a model modern American, Huey railed at the enormous waste of time and energy spent on the "mere mechanics" of spelling. Whole years were being devoted to this pointless and counterproductive labor! Current teaching generated "mechanical, stumbling, expressionless readers," and it separated "poor thought-getters from what is read." This, then, was where the country urgently needed the scientific revolution in reading to take effect.[46]

In the end, Huey ended up advocating that the early years of childhood should not be spent in learning to read at all. That would be better deferred until eight to ten years old, as Dewey believed. Instead, young children should spend their time acquiring the experiences and desires that would make reading a "natural" and "meaningful" acquirement in due course. Once again, it was a matter of evolution repeating itself: "a correspondence with, if not a direct recapitulation of, the life of the race." At an early age, a child had not yet developed the "logical and ideational habits" that reading required, "any more than had primitive man," so the child too should remain "in the oral stage." Huey reassured his own readers that all laboratory research confirmed that "reflection, reading, writing, reasoning, voluntary attention, etc., came late to the race and should not be hurried in the child." If a child really *had* to learn to read early, then it should be done at home, where it might be made a matter of natural curiosity about the typographical environment of a modern household. There the "natural method" could be followed of learning by imitation at the level of whole sentences.[47] Schools, with their artificial and counterproductive methods, should simply hold off.

Huey consequently made two radical proposals. The first was that books should be banished completely from primary and elementary schools. And the second was that teachers had to transform their concept of reading to embrace the centrality of creative autonomy. So, for example, they should accept that if a student faced with an unfamiliar passage grasped "approximately, the total meaning of the sentence," then that counted as having read it. "Even if the child substitutes words of his own for some that are on the page," he insisted, "provided that these express the meaning, it is an encouraging sign that the reading has been real." He acknowledged that this suggestion would cause "shock." But that was simply an indication of how strongly the "false idea" of reading had taken hold, according to which "to read is to say just what is on the page, instead of to *think*, each in his own way, the meaning that the page suggests." Indeed, Huey went further still, and suggested that for true reading to occur it was even *necessary* that the student should use words "somewhat variant" from those on the page. "Reading is always of the nature of translation," he pointed out, so "to be truthful, [it] must be free." The point was that real reading was "*thought-getting*."[48] Literal regurgitation of characters did not count. To the endlessly vexed question of how to teach reading, then, in the end Huey's answer was that we should not be teaching it at all. On the other hand, though, the identification of reading with thinking reinforced the point that when people *did* finally learn the skill, it was very important that they should read at high rates of speed. Reading should converge with imaginative thought. And this rapid reading-as-thinking must itself be pursued purposefully, with the reader being made ready to skim, trim, and digest. In a modern age, citizen readers must know how to find texts and cull information from them "without dissipating energy." Schools must therefore use the time they currently wasted on drills and primers to train their students in the creative use of bibliographies, catalogs, and indexes, and also in efficient note-taking practices. Only with these skills in hand could reading become "one of our most effective means of mental discipline." This was true efficiency. "The library is the reading laboratory," Huey noted—"and reading is a laboratory subject."[49]

The Hygiene of Reading

These reforms would surely help, but Huey cautioned that they were not sufficient. The fact remained that the eye, the mind, and the brain were all being called upon to perform tasks that were fundamentally unnatural.

In time, the human organism would surely evolve to cope with the complexities of modernity—evolution never ceased—but in the short term
readers risked "disaster." There was simply no natural precedent for the
kind of attention that a reading mind had to sustain as the eyes skittered
rapidly from sentence to sentence. "Not only must a really vast amount
of mental machinery function in the mental construction of the words
upon any page," Huey observed, "but a still larger number of words must
be sub-aroused, almost to the point of actual construction, as associative
expectancy points." Reading rapidly and smoothly certainly reduced fatigue, and the mind, like the body, could be kept in shape by a regime of
exercise and good diet. But nothing could eliminate the issue altogether.
"The general stock of nerve energy" simply had to be depleted by such
work. "We cannot get something for nothing," Huey pointed out, "even in
psychic economics." There was no such thing as a free book.[50]

It had been fatigue that had initially sparked Huey's interest in reading,
and so it was fitting that he ended with fatigue once again. And in doing
so he once more aligned the science of reading with a major fault line of
American culture at the time. Fatigue was the great problem of industrial
society, or so Americans were constantly told by experts in industrial efficiency. It emerged from inefficiency and it led to degeneration.[51] We
have seen already that early scientists of reading like Cattell and Dodge
were preoccupied with the concept, and in the rest of this book we shall
see that for the next half-century fatigue would continue to be a shibbo-
leth for their successors, even as the eugenic contentions of Cattell and
the evolutionary themes of Huey receded from explicit view. But it was
Huey, once again, who made the most authoritative statement of its importance and of the vital need to remedy it.

In the case of reading, the symptom of fatigue was disability. In particular, the "tremendous" rise in myopia witnessed by Huey's generation
was widely attributed to reading fatigue. Myopia was incurable, as Huey
warned; it could lead to blindness, and, he hinted, it might even be inherited by one's children—as might bad reading habits themselves.[52] But
although the problem was at root insoluble—it arose because reading
placed demands on the human organism that evolution had not equipped
it for—Huey indicated that such effects could be mitigated. The measures
needed to achieve that end were collectively known as "the hygiene of
reading." They included personal habits of diet, comportment, and the

like, but they also extended beyond the person to matters of furniture design, lighting, and even architecture. For example, seating should encourage sound posture. A reader's desktop must be at an appropriate angle and height. Lighting must be situated behind the reader and off to the side, and preferably powered by electricity because oil or gas would dry out the cornea. The time management of the workplace mattered too: spells of reading should always be interrupted by regular rest periods, during which it was a good idea to do some gymnastics.[53] And the hygiene of reading suggested a range of reforms to books, too. Scientists of reading had already spent a lot of time investigating the legibility of different typefaces and arguing interminably about font size, the thickness of strokes, spacing, leading, and other characteristics. Huey summarized them and added his own recommendations. Fatigue could be countered by optimizing the length of a line of print, for example.[54] In this respect the forces of evolution had clearly acted more powerfully on newspapers than on books, for the dailies had arrived at what eye-movement traces proved to be the near-ideal width of roughly 90 mm. Books should in future be printed in two uniformly justified columns to adhere to this ideal, with regular minor indentations on the left to help align the eye. Volumes should be small enough to be held in the hand for long periods for viewing at the optimum angle. Paper should be pure white, and matte—not the yellow that Javal had favored, and before him Charles Babbage too. Ink should be deep black. Huey decreed that schoolteachers across the country must examine their students' books for such issues with ruler and magnifying class, so that those with formats likely to cause injury could be "mercilessly discriminated against."

Huey was in good company when he made this case. Long before he even began his work, in 1884 the president of the American Public Health Association had declared dire reading conditions at schools to be a health problem akin to their awful sanitation. Children's "undeveloped plastic bodies" were, he declared, "distorted on uncomfortable seats, at uncomfortable desks," and "their eyesight [was] progressively deteriorated by glaring windows and poor type." In the 1900s, good reading pedagogy became as integral to the hugely influential school hygiene movement as cleanliness, well-ventilated classrooms, and physical exercise. Ophthalmologists and medical practitioners constantly warned of the physical and mental damage caused by the conjunction of poorly designed desks and small typefaces. Well-illustrated jeremiads on the topic were commonplace for decades. Such books included advertisements for the latest desk and "reading and writing slope" designs, which were also displayed at public exhibitions and publicized in the newspaper press. Schools bought and installed them by the tens of thousands, so that by the 1920s the physical habits of America's young readers, it may be inferred, had been radically changed.[55]

But Huey went much further. Expanding on the theme, he called for a revolution in typography itself. Society had been rationalizing every other aspect of life apace, from transport to communications, and it was past time for the most fundamental means by which modernity worked to be rationalized too. New fonts were "cryingly needed," for example, as a couple of decades of laboratory work by Cattell and others proved. A "total rearrangement of our printed symbols" must happen, and soon, "in the interest of economy of time, energy, and effectiveness in getting thought from the page." It was not yet clear what the new characters should look like. Maybe the various modest reforms advocated by the likes of Cattell would be enough. But Huey also raised the possibility of a truly radical shift, away from linear alphabetics altogether and toward a new kind of compositing idiom based on the principle of capturing whole ideas in a single word-sized pattern, which would presumably accord better with the psychophysical reality of reading. It would be rather like an efficient, modern, industrial counterpart to ancient Egyptian hieroglyphics. Such a system would reduce the fatigue of the eye by allowing for much more information to be taken in at a glance. Failing that, perhaps society might move to reading lines vertically, as in Japanese, which he thought would save "at least three-fourths of the fatiguing eye-movements."[56] And perhaps written texts would ultimately be superseded altogether by some kind of "graphophone-film book," with listening replacing reading at a massive saving of "energy."[57]

Exciting times, evidently. Huey recommended a systematic, collective scientific program—perhaps the Carnegie Institution might like to fund it—to resolve these questions, based not only on experiments in the science of reading, but also on the histories of typography, aesthetics, and culture. "When the proposed reforms are shown to mean definite savings of time, money, and health, and definite improvement in mental habits as well, practical sense will sooner or later see to it that they are duly installed," he affirmed. Thanks to this new science, "waste" would be radically reduced. A modern economy of reading could finally arise.[58]

Principle into Practice?

The Psychology and Pedagogy of Reading was certainly the most compelling work on the science of reading to have appeared to date. There would not be another work in the field to compare with it until at least the 1960s—when the competition would be represented by a newly issued reprint of the same book, seen now as a masterpiece of cognitive science. It not only summarized the research done by that time but supplied it with an authoritative evolutionary rationale and a practical agenda for future efforts. In effect, it cast the science of reading as the key to understanding the history, character, and future of human civilization. That is why it merits the kind of attention that I have given to it here. Every aspect of reading in modern America was now in line to be transformed on the basis of this sweepingly ambitious scientific program.

But Huey did not go to work to realize the agenda expressed in his book, at least in any evident sense. Instead he moved on immediately and dedicated himself to what looked like quite different interests. He resigned from Pittsburgh abruptly and left his academic career behind, instead devoting himself to the clinical psychology of mentally ill children. Huey first returned to France, where he worked for a year alongside the great psychologist of trauma and mental disorders, Pierre Janet. Then he took up a position as a clinical psychologist in Lincoln, Illinois, at the state's Asylum for Feeble-Minded Children—an institution later to become notorious for abuse. There he was placed in charge of the first psychology department in any American state institution of the kind. He seems to have acquitted himself well, but by 1911 he had moved on again. Now he was to be found working on clinical research at Johns Hopkins alongside the renowned mental hygienist Adolf Meyer. A second book devoted to "the psychology of defectives" emerged from these experiences in 1912, and a third would have appeared if the manuscript had not been destroyed in a fire. But Huey had long been in ill health, and around this time he contracted tuberculosis. A move to Washington State in search of a better climate did little good, and he died there in 1913, aged about forty-two.

In one sense, Huey did clearly move away from the science of reading after 1908. His successors in the field occasionally expressed a rather sad bafflement at his decision. But in another sense it may have been his only possible move. It is conceivable, perhaps, that the departure reflected precisely the degree of importance that that science held for him. He was already convinced that linguistic disorders were largely attributable to pedagogical malpractice, and in moving to engage with psychiatric patients themselves he was extending scientific practices to a domain where their promise could be made good. All that talk of hygiene and of the psychophysical effects of poor pedagogy had been meant in absolute seriousness. After his book appeared, perhaps, there was nothing left to say. What remained was to do.[59]

3

Reading, Looking, and Learning in Chicago

The appearance of Edmund Burke Huey's *Psychology and Pedagogy of Reading* in 1908 can be taken as one of two events occurring close together that marked the decisive emergence of the science of reading as a major enterprise. The other was the University of Chicago's hiring of Yale experimental psychologist Charles Hubbard Judd to head its Department of Education. Almost as soon as he arrived, Judd made an impact. A dedicated and rather stern figure, likened in a later tribute to an academic Caesar, he was exactly the kind of austerely self-disciplined scientist on whom the university would come to model its institutional identity. Judd's intense stare, barely mitigated by steel-rimmed spectacles, intimidated many a coworker and colleague. Reputedly, students who failed to live up to his standards were exiled so routinely that they ended up forming a society they designated the FBJ Club, the acronym standing for "fired by Judd." Under his rule, Chicago's School of Education became America's leading site for the science of reading. It was far from the only place where the science was done—other sites included Stanford, Minnesota, Iowa, and, most important of all, Columbia University, where Teachers College connected laboratory and school in a similar way. But Chicago would retain its prime position for decades.[1]

If Judd founded what may be called one "Chicago school" in the science of reading, however, it was not to be the only one at his university. Reading has always been a protean activity, and there were other ways to tackle it than by means of the intricate and time-consuming laboratory experiments over which he and his students labored. At the very moment when his approach was at the peak of its success—and when, as we shall see, it was set to generate possibly the most consequential set of publications to emerge from any of the human sciences in the twentieth century—a quite different approach took up residence at the university. In 1928 the Carnegie Corporation sponsored the creation of America's first graduate library school at Chicago. The new dean immediately sought out an expert in the social science of media for his new school, and found one in a colleague of Judd's named Douglas Waples. Waples accepted a promotion to leave Judd's department behind and become the new venture's "professor of educational method." He felt strongly that "library science," as he always made a point of calling it, should be a genuine social science, standing alongside the disciplines of sociology and anthropology in which Chicago already enjoyed global renown. His nascent discipline needed a secure methodological and statistical foundation, and to that extent he retained the predilections of the science of reading. But at the same time he wanted it to leave the laboratory behind and ask questions of the broader urban and international world—questions about the social and political ramifications of reading communities and their choices. Waples proved himself as energetic in pursuit of this vision as Judd was in pursuit of his, and by the mid-1930s he had cultivated a substantial canon of literature. In effect, he fostered a Chicago school of the *social* science of reading, to stand alongside Judd's Chicago school of the *laboratory* science of reading. This social science was strongly aligned with Roosevelt-era government initiatives in researching contemporary domestic and international problems, including those of race and authoritarianism. And it increasingly made common cause with investigations of the other communicative media that defined the age: radio, cinema, and early television. Together, the two enterprises highlighted urgent questions about the place of "media" in a democratic world that everyone now recognized was in mortal danger.

In short, it was in Chicago that the science of reading responded to the great crisis of modern democracy. From that response, as we shall see, emerged a concept of communication that would help shape the informational world in the next century.

Lab Work and School Work

Well before he moved to Chicago in 1909, Judd had orchestrated a so-phisticated and profound set of experimental researches that extended beyond reading sensu stricto to embrace matters of aesthetics, habitua-tion, mental action, and the very nature of experimental knowledge. His *Genetic Psychology for Teachers*, published in 1903, made some of the im-plications accessible to working teachers. It argued, among other things, that they could learn about children's acquisition of reading skills by watching the training of telegraph operators in Morse code; and his text-book, *Psychology*, reiterated the idea for student psychologists in 1907. The modern notion of a "learning curve," incidentally, arose in the con-text of the studies into the reading performances of telegraphists that Judd was invoking here, and those studies would later become pivotal for early information science too.[2] At any rate, by this point the twenty-six-year-old Judd was already among the country's top practitioners of the new science.

Judd ended up spending the rest of his career at Chicago. During that period of almost thirty years, the university became the principal home of the science of reading in the United States. One factor that gave Chi-cago the lead was the closeness of its relationship with an active school— Dewey and Parker's so-called Laboratory Schools, situated next door to the university. (One of the buildings that the school occupies, incidentally, is now called Judd Hall.) This was the institution that Huey had lauded. It gave Judd and his group the opportunity for constant reciprocal com-munication with practicing teachers and trainee readers. But he also maintained ample laboratory facilities, of course, including a workshop regularly called upon to construct new instruments. There he cultivated America's first stream of PhD students in the field, and encouraged gen-erations of researchers skilled in his experimental techniques. Several of them then took up positions in other universities, spreading the word (and the instrumentation) of the Chicago approach across the country.

Above all, perhaps, Judd produced a way for his protégés to get their work into print. Fundamental to the success of his enterprise was the strong relationship that he forged with the University of Chicago Press.

The press was unusual in American academic publishing for its close integration into the administrative structure of the university, and Judd made the most of that. He became chair of an editorial committee that oversaw the journals *School Review* and *Elementary School Journal*, both of which were dedicated to progressive education (they included much material emanating from experiences at the Laboratory Schools) and more particularly a long series of monographs issued under those journals' joint imprimatur. These books, collectively entitled Supplementary Educational Monographs, began to appear in 1917. They constituted a series that eventually lasted into the 1960s and totaled ninety-seven volumes. They were devoted to a range of themes, but in the early years reading was dominant among them. Three of the first five volumes were doctoral dissertations on reading by students of Judd. The first of all was William S. Gray's *Studies of Elementary-School Reading Through Standardized Tests* (1917); it was immediately followed by William Anton Schmidt's *Experimental Study in the Psychology of Reading*, and soon after by Clarence T. Gray's *Types of Reading Ability as Exhibited Through Tests and Laboratory Experiments*. Before long the series added Judd's own *Reading: Its Nature and Development* (1918). All four of these appeared within six months, right at the outset of the venture. They effectively defined what a monograph series in educational research would be. And all of them made full use of the eye-movement instrument that, as we saw in the last chapter, Judd had developed after seeing the early version by Dodge. Clarence Gray had his own improved version of the instrument constructed, and this then became the standard both within the Chicago laboratory and beyond; Gray subsequently introduced the device at what became his long-term home, the University of Texas. Well into the 1930s, his account of the machine in his *Types of Reading Ability* would be the customary reference for later scientists wanting to convey the nature of the instruments used in the study of reading to outsiders. Finally, all these researches also used the graphical conventions that Judd had invented at Yale to represent the motions of the eye along a line of text.

In these publications commencing in 1917 we can therefore see the consolidation of a scientific discipline, sharing methods, instruments, graphical conventions, and communication channels. The results were striking. There had been only thirty to forty papers published in English on the science of reading before Judd arrived at Chicago, but in the ensuing fifteen years more than 430 saw print. Moreover, the works increasingly appeared in venues where schoolteachers and principals themselves would encounter them. They were encouraged to assimilate the scientists' arguments and revise their own teaching practices accordingly. If the science were to realize its long-stated ambition to change how readers were made, then this was a critically important step toward doing that.[3]

As a result of these efforts, it would not be an exaggeration to say that in the 1910s and 1920s the Chicago school of the science of reading was the primary agent in a transformation of the practice of reading in the United States that would last generations. What characterized that school were five key elements. First and foremost, it was a laboratory practice using (and constantly refining) standardized tests, tachistoscopes, and, above all, eye-movement instruments. As a branch of experimental psychology in the Judd style, it regarded reading as an experience involving the entire body—the "total organic situation," centered on the self at the core of the central nervous system—and eye movements were the trace and symptom of that.[4] But it was never *only* a laboratory practice: the need to bridge the gap between lab and world was a constant preoccupation. So, second, the Chicago-school scientists of reading built networks of sociability, correspondence, and expertise that extended across the country, and these networks were channels for reciprocation, not simply for top-down dissemination. Another reason why this had to be so was that, third, however useful standard instruments were, the fact that reading was an interpretive act reliant on prior experiences meant that problems were inevitably particular to individual settings and readers. So diagnosis and remediation had to be tailored to specific cases, and case-knowledge was as indispensable as laboratory experimentation. Fourth, the science of reading must respect the differences between age groups too. One could not understand childhood reading by measuring adults, and vice-versa, and the applications of the science would therefore differ among children, adolescents, and adults. The Chicagoans insisted that their science was vital for elementary schooling, then, but they also argued fervently that adult workers in cities like Chicago, Toledo, and Detroit desperately needed their help—and that America's industrial and political culture depended on their getting that help. Fifth, and finally, Judd, Gray, and their colleagues placed great emphasis on silent reading rather than vocalized. For these experts, the two kinds of reading were ultimately different practices altogether. Silence, as Huey and other pioneers had argued, was the only way to make reading proceed at the speed of thought, and true reading must be a kind of active thinking. And it was this, more than anything else, that made manifest the influence of the science of reading at large. By the late 1920s the old emphasis on reading aloud had almost disappeared in schools, and what had once been a thriving trade in elocution for public reading was being deprecated generally. The Chicagoans could pride themselves on having wrought a transformation in the practice of reading that was widely acknowledged to be revolutionary.[5]

Of all the students Judd oversaw, the most important and influential was William Gray. Gray (1885–1960) was an Illinois native who had become a public-school teacher as soon as he left high school. He then attended Illinois State Normal University—a teacher-training institution—for two years, before a chance meeting with Judd led to his matriculating at the University of Chicago for an undergraduate degree. He then acquired a master's from Columbia, learning from Thorndike about his enthusiasm for standardized reading tests, before returning to Chicago for doctoral work. He, Schmidt, and C. T. Gray seem to have been the very first American students to gain PhDs on the science of reading, and all did so under Judd in 1917 with work that promptly appeared in Judd's new monograph series. By that time Gray was already the university's Dean of Education; he remained at Chicago for the rest of his career, becoming Director of Research in Reading at the Graduate School of Education and leading the teacher education program until 1945. He also became a lynchpin of the campaign to bring the science of reading to national attention. Starting in 1925, Gray published annual surveys of achievements in the expanding field. And in 1939 he launched a series of annual conferences devoted to the subject, which continued for decades after his own retirement. As well as publishing some hundreds of papers in his own right, he also authored a number of books that brought the science of reading to public attention. Much later, a committee charged with identifying the ten most important papers in the field ended up with a list in which Gray had authored one of the papers and edited five more.

Gray was a pivotal figure in defining how the science of reading affected the everyday experiences of readers nationwide. He found ways to connect the laboratory in Chicago with classrooms and libraries across the country. Perhaps the most significant early instance of this occurred in the summer of 1920, when he was introduced to a boy who had fallen so seriously behind in his school work that his parents—professionals both—were worried that he might drop out altogether. We do not know the name of this ten-year-old, but he had entered public school late because of prior illness and then charged rapidly through the first few grades before hitting problems. Gray took him to the laboratory and subjected him to three kinds of appraisal. First came a battery of intelligence, reading, vocabulary, and memory tests—the combined repertoire that Thorndike and others had started to develop in the previous decade and, as we shall see, were just attaining general credibility. He scored very highly on the intelligence test but otherwise displayed major problems. The boy had a habit of reading words individually rather than as groups, for example, and another of reading words out of sequence. Gray then turned to the tachistoscope to assess how well he was able to recognize letters, words, and groups of words at a single fixation. He made twice as many errors as four other children brought in to calibrate the experiment, and the mistakes increased with the number of characters displayed. And in the third place Gray used an eye-movement instrument to photograph the boy's eye motions while reading silently. This revealed that he was stopping at every word to take it in individually, and that the movements were generally haphazard, sometimes leaping ahead and sometimes regressing backward. This was a reader, then, who had never developed rhythm or fluency. These problems would need to be solved before the child could read "economically and effectively." Over an eight-week period Gray worked with a teacher at the Laboratory Schools to retrain his habits. They used typewritten selections, adding spaces between words to aid in building regular eye movements across a line and wider separations between lines to help train an accurate return sweep. In a separate exercise, words were grouped together as "thought units" to encourage his attention to grasp such larger patterns. Over time, the "regularity" of the boy's eye movements increased. As it did so, so did the "fluency" of his reading. By December he had made a year's progress in one term. Above all, his rate of reading had more than doubled, and he was far better equipped to read for meaning too. Gray summed it all up as "highly gratifying."

This case became the occasion for a national reflection on how the science of reading affected teaching. Gray published a report on the boy's

progress and sent it to some eight hundred principals and superintendents in schools across the United States, with an invitation to respond
with their own reports. He garnered from them fifty new case reports,
originating in thirty school systems ranging from rural Michigan, through
Minnesota and Oklahoma, to Seattle. He added to the accumulating heap
of information by persuading the city of Toledo to allow similar studies of
twenty-six slow readers in the school system there. Overall, the venture
involved Gray not only collecting information from across the country
and imparting his own strategies to teachers, but accepting those teachers' own testimonies—as he had to, given what was assumed to be the
irreducibly individual character of the problems they addressed. The
project seems to have been a genuinely mutual conversation spanning
research and implementation. Published in 1922, *Remedial Cases in Reading* was not only one of the earliest books to draw professional educators'
attention to the issue of "congenital word blindness, or dyslexia." In some
ways it was also a pioneering example of what is today known as citizen
science.

That said, however, Gray's initiative—and this was true of the science
of reading in general—also exemplified the stark social inequities then
characterized science (and society writ large), especially those of gender.
Schoolteachers were almost invariably women, while the researchers in
the field were generally men. Authorship was typically assigned to the researchers alone. *Remedial Cases in Reading*, for example, was "by" Gray,
but "with the co-operation of" Delia Kibbe and Laura Lucas, two teachers at the Laboratory Schools, and graduate student Lawrence William
Miller. And that was only on the title page: inside the book, the introduction named four more teachers and supervisors in Toledo, three of them
women: Superintendent Charles S. Meek, Assistant Superintendent Miss
Estaline Wilson, Primary Supervisor Miss Florence E. Hawkins, and In-
termediate Grade Supervisor Miss Flora Nettleman. And Gray added that
the remedial work there (as opposed to the research, presumably) was in
fact done by ten unnamed "special teachers," acting under Wilson's direction. The hierarchy was clear. This was not an absolutely fixed convention, of course, because different researchers handled the question of ac-
creditation differently. And customs also changed over time. In 1930, *The
Reading Interests and Habits of Adults* was "by" Gray and Ruth Munroe, a
one-time librarian in Cleveland, and in 1935 *What Makes a Book Readable*

was "by" Gray and Bernice Leary, but it was largely based on Leary's PhD dissertation in Gray's department. In 1956 *Maturity in Reading* was "by" Gray and Bernice Rogers, "Research Associate in Reading at the University of Chicago." By then Gray himself was emeritus "Director of Research in Reading" and the author of more than a dozen books. But during that time women had also become authors in their own right in the science. As we shall see, by the mid-1940s several of the most influential studies highlighting the broader social and cultural implications of the science of reading were the work of women—such as Clara Schmitt, to be encountered later in this chapter, or Eliza Gleason and Elizabeth Cleveland Morriss, whom we meet in the next—who thereby took on an authorial prominence in this field that was still rare across the sciences in general.[6]

Silent and Other Readings

If Gray was the most publicly influential face of the science of reading à la Chicago, Guy Buswell (fig. 3.1) was its most imaginative researcher. Trim, energetic, and witty, Buswell mastered all the tedious methods of the laboratory scientists, but he also spotted questions that others had not thought to ask, and saw how the techniques of the science of reading could be applied to answer them.

Buswell hailed originally from an itinerant minister's family in Nebraska. He had paid his way through college by working long enough hours in a pharmacy that he became a registered pharmacist himself. He first arrived at the University of Chicago in 1914 hoping to study psychology, but found his ambitions frustrated, first by poverty and then by service in World War I. But by good fortune he was assigned to the signal corps, and Buswell spent his military years at Kansas State Agricultural College learning about the communications technologies of radio, telephony, and telegraphy. As Judd had argued, telegraphy training was a fine way to come to appreciate the finer issues involved in learning to read, and Buswell returned, via York College in Nebraska, to the University of Chicago to pursue that subject. He gained his PhD in 1920 under Judd for a dissertation on what he called "the eye-voice span"—that is, the space opened up as the eye raced ahead of the voice when reading aloud.[7] He found that he could measure the span by combining eye-camera film with Dictaphone voice recordings, tying the two together by an ingenious

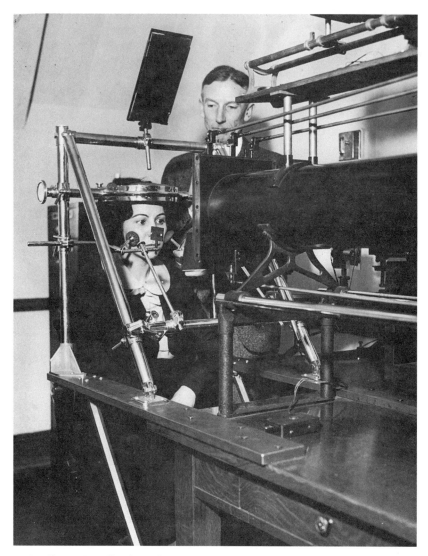

3.1 Guy Thomas Buswell and a reader using an eye-motion instrument. The identity of the reader is unknown. University of Chicago Photographic Archive, apf1–02381, Hanna Holborn Gray Special Collections Research Center, University of Chicago Library.

mechanical system that allowed the experimenter to mark the moment of a particular sound visually on the film. The eye-voice span was wider for better readers, he discovered. He suggested that this was because the gap allowed the mind to grasp the sequence of material being read and decide promptly on how to pronounce an otherwise ambiguous word (like *read* in this sentence) in the context of its surroundings. A method of measuring this span provided a gauge of ability, then, and suggested a diagnosis of reading problems; and it implied that a method to train a reader into increasing the eye-voice span might constitute a treatment. Judd was impressed enough that he engineered Buswell's appointment to the faculty, where he would remain for nearly thirty years. And the eye-voice span became a standard measurement, made routinely in diagnoses of troubled readers by people trained in the Chicago tradition.

Like Gray and others in the Judd orbit, Buswell worked to extend the use of eye-movement measurements into the design of new ways to improve adults' reading abilities. Scarcely had his doctoral work been completed when he and Judd began collaborating on a survey of the different kinds of silent reading. Here, they realized, was a subject where the new science had already had a clear impact. "The discovery of silent reading," they declared, was "directly traceable" to the work of laboratory scientists in this tradition. It had been almost ignored in schooling in favor of reading aloud, until their work made its importance evident. That schools now focused overwhelmingly on it was perhaps the science's proudest achievement. But Judd and Buswell wanted them to go further and embrace the notion that there were different *kinds* of silent reading, and that schools' existing practices were inculcating the bad kinds. The view that individual words possessed distinct meanings and that the act of reading was a matter of tracing them out would not hold water. A printed page was in fact "a source of a mass of impressions," and it required "the active mind" to "organize and arrange [them] with reference to some patterns which it is trained to work out." Readers came to pages on different occasions with distinct purposes in mind, such that the outcomes would be different. To explore this, Judd and Buswell measured eye fixations both for various kinds of texts (from blank verse to an algebra textbook) and for various kinds of reading (casual, studious, rapid, and more). At the extreme, their instruments told them that the kind of analytical work typically characterized as reading in schoolrooms—and above all in Latin classes, steeped as they were in traditional methods of construal—should not be called a form of *reading* at all. It was something else entirely. Indeed, in their terms the skill of parsing acquired in Latin lessons was akin to a virus. Once infected with it, people grew up to think that it was "immoral" to deal with texts in any way except slowly, laboriously, and pedantically.[8]

In essence, Judd and Buswell argued that the very practice that was central to educational experience was directly antithetical to modern culture. Schools, accordingly, must retool once again. They must train students in a variety of reading practices, suited to different circumstances: there should be classes to teach "methods of reading science," for example, and others to teach students "to read algebra fluently and intelligently."[9] Moreover, concerted action must be taken to combat what was currently "the most dangerous obstruction to a rational reading program," namely the textbook industry. Textbooks were designed not to be read at all, they argued. Such volumes were monstrosities—compilations suited only to be analyzed, paraphrased, and excerpted in an endless ordeal of anticreative tedium. That kind of training was liable to be personally and nationally devastating. Students should be told to discard those books forthwith, and instead go to libraries and ferret out knowledge for themselves.[10]

The Chicago scientists of reading excelled at this kind of argument. It was presented as based in rigorous and "objective" (a word practitioners liked) laboratory work. That work focused on tiny, arcane details, such as the movements of the eye or the times taken to read different combinations of characters. The findings were presented in dry, even pedantic scientific prose. And it was, at least in many statements, distinctly acontextual: a student told to ditch her textbook and go to a library instead might find the advice difficult to follow if she lived in a rural area with no library within a hundred miles. Yet the ambitions were nevertheless lofty, sweeping, and progressive. They were also radically relativist, in a sense, at the same time as they proclaimed their objectivity. Reading was not one thing, but many, and the types could be discerned in the laboratory. It depended on both reader and text. It had been shaped by history and evolution. An act of reading, accordingly, could never be simply a matter of pure information transfer. Perhaps (this was implied, though never announced explicitly) such an act could never be repeated even by the same person, because the self was never the same the second time around. Books must be redesigned to cope with this reality. And anyone hoping to teach or learn reading—which is to say, anyone hoping to flourish in modern society—had, they said, to accommodate themselves to the reality they revealed too.[11]

The implications became all the starker after the stock-market crash of

October 1929 and during the decade-long Depression that followed. With mass unemployment plaguing America, and the rise to power of totalitarian parties in Germany, Italy, the Soviet Union, and Japan, the fitness of democratic politics for a modern technological society came into question as never before. If democracy were to survive—if Americans were to have a reason to want it to survive—then a convincing account was needed of how public opinion related to politics, and that in turn required public opinion to be based in citizens' experiences, reading experiences being prominent among them. So it was a matter of profound unease, as Buswell noted, when a census of Chicago found that fully 60 percent of adults had an eighth-grade reading education or less. Not only industrial modernity, now, but the political order itself was at risk in such a population. For industry and democracy alike, the sine qua non was that citizens be able to read, and read well, not least because all other efforts to uphold a responsible public opinion depended on that. Without a populace able to read effectively, Buswell bluntly insisted, "the entire program of adult education breaks down."[12] We shall see in the next chapter how resonant such a warning could be.

So the stakes for a science of reading were suddenly both enormous and clear. If America could reach the point where most citizens could "read intelligently" sources like newspapers and magazines, then, Buswell said, "the social implications of the improvement would be impressive." Like the social scientists we shall encounter in a moment, he recognized the dependence of social democracy on "that independence of thinking which results from wide and critical reading."[13]

In the 1930s, then, the science of reading took its eye-movement tools and went to work for American democracy. Buswell in particular launched himself into a prodigious endeavor to take and analyze eye-movement traces for a total of 1,120 readers—1,020 adults (more than half of them women) and one hundred children. First he administered tests. Even though their average level of education was higher than that of the general population, he found that 83 percent of his adult pool scored lower on reading tests than high-school seniors. People who had recently left school actually did measurably worse than those who had left school twenty years before—a finding that Buswell laconically suggested schoolteachers might not want to pass over too casually. Next he measured the eye motions of every single person. Using a newly constructed version of the eye-movement camera, along with a disk-cutting sound recorder specially made by an audio company, and an oscillograph that rendered a representation of speech onto the same film as the eye-movement traces, Buswell captured 4,000 records in a total of 22,000 feet of film. Every single frame had then to be examined individually to correlate fixations to the various texts he had had his subjects read, which varied from blood-and-thunder passages extracted from newspapers to articles in the *New Republic* and *Short Wave Radio Magazine.*

This is perhaps the place to detail what a scientist of reading like Buswell actually did when faced with a record of a reader's eye movements. Luckily for us, Buswell himself recounted the procedure at some length. In this example (fig. 3.2), he had his volunteers read silently a twelve-line selection from an article in the *New Republic* by the southern writer Stark Young with a simple enough vocabulary that it should not be hard for any adult reader to get through. All 1,120 people read it "in their natural manner," each generating a photographic trace, and about half then took an oral comprehension test. To make a reference point on the film, each reader was asked to fixate first on two dots marking the length of a line of text at the top of the paragraph, which allowed Buswell to plot

At the Battle of King's Mountain a young Mr. Black was badly hurt. His colored servant saw that his master had been left to die and carried him to the nearest town. There he hunted until he found a doctor. The young soldier lived. After the war he gave the Negro a little house on his estate. Always on the day of the Battle of King's Mountain a fine dinner was prepared for the old servant. He sat alone at the head of the master's table and was served by the two daughters of the family. This was done every year that the old man lived.

PLATE VI.—Selection 1. Read silently

3.2 How to analyze reading, 1: G. T. Buswell, *How Adults Read*, Supplementary Educational Monographs 45 (Chicago: University of Chicago, August 1937), plate 6, p. 43.

the resulting trace onto the printed text itself. Figure 3.3 shows three of the resulting records, and figure 3.4 the plot of that produced by reader 983 on lines two through eleven. In each case, the small figures written on the plot of eye fixations show the number of each pause in the sequence above the line and the duration of each pause below it. This female reader had the most regular rhythm and the broadest span of recognition of all Buswell's subjects, and was thus measurably the best reader: you can see the regularity of her fixations on the trace (the leftmost one in the photograph). She averaged 3.7 fixations per line at 6.2 thirtieths of a second each time, and made no regressions at all. "The perceptual habits of this subject had been developed to a point where there was evidently no difficulty in grasping the thought of the selection," Buswell remarked. "Her eye-movements gave no indication of any interruption in the process of securing meaning." Subject 567 was roughly average, with 8.7 fixations per line at 7.4 thirtieths of a second each, and 1.5 regressions per line (fig. 3.5). The third trace was from subject 832, the poorest reader in the entire pool. He produced an average of 22 fixations per line at 9.4 thirtieths of a second each, and 5.9 regressions (fig. 3.6). And he was unable to answer a single comprehension question. It turned out that he had dropped out of school at grade four. "One might say that he had failed utterly to master the type of response to a printed page which is called reading," Buswell concluded. Such records, he declared, were "excellent data" for analyzing the reading process. They showed what the task of remediation was: it was to take a reader like 832 and retrain him so that he developed the habit of reading in a way that would produce a trace like that of 983. And repeated eye-movement traces made during that training process would allow progress to be monitored. But Buswell was nevertheless emphatic that these movements were symptoms, not causes. Poor reading manifested itself as eye motions of particular kinds, to be sure, but what actually caused it remained to be determined. So retraining the eye was a therapeutic practice, but it was not necessarily the ultimate solution in all cases.[14]

In analyzing this daunting array of records, Buswell looked for the standard Chicago-school panoply of signs: span of recognition, regularity (or lack of it) of movements, excessive regressions, inaccurate return sweeps, and long fixation durations. And he found meaning in it all. For example, almost three-quarters of his readers displayed smaller spans of

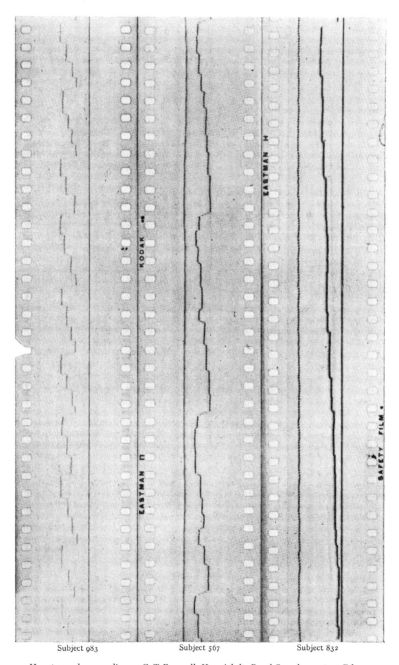

Subject 983   Subject 567   Subject 832

3.3 How to analyze reading, 2: G. T. Buswell, *How Adults Read*, Supplementary Educational Monographs 45 (Chicago: University of Chicago, August 1937), plate 12, p. 48.

Mr. Black was badly hurt. His colored servant

saw that his master. had been left to die and

carried him to the nearest town. There he hunted

until he found a doctor. The young soldier lived.

After the war he gave the Negro a little house on

his estate. Always on the day of the Battle of

King's Mountain a fine dinner was prepared for

the old servant. He sat alone at the head of the

3.4 How to analyze reading, 3: G. T. Buswell, *How Adults Read*, Supplementary Educational Monographs 45 (Chicago: University of Chicago, August 1937), plate 13, p. 49.

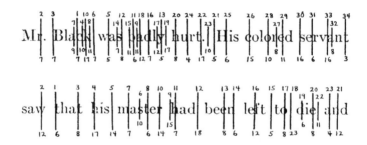

3.5 How to analyze reading, 4: G. T. Buswell, *How Adults Read*, Supplementary Educational Monographs 45 (Chicago: University of Chicago, August 1937), plate 14, p. 50.

3.6 How to analyze reading, 5: G. T. Buswell, *How Adults Read*, Supplementary Educational Monographs 45 (Chicago: University of Chicago, August 1937), plate 15, p. 51.

recognition than high-schoolers. "Obviously," he reasoned, "the independence of thinking that comes from wide reading is beyond the power of large sections of the adult population."[15] The situation was clearly dire, but how could it be improved? To begin with, Buswell dismissed the standard strategy of simply calling for more books to be made available and for people to be urged to read more. For one thing, there was no lack of materials already—over 90 percent of his pool said that they regularly read newspapers. For another, given the nature of the problem, simply reading more might well make things worse. Many of his subjects, as if to confirm his earlier prediction about school experience, believed that reading *ought* to be laborious, so merely reading more would reinforce a bad habit—a habit, indeed, so bad as not to qualify as reading at all.[16] Instead Buswell came up with a characteristically Chicagoan remediation strategy, designed to bring results quickly and effectively.

Buswell began by using a stereopticon (a projector that could cast two still images at once) to project sections of rhythmical poetry onto a screen. A group of people would then read them together, as a flashlight highlighted the words being uttered. The speed with which the light moved forward increased as the readers improved, encouraging smooth fluency; once a good rhythm had been established, a metronome could be used instead. At that point, Buswell switched to prose and to a movie projector. He cast up an image of a page of text, with most of the lines only dimly visible but one-third of a line at a time being highlighted. As with the poetry, the highlighted section would move forward as it was read and with increasing speed as the group improved. This projection technique had the great advantage of making eye-regressions impossible, so it trained readers to use fewer of them. Moreover, he would sometimes switch off the highlighting lamp and ask which words came next, revealing something of a reader's span of recognition and presumably bolstering it as he did so. And he also found that if he reversed the film then it produced "an excellent tachistoscopic situation," because words became disconnected from phrases and had to be perceived in themselves. It all sounds laborious. Yet his participants were apparently enthusiastic, and they grew more so as they improved. Some even took to devising their own experiments after hours, using home-made pendula to accelerate their reading in bungalows and apartments across the city.

Among the readers who came into Buswell's laboratory were 145 African Americans, and he devoted a discrete chapter of his book to this group. It was probably the first real attempt to apply the laboratory science of reading to questions of racial difference. When tested, they were found to be much worse readers than whites of comparable education, and their eye movements were halting. Buswell found that they had been educated in the period 1905–26, to varying grades, but that seventy of them—almost half—had not gone beyond sixth grade. They reported reading magazines (but not other matter) more frequently than whites, but their range was much narrower and mainly comprised publications of what Buswell called lower "quality"—"love and adventure magazines" rather than those devoted to news, "serious" discussions, and the like. Buswell was keen to point out, however, that none of this counted as evidence for "innate" inequalities. His preferred explanation appealed to what was then known as *environment*, which here meant the geography and political culture of educational institutions. Some 75 percent of his group had attended school in the southern states, he noted, where the poor quality of schools that African Americans could attend was "common knowledge." Reading abilities and eye movements both confirmed this, as they indicated that those in the group who had attended schools in the North enjoyed major advantages. But even though schools in the North were mixed-race, the Black population there was still commonly segregated in poor districts, so the schools they attended would still be substandard compared to those for many whites, and that explained the differences between those populations. In short, the science of reading, if correctly construed, contributed not to commonplace beliefs about racial difference but rather to "the accumulating mass of evidence showing the results of unequal educational opportunity." Of course, an even casually attentive reader would notice that Buswell had quietly demonstrated the high political stakes of American schooling as a whole—a theme that others would pursue much more earnestly, as the following chapters will show.[17]

In the 1930s, Chicago-school initiatives like Gray's continued to target schools. But, as in Buswell's case, they also focused increasingly on the need to improve the reading of adults. Typically, they sought to arrive at a diagnosis using eye-movement traces, and then to improve matters by means of a set of exercises that could be undertaken in a relatively short period. The main objective tended to be to speed up the *rate* of reading. This was not pursued simply for its own sake. The point was that ideally, or so the Chicagoans believed, reading should be a close adjunct to thinking, and as such should take place at roughly comparable speed. They were good followers of Huey. To that end, Buswell's own priority was to eliminate vocalization, broaden the span of attention, and inculcate a rhythmic regularity of eye motions (or, as he put it, reduce their "spasmodic fluctuations").[18] But there was a problem with this, in that the machinery of the science was so cumbersome. Using eye-movement cameras was vital for research, Buswell declared repeatedly, because they alone provided a "completely objective" record of a reader's abilities and problems. But he accepted that it was impractical to expect every school to have a device like the elaborate one he used in his own laboratory. Generalized strategies would have to be followed, even though reading did differ from individual to individual. This created what Buswell immediately recognized as an opportunity, and he was far from alone to see the potential. The very success of the science of reading meant that by the 1930s entrepreneurs and corporations were eager to make and sell machines purporting to realize its aims far beyond the confines of the lab.

The experimental instruments of the science of reading prior to the 1930s tended to be bespoke pieces of equipment. They were designed to stay in one place and have readers come to them. That was especially true of eye-movement machines, which were extremely cumbersome and delicate, and required extraordinary efforts by skilled operators if they were to produce usable traces. They were also extremely expensive: the famous information pioneer Paul Otlet's claim that one Chicago instrument cost $600,000 seems implausible unless it included the costs of the entire operation, but his general point was valid.[19] If the science's move out into the schools and cities was to succeed, this had to change. So it was that Buswell himself constructed the first portable eye-movement device in 1935 (fig. 3.7). Immediately, private companies saw a chance to get in on the act. The American Optical Company of Massachusetts—the largest maker of spectacles in the country—spotted the opportunity first. It marketed its own eye-movement device called the Ophthalm-O-Graph, modeled on Buswell's, from about the same time (fig. 3.8). And it also worked with the multi-generational Taylor family of Educational Laboratories, Inc., in Brownwood, Texas to develop and market their "triple-shutter tachistoscope," the Metron-O-Scope (fig. 3.9).[20] Both of these were portable, relatively rugged devices, built for classroom use. The first could be used to diagnose problems, and the second to remedy them by training readers to take in more details every time their eyes paused. And they were accompanied, importantly, by publications both large and small—most notably a fully footnoted but accessible volume on *Controlled Reading*, authored by Earl Taylor of the American Optical Company's "Bureau of Visual Science" and published by the University of Chicago Press, which made it seem part and parcel of the Chicago school. Taylor provided a full, illustrated history of reading research dating back to Cattell's first forays, along with a complete bibliography, and he acknowledged the

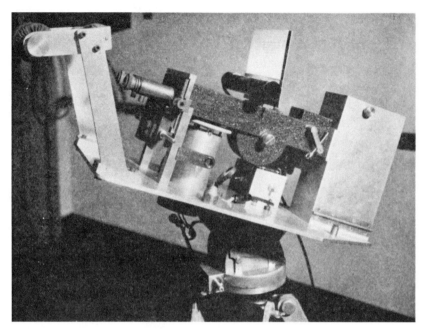

3.7 Guy Buswell's portable eye-movement device, invented in 1935. E. A. Taylor, *Controlled Reading: A Correlation of Diagnostic, Teaching, and Corrective Techniques* (Chicago: University of Chicago Press, 1937), 104.

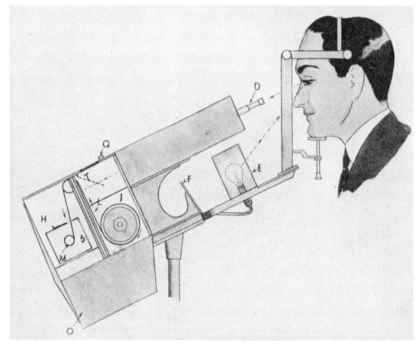

3.8 The ophthalmograph, marketed by the American Optical Company after Buswell's design. E. A. Taylor, *Controlled Reading: A Correlation of Diagnostic, Teaching, and Corrective Techniques* (Chicago: University of Chicago Press, 1937), 97.

3.9 The Metron-O-Scope, a classroom tachistoscopic machine, being used to help students with reading problems in a reading clinic in Evanston, Illinois, in about 1936. E. A. Taylor, *Controlled Reading: A Correlation of Diagnostic, Teaching, and Corrective Techniques* (Chicago: University of Chicago Press, 1937), 289.

contributions of educationalists, optometrists, librarians, and psychologists across the country. He included every single principal figure in the science of reading, from Cattell (now listed as an editor), through Dodge and Dearborn, to Judd, Gates, both Grays, Thorndike, and, indeed, Buswell himself. The last of these was given pride of place in Taylor's listing of major figures in the field. But the real purpose of the monograph was to introduce the company's machines. And this Taylor did by invoking a research study based on the photographed eye movements of 2,500 readers ranging in age from six to sixty—which probably made it the largest scientific study of reading ever undertaken to that time. The Ophthalm-O-Graph and Metron-O-Scope were presented as the keys to a future in which America's schools, equipped with these marvelous machines, would turn out generations of democratic and entrepreneurial creative citizens. Industry would bring the science of reading to the people.[21]

Buswell was, if not horrified, then at least perturbed. Many practicing teachers (then as now) were leery of technology companies that emerged from nowhere bearing gizmos, and he shared their skepticism. He was not entirely opposed to the use of technological devices, of course, and in fact he recommended that schools use projectors, if they could afford them, to control classroom reading in the way that he had. And he himself devised bits of apparatus that might meet their needs—not only the first portable eye-tracking machine, but a tachistoscope-like projector using 35 mm film that would cost a third as much as commercial "mechanically operated flash devices" such as the Metron-O-Scope and would allow both the grouping of letters or words and the rate of reading them to be "completely controlled." In fact, it seems likely that Buswell's invention was the inspiration for the commercial device (fig. 3.10). It would show

3.10. Guy Buswell's camera and reading materials used for his projecting tachistoscope, the inspiration for the Metron-O-Scope. G. T. Buswell, *Remedial Reading at the College and Adult Levels: An Experimental Study*, Supplementary Educational Monographs 50 (Chicago: University of Chicago, 1939), 24.

3.11 Guy Buswell's "apparatus for covering reading material at a controlled rate"—the prototype for "reading accelerators." G. T. Buswell, *Remedial Reading at the College and Adult Levels: An Experimental Study*, Supplementary Educational Monographs 50 (Chicago: University of Chicago, 1939), 68.

passages of text of various lengths for a specific duration, which could be decreased over successive sessions. But he accepted that even his simplified machine might be too extravagant for many school districts. So he proceeded to think up an even simpler invention. This was a clockwork machine that lowered an aluminum sheet over a page of print at a preassigned rate, thus forcing the reader to proceed at a certain speed (fig. 3.11). No projector or film was required. Buswell thought the gadget could be marketed for about fifteen cents. Finally, it seemed, a machine existed that could bring the science of reading to the people. And it did. Before long, as we shall see, this simple invention would give rise to a whole class of instruments collectively known as *reading accelerators*, which would be circulated in the hundreds of thousands and become the lynchpin of a huge national market in reading improvement.

Buswell was clearly no skeptic about machines as such, although he certainly questioned their expense and their utility in a classroom. But he was just as quizzical about the theories peddled by means of such machines. The fact was that *any* apparatus was merely a means to an end—"a means," he specified, "for implementing a modern theory of read-ing."[22] But the true science of reading—the kind carried on in laborato-ries like his—was a nuanced affair, precise, attentive to individuality, and multivalent. Such niceties risked getting lost in the gee-whiz publicity for educational technologies like the Metron-O-Scope, which implied a one-size-fits-all strategy. Advocates like Taylor tended to imply that the point was simply to train people to move their eyes faster, but to Buswell this was hopelessly simplistic. In his phrase, it was "mechanistic folly"—and we shall see that he was right to be nervous, because such simplistic ac-counts *did* gain sway in later years. What he was really after in using his techniques were "changes in perceptual habits," and eye movements were no more than an indication and a measure of those habits. By 1939, then, after a decade of efforts to refound schooling on the basis of the science of reading, Buswell, like Judd before him, and Gray too, had become some-what fearful of what success had led to. Dozens of conferences were tak-ing place annually on the improvement of reading skills, and nine hun-dred people had just showed up at the University of Chicago itself for one of them, astounding the local organizers.[23] But the major result of all this activity was that schools everywhere were getting ever more anxious about the problems the science highlighted. In the absence of clear and consensual recommendations, such concerns left them vulnerable to the sales pitches of corporations keen to market gimcrack devices backed by "quasi scientific or pseudoscientific literature."[24]

Yet when it came to the laboratory, Buswell's imagination and appetite for new instrumentation did not flag. At the same time as he was warning darkly of the exploitative maneuverings of pseudoscientific instrument-mongers, he was also advocating for his own institution to embrace the most ambitious device of all. In 1939, Buswell began experimenting with what he called "brain-wave apparatus." By this he meant the radically new technology of electroencephalography. He corralled Chicago's De-partment of Physiology, where Ralph Gerard was pioneering such ma-chines, into helping him to build an apparatus suitable for reading re-search. Buswell's initial hope was to detect a difference in alpha-wave patterns between silent and vocalized reading. But he looked beyond that, too, and envisaged a time when such traces might provide, finally, the long-sought objective way of distinguishing between species of read-ing overall—species that could then properly be described as fundamental. This "objective distinction of degrees of mental effort," he predicted, would create "a wide field for study" that had scarcely been imaginable previously. "A new technique," he affirmed, was "the key to a new body of understanding." Prescient words, perhaps. But just building the device proved a troublesome and arduous undertaking. It took so long that the machine arrived too late to conduct real experiments before the war threw all of American science into turmoil.[25]

Buswell's distinctly visionary proposal for a new epoch in the science of reading came just as the Judd tradition was losing its sheen. Buswell's portable device notwithstanding, eye-movement photography in general had remained expensive, demanding, and time-consuming. By the early 1930s, voices could be heard expressing skepticism that such machinery had much of a role to play outside highly specialized research laboratories. Even if it could be made practicable, critics began to ask whether eye-movement photography were really useful in correcting disabilities. After all, poor eye movements were symptoms, not causes, of poor reading, as the indefatigable psychologist of typography Miles Tinker pointed out (and as Buswell, for one, fully acknowledged).[26] In that light, there were other signs of difficulty that were much easier to spot—starting with the very fact that someone had shown up at a reading clinic in the first place. In the years around 1930, Tinker published a series of responses along these lines to the more extravagant claims that Judd's enterprise had incited. Many of them, to be sure, confirmed points that Judd, Buswell, and Gray had themselves endorsed—the point about movements being symptoms rather than causes being one—and in his own trials Tinker found the records produced in laboratory eye-movement experiments to be reliable and robust. (Those trials were heroic: he advertised on the title page of his book *How to Make Type Readable* that he had conducted precisely "twelve years of research involving speed of reading tests given to 33,031 persons.") But he contended that outside practitioners who rushed to adopt Chicago techniques failed to appreciate their subtleties, not noticing, for example, that a diagram seeming to locate fixations in a line of prose should be taken as identifying fields of perception rather than dimensionless points. By the late 1930s a broader retreat was at hand, not from the laboratory science of reading itself, but from the more enthusiastically scientistic hopes it had fostered for an educational revolution capable of saving American modernity.[27]

How to Read a Picture

On the face of it, the most remarkable publication to arise from the science of reading between the wars was not about reading at all. It was about the viewing of images. Buswell had been at Chicago for about fifteen years when he produced it, and it was based in his intimate familiarity with the peculiarities of the Chicago laboratory's instrumental repertory. In the early 1930s he took advantage of this repertory to return to the field that had initially inspired Judd himself. He had the Education Department's resident mechanic, Ralph Larsen, build a new version of the old Judd eye-movement instrument, and used it to investigate the viewing of pictures of various kinds. The result was *How People Look at Pictures: A Study of the Psychology of Perception in Art*, published in 1935. It was, as Buswell himself declared, the first thoroughgoing experimental study ever published of the ways in which human spectators experienced artistic images.

When art critics and connoisseurs described their experiences of pictures, Buswell noticed, they often adopted a style akin to that of the physiological aesthetics that had exercised Stratton and Judd back in around 1900. For example, in Eugen Neuhaus's *The Appreciation of Art* (1924), Neuhaus referred to how patterns in a painting would "draw the eye from the lightly developed part toward the more expressive." Likewise, Carleton Eldredge Noyes in *The Gate of Appreciation* (1907) spoke of how a beholder's eye would rest "lovingly" at the balance point of a composition, and follow smoothly "the rhythmic flow of line." Buswell quoted about ten different experts using language like this. The trouble with such evocations, he pointed out, was that they were based exclusively on "introspective and subjective evidence"—if, indeed, they were based even on that. Critics, he remarked, constantly evoked eye movements (and hence perceptions) that they said were experienced by an ideal viewer. But were they? What about real viewers? He decided to find out, by using an instrument proven in the science of reading to furnish data that were "entirely objective."[28]

So Buswell followed Judd's lead of thirty years before and extended the science of reading into two dimensions. His machine had to be specially built, however, because in order to view paintings his viewers would need an unimpeded forward field of vision (fig. 3.12). The resulting apparatus

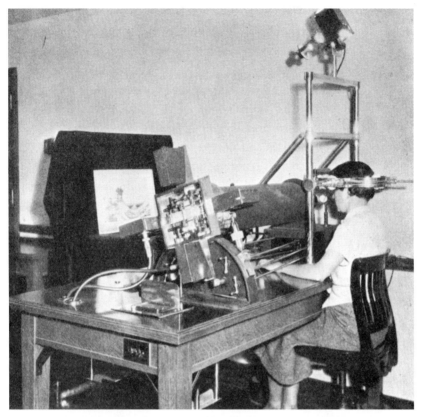

3.12 Guy Buswell's apparatus for extending the experimental study of reading into that of the viewing of images. G. T. Buswell, *How People Look at Pictures: A Study of the Psychology of Perception in Art* (Chicago: University of Chicago Press, 1935), 12.

was an ingenious contraption of lenses, mirrors, and prisms. It channeled a ray of light up from a lamp housed beneath a desk onto the subject's eye (the device was monocular), whence it was reflected sideways onto two moving kinetoscope films, one oriented vertically, the other horizontally. A rotating fan blade interrupted the light beam every thirtieth of a second, so that what was recorded on each film was a series of dots, each of which marked the passage of that much time. This allowed Buswell to measure the duration of the eye's movements and pauses simply by counting up the number of dots in a given cluster. As with Judd's earlier machines, the viewer sat in a specially constructed chair, with calibration of head movements—"absolutely necessary," Buswell noted—being provided by using a spectacle frame holding a reflective bead to create a trace on the films. (He does not seem to have used a jaw retainer.) Each viewer was asked first to fixate on four dots at the corners of the image to provide points of reference, and then, in most cases at least, to view the image simply as they normally would. The instrument would then record the eye's motions as the viewer took in what they saw.

The result from each camera was a pair of traces, showing respectively the horizontal and vertical movements. Having obtained the two films from a session, Buswell then projected them through a stereopticon onto a copy of the picture itself, allowing the horizontal and vertical locations to be plotted, with the four reference points providing calibration. In tests with almost fifty people, Buswell found that the resulting traces were accurate to within five millimeters in about 90 percent of cases—which, given that the eye was never quite stationary anyway and that each fixation point was just the center of a broader area of perception, was certainly sufficient to draw firm conclusions. He and his helpers could then chart the results in various ways. For example, they could create what he called a "density plot" showing the fixation points for a population of viewers; or they could generate a connected series of lines showing how one individual viewer's eyes had moved around a picture. Both gave rise to compelling graphical displays. Impressive as it was, however, the technique was essentially that familiar from the science of reading, and, as Buswell conceded, it was "new only in its application to the field of art."[29]

So the science of reading here became a science of visual attention. Its grounding assumption was simple: "that in a visual experience the center of fixation of the eyes is the center of attention at a given time." The assumption was not an inevitable one, but it came naturally to a scientist of reading because of the axiom in that field that perceptions happened only at fixations. "If this is true," Buswell argued, "then the record of eye movements in looking at a picture supplies objective evidence of the pattern of perception during that experience." He was reluctant to make any further inferences about mental processes such as appreciation or understanding, but he insisted that if a viewer's eye paused over a certain element in a picture then the viewer must be "interested in" that element. And this could now be shown "objectively" for the first time. His science had implications not only for poets, newspapermen, and teachers, now, but for artists, architects, illustrators, and all other creators of graphic work. At hand was a radical extension of the scope and scale of the science of reading.

Buswell put together a portfolio of fifty-five different images to conduct his trials. Some were in color, others monochromatic; some were of paintings (mostly from the Art Institute of Chicago), others images of pots, tapestries, buildings of various kinds, furniture, and decorative motifs used in interior design. One or two were specially drawn geometrical designs exemplifying symmetry or balance. He classified them all into eight groups:

paintings (color and monochrome)
vases and dishes
furniture and design
photographs of statuary and museum pieces
tapestries and designs on cloth
architecture and interior design
posters
outlines, silhouettes, and geometric figures

This done, he sat people in his machine and asked them to view differing numbers and sequences of images while he recorded the movements of their eyes. Some two hundred volunteers took part. Twelve were elementary-school pupils and another forty-four high-school students—all these probably came from the University of Chicago Laboratory Schools. Of the 144 adults who participated, forty-seven were students who had trained in art at the School of the Art Institute of Chicago, and another fourteen had sufficient training to be classified as art students. Most of the rest were college or graduate students, again probably hailing from the University of Chicago. In all, Buswell and his assistant exposed about 18,000 feet of photographic film, recording 1,877 usable traces. That was on a par with the prodigious numbers that one often saw in scientific experiments on reading, but it is worth noting that Buswell found it barely sufficient. He lamented that his subjects could not sit still for long enough to yield 2,000 fixations in a single session.

Buswell found, as Judd had, that viewing pictures was similar in important ways to reading texts. In particular, the eye never moved smoothly, even though subjectively it might seem to, but proceeded "in a series of quick jerks and pauses." That meant that in principle the experimental and conceptual repertoire of the science of reading was entirely applicable. First, then, Buswell looked at the "centers of interest" or "centers of attention"—those places on a given image where fixations congregated. He could easily render these visible by making a density plot of the fixation points of multiple readers. For example, figure 3.13 shows the "distribution of interest" for sixty-eight people who looked at Winslow Homer's painting *Stowing the Sail*, making 3,267 fixations in all. The higher the density of dots, Buswell reasoned, the greater the attention devoted to that small region, and therefore the more meaning acquired. In the Homer case the centers were fairly clearly visible and somewhat predictable, and that was true for many kinds of pictures, portraits in particular. But it was not always so. In a Japanese image called *The Wave*, Buswell found that viewers' fixations did not congregate so clearly, and they sometimes seemed to gather around unpredictable elements such as the artist's signature. But he then isolated the first few fixations in order to show what had *immediately* captured viewers' attention, and in the case of *The Wave* that was the gigantic wave itself (fig. 3.14). Some laborious statistical

3.13 Density plot showing the position of all fixations for sixty-eight viewers of Winslow Homer's painting "Stowing the Sail." G. T. Buswell, *How People Look at Pictures: A Study of the Psychology of Perception in Art* (Chicago: University of Chicago Press, 1935), 19.

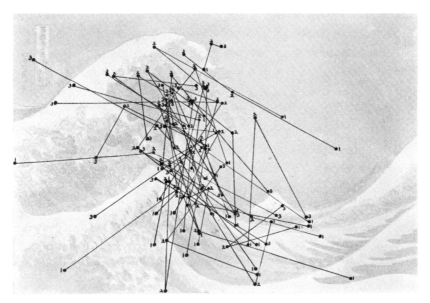

3.14 The first three fixations for forty viewers of the Hokusai image *The Wave*. G. T. Buswell, *How People Look at Pictures: A Study of the Psychology of Perception in Art* (Chicago: University of Chicago Press, 1935), 28.

work led him to the further conclusion that the sequence of viewers' pauses had generally followed the direction of the wave itself, as though it were in motion.[30] And Buswell also turned to images with less evident meanings, like Duchamp's *Le Roi et La Reine*—which, he noted, few of his subjects could make head or tail of—and tapestry designs, in a further quest to find significance in what people initially looked at. For many images this showed nothing very much. But in certain cases—*Stowing the Sail* being one—Buswell was able to affirm that a feature that critics routinely claimed to be particularly compelling (here, the red flag in front of the mast) in fact hardly registered with viewers at all.[31]

These were all collective traits, shown by groups of people considered together. Buswell also highlighted certain individuals, showing how their eyes moved across a picture in a sort of aesthetic exploration. Some, he found, took quite simple paths; others ranged back and forth elaborately, pausing over different spots and revisiting some of them. These plots— for which Buswell linked successive fixations by straight lines, marking out a map of the perceptual journey—revealed something of the viewing habits of the individual concerned. Comparing the traces of two viewers of a painting called *The Grief of the Pasha*, for example, he found that one focused on the dead tiger and the bereft Pasha, while another concentrated on the elaborately decorated architecture. That seemed to speak against conventional assumptions about an artist's control over a picture's reception. On the other hand, though, one viewer of *The Wave* left a trace that followed the breaking wave closely and thereby, Buswell commented, provided "an excellent example of how an artist is able to control the perceptual process by the composition of his picture."[32] Similarly, images of the interiors of cathedrals did tend to confirm critics' statements that vertical lines encouraged a vertical sweep of the viewer's eye. But whereas those critics always leaped to the assumption that this was an *upward* sweep, associated with spiritual uplift, Buswell's traces revealed the "disturbing" fact that, faced with a photograph of the interior of St. Patrick's cathedral, viewers' eyes were more likely to sweep downward— falling from heaven to earth, as it were.[33]

When viewers paused over a particular feature of an image, what were they doing? This too was a question inherited from the standard repertoire of the science of reading, in which it was taken for granted that nothing very much happened while the eye was in motion. It was well known that readers of texts were not engaged in "the simple recognition of words," but in a complex fusing of words together into meaningful "units of thought." That was why the fluency and ease of better readers corresponded to shorter pauses. Buswell assumed that something similar must be happening with images. And yet, the cases were not directly comparable. Here, pauses varied from one-tenth of a second to two-thirds of a second or even longer. The durations of fixations in image-viewing did not decrease with maturity, as they tended to do in reading; and there was no evident way to correlate them with difficulty, whatever "difficulty" might mean in this context. Puzzled, Buswell fell back on another practice that he had investigated using eye cameras, namely arithmetic. Someone asked to add a column of figures generally showed eye fixations a lot longer than those in regular reading, he knew, because the eyes were awaiting the completion of a mental process. Long durations presumably implied the presence of discrete thought processes, then. If this were happening in the viewing of images, he reasoned, then fixations should behave in almost the opposite fashion to what one found in reading. They should get longer as a viewing proceeded, because an initial glance would give way to a more detailed appraisal. And a statistical analysis showed that this was indeed the case. So pauses looked very much like symptoms of thought itself.[34]

Then Buswell looked for another parallel to reading. He had viewers look at regularly repeated horizontal patterns, of the kind that were often used to highlight a coving, picture rail, or other linear feature in a building (fig. 3.15). Here, again, aesthetic critics generally asserted that a design controlled how fast one's eyes moved along the pattern. They tended to assert that certain characteristics encouraged certain eye motions: that flowing lines, for example, encouraged rapid, smooth, and easy motions, while "static" designs did the opposite. Repetitive elements in a design were commonly said to produce a kind of "rhythm" that induced a pleasing effect by virtue of the corresponding "muscular rhythm in the movements of the eyes." Buswell put all this to the test and found it baseless. His subjects showed no trace whatsoever of such a rhythm. Instead, they tended to look in zig-zag fashion back and forth along the line, and if anything they paused longer over patterns that seemed to facilitate speed than over those said to induce delays. But they also showed a general tendency that was quite unexpected: they would start in the middle, make one sweep to the left, and then move across the pattern from left to right. Here, it seemed, was a telling manifestation of how powerful the habituated skill of reading could be. He suggested that his instrument was revealing "much the same" phenomenon of regressions as occurred when someone read a line of Latin prose: in each case the eye had to move back and forth to garner the elements of an interpretation that felt, subjectively, to have been acquired by a smooth linear motion.[35]

Altogether, a dismaying number of the issues that art critics ordinarily reckoned important turned out to leave no trace in Buswell's films. Colored images, for example, seemed to be viewed in more or less the same ways as monochrome ones. The aesthetically critical issue of symmetry, or "balance," made no impact whatsoever. "Oriental" people seemed to perceive Asian art no differently from Caucasians.[36] And he found no difference between skilled subjects (experienced art students) and unskilled ones. That last result stood in such stark contrast with what happened in reading, however, that even Buswell balked, declaring it "in no sense satisfactory." Perhaps, he cautioned, it was really a consequence of the fact that children could not sit still for long enough. Even in the one

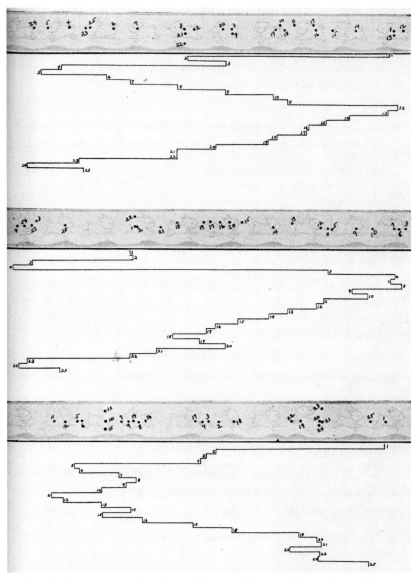

3.15 Record of eye movements for a viewer of a horizontal decorated border design. G. T. Buswell, *How People Look at Pictures: A Study of the Psychology of Perception in Art* (Chicago, IL: University of Chicago Press, 1935), 111.

case that produced numbers he thought worth recording, an eight-year-old boy spent half the session in an apparent state of boredom. His gaze ranged so far away from the picture that the trace went off the edge of the photograph.[37]

It may have been frustrations of this kind that impelled Buswell into what proved to be perhaps the most consequential aspect of the whole work. He became acutely suspicious of the effects that his own instructions might be having on the people whose eye movements he was recording. So he addressed those effects explicitly. Unusually for experimental psychologists of the period—but, as we have seen, less so for the scientists of reading in Judd's orbit—he came to think of his experiment in reflexive terms, as a two-way, constructive process of communication between experimenter and subject. Moreover, this was not merely a matter of noting *that* the framing of the experiment might affect its results: Buswell sought to understand *in what ways* it did so. For the most part he had simply asked people to "look at the pictures in their normal manner, whatever that was," but he now gave instructions that were more particular. After having a viewer look at an image of Chicago's Tribune Tower under that general instruction, for example, he then asked the same viewer to find a person looking out of one of the windows and had him look at the image again. The result was a markedly different set of movements and pauses, mapping out a visual search operation. In other cases, he had viewers return to an image after reading a page of commentary on the picture concerned. Or he asked a viewer to choose a favorite among several panels of an image. Simple as they may seem, the implications of these moves were momentous. They proved, "objectively" as Buswell would put it, that what you saw when you looked at something depended profoundly on what questions you were asking. And those questions could be assigned ones.

That point took on an added *frisson* in the context of the three images that formed the climax of Buswell's book. These images were of a very different kind from all those he had hitherto used, in that their purpose was expressly to manipulate their viewers. Buswell introduced a trio of advertising posters, for Chevrolet, Pontiac, and Wrigley's Gum, each of which exemplified the graphic style of the time. He found that viewers tended to focus on specific pictorial elements in these posters—especially the female figure in the Chevrolet ad—rather than the textual elements. Preliminary and sketchy as it was, Buswell noted the potential that this experiment held. "In terms of the extensive use to which a single poster is often put," he remarked, "the reactions of a sample group of subjects, selected from the class of individuals to whom the advertisement is supposed to appeal, would be of great value in the selection of picture advertising." So the science of reading gave rise to a proposal for an experimental discipline of graphic persuasion, and this happened a decade before the first "focus groups" were corralled together at Columbia University in what is conventionally regarded as a founding moment of modern media studies and market research. Entirely ignored at the time, for a later generation this insight would become central to the commercial display of information.[38]

Learning to Read, Scientifically

The vast majority of modern Americans learn to read in the company of elementary books. In the last two centuries, these books have been successively called *primers*, then *readers*, and later *basal readers*. As Carl Kaestle and Richard Venezky point out, they are perhaps "the most ubiquitous pieces of print culture in our history." Through the early twentieth century, thanks to the consolidation of the textbook publishing industry, they came in increasingly formalized series, ostensibly designed for sequential use across years of classes, and accompanied by teachers' guides, wall-posters, flash cards, and other paraphernalia. In earlier periods these books were often written by individual authors, the most famous of whom was William Holmes McGuffey; later series were more like industrial products, being concocted by teams working under the leadership of some authoritative individual. McGuffey died in 1879, but the McGuffey's Readers books were still in use in some schools well into the twentieth century. Through the early 1900s, such volumes generally espoused some variant of an approach in which students learned by the accretion of letter-sounds (what would later be popularly known as *phonics*). A central aspect of this was the assumption, common through the later nineteenth century, that reading was fundamentally oral and that it was best learned through sounds. The point was not to master meaning, in fact, but to inculcate elocution. A strong secondary mission was to convey the elements of a very conventional kind of morality. Texts in these books were therefore short, moralistic, and intended to be read aloud.[39]

After 1900, this genre underwent radical and lasting changes. They arose largely because of the centrality of the science of reading to ideals of progressive education, but they were relatively coherent because of the coalescence of the textbook industry itself into something approximating a "trust," the characteristic corporate form of the time. Given that the industry was not a conventional one, such agglomeration may have been all the more inevitable. It had to sell huge numbers of low-profit items, not through retail stores but directly to schools and school districts, in a short annual season. It also often had to tweak volumes to meet states' differing requirements. At any rate, in 1890 what had been a relatively diverse industry regrouped into the American Book Company (ABC for short), and over the next decade the conglomerate swallowed up another ten rivals. Although it was never technically a monopoly, ABC came to enjoy a position of dominance when competing before the all-important textbook adoption committees that by 1900 operated in many states.[40] It acquired a reputation for ruthlessness and corruption—and there is considerable evidence that it deserved that reputation. By the 1920s it had attracted the attention of Upton Sinclair, whose *The Goslings* tried to do for the textbook trade what *The Jungle* had done for meatpacking. Sinclair's book offered a stinging exposé of the industry's practice of "corrupting school officials and making away with school money in every section of the United States."[41] Such bribery and strong-arming did help the textbook business maintain a certain homogeneity, however, and it meant that changes in one area were likely to produce changes elsewhere too.

After about 1910, books to teach new readers therefore departed from their old formats in a relatively coherent fashion. The previous emphasis on reading aloud was dumped: the science of reading had shown that silent reading was both more efficient and more conducive to active understanding. Also discarded was the dedication to phonics. Now, students were to learn to read by apprehending whole words, as Cattell and Huey recommended. Alongside this, the old moralism faded away, a little more gradually, to be replaced by a more neutral (if boring) blandness. The shifts were explicitly inspired by the educational philosophies of Dewey and Edward Thorndike of Columbia, alongside the scientific interpretations of reading advanced by Huey and his peers. Thorndike's own *Reading as Reasoning* (1917), a nontechnical outcrop from the science of reading, was to some extent the manifesto of the transformation.[42] It was perfectly possible, Thorndike indicated, to read letters, words, or sentences correctly in a mechanical sense, and yet to come to ridiculously false accounts of their meaning if one did not place them in their paragraph context. To read a whole paragraph with understanding, on the other hand, was "like solving a problem in mathematics." It could not be done without taking into account the whole section, while resisting the pressure of any one element in order to select what was significant for the mind's purpose. So, as Huey had insisted, reading must never be treated as a "mechanical, passive" task, but as an active, interpretative one. It was in order to foster this kind of ability that reading must be taught as a silent practice, dedicated to meaning, not speech. This would both fit the next generation for its place in industrial modernity—which was Thorndike's own priority—and, as Dewey and Huey urged, equip students to become truly thinking humans.[43]

By the 1920s, mass-produced elementary books had thoroughly embraced silent reading by whole words (or even larger units). That priority would last at least until the 1960s and the eruption of the "reading wars," which we shall encounter later. Not only was it influenced by the science of reading in general: some of the foremost series of primers, including the most influential of all, were in fact authored by the scientists of reading themselves, who designed them explicitly according to their scientific principles. Thus Guy Buswell, the inventor of the portable eye-reader, was also behind a number of elementary textbook series. In particular, he masterminded the Silent Reading Hour series issued from 1923, in which stories were printed at length, in perfectly justified layouts of eminently legible type. And Buswell made the foundations of the series explicit at the very outset. In sections printed at the start of special teachers' copies, he represented each book as "an instrument for teaching silent reading," and warned that its functioning would be "seriously impaired" if a child were allowed to read aloud from it. He insisted that silent reading was "the complex process of getting thought from the printed page," and that as such it required "an entirely new pedagogy" that must be based in eye-movement studies of saccades and fixations. Only such studies could provide "an exact and objective record of one's reading habits." Illustrating his

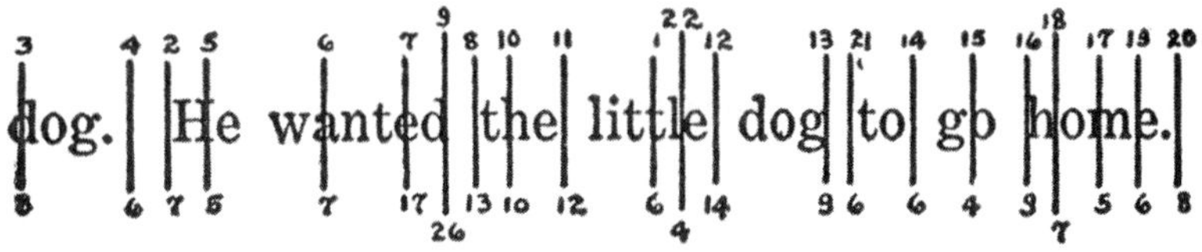

FIG. I

3.16 Guy Buswell's instructions to elementary-school teachers for the use of his primer for beginning readers. The upper line records the fixations for a typical first-grade pupil, the lower those for a typical college student. G. T. Buswell and W. H. Wheeler, *The Silent Reading Hour, Second Reader* (Chicago: Wheeler, 1923), iii.

point with two traces of saccades, he warned that it was "the task of the school" to "change the crude reading habits disclosed in Figure I to those of the mature type of reading shown in Figure II" (fig. 3.16). The prime importance of this was now "fully established" and beyond dispute. So it was that Buswell included whole stories (not snippets), newly written (by female authors) and designed to excite active interest in order to foster a habit of silent reading. They were reflective of real life, not fantasy, so as to help train "constructive thinking." And he presented them in type the sizes, faces, spacings, and leadings of which were all carefully designed, and arranged in short, uniform lines so as to reduce eye fatigue. Even his paragraphs had proper indentation and justification, "in order that the child shall develop from the very beginning the rhythmic eye-movement habits which will be required in mature reading." All of this was invisible to child readers and their parents, but any teacher who read the special introductory section to a volume knew all about it.[44]

As these changes were taking place, so schools of education were coming into being across the country to train future teachers in such techniques. The new field of "educational psychology" was the central attraction for these institutions, which were dedicated to creating the population of teachers that would be needed to put these books to use and create the reading population of a bright future. Judd and Thorndike

were towering figures in this environment. The science of reading consequently became a mainstay in the cultural formation of this population,
not as a laboratory science, but as a foundation for everyday, common-
sense classroom exercises. The teachers of reading who emerged from
the new schools of education would see themselves as teaching *this* kind
of activity. And authors of basal readers, where they were not actual scientists of reading like Buswell, were often those teachers themselves, or
their teachers—or, sometimes, their students.

The undisputed star of this population of scientist-authors—and, significantly, almost the only person to move freely between the two Chicago schools of reading—was in the end not Buswell, but William S. Gray
(fig. 3.17). From about 1930 until his retirement in 1950, Gray worked to
produce basal readers, incorporating the contentions of the science of
reading in very much the way Buswell did. Having collaborated with Chicago publishers Scott, Foresman, and Company since the early 1920s, in
1927 he started working extensively with an elementary-school principal
and consultant for the firm named Zerna Sharp. Sharp and Gray resolved
to create a new generation of books, designed around the wholesome ad-
ventures of two model characters that they would mold into the ideal vehicles to carry Chicago's science of reading into schoolrooms and houses
across the land. After puzzling for a while, they settled on names for these
characters: Dick and Jane. As far as I have been able to tell, they did not
include saccade diagrams like Buswell's in the extensive aids for teachers
that Scott, Foreman produced—an omission that perhaps reveals something about his changing ideals for the science. But in other respects the
commitment to the science of reading was explicit and complete. And
whereas Buswell's series were moderately successful, those of Gray and
Sharp were stupendously so. They became some of the most culturally
influential objects of the entire century.

The consequences hardly need to be stated explicitly.[45] For millions
of Americans in midcentury and beyond, reading was a habit acquired
through encountering and following these two characters, Dick and Jane.
They stayed with them throughout their childhoods, for years on end.
They did not know, of course, that the habit they were developing was
being formed through familiarity with formats, type, and concepts all expressly shaped by the principles of a science of reading. But it was so. By
the middle decades of the twentieth century, it would not an exaggeration

3.17 William Scott Gray. University of Chicago Photographic Archive, apf1–06533, Hanna Holborn Gray Special Collections Research Center, University of Chicago Library.

to say that the overwhelming majority of Americans had been forged as readers by means of this science.

Strephosymbolia and the Limits of Science

The science of reading pursued by Judd, Buswell, and their peers underpinned what the practice of reading itself was for Americans in the interwar years. But its very success generated a substantial problem. Illiteracy as traditionally defined—the inability to read at all—might well be on the decline, and this was indeed a great achievement. But the retreat of old-style illiteracy, coupled with the intensity of research that characterized the science of reading, made visible a persistent problem that threatened to be even harder to solve. It became clear that there were still some people who defeated the best measures that this science could suggest—people like the ten-year-old who had fascinated Gray in 1920. As their numbers rose, so their existence became a matter of increasing embarrassment, almost scandal. If a true science dealt in the laws of nature, what should be made of a professedly scientific discipline whose laws were apparently breakable?

It was known almost from the outset of the science of reading that even with the latest teaching methods, well-designed books, and ample resources, individuals existed who never seemed to acquire fluency in reading. A few of them never mastered the basic ability at all. It was tempting to attribute this to a lack of "intelligence," especially in the 1920s, when IQ testing was all the rage. But there were always intractable cases, like that of Gray's boy, that cast doubt on any such assertion. Puzzled psychologists heard now and again of a child or adult who was in every other respect bright and diligent, but who just could not learn to read well. This problem loomed increasingly large as the science of reading spread and its influence grew. In the end, it gave rise to a fundamental challenge to the very enterprise and its central claims.

The condition was first named in 1877—at exactly the time Javal was doing his foundational experiments in Paris—by a Strasburg physician named Adolf Kussmaul. Kussmaul called it *word blindness*. It was a slightly unfortunate term, because, as he himself stressed, sufferers could in fact see perfectly well, and might even be able to identify individual numerals and characters. They simply could not construe words and phrases in the way that readers did. Many people who experienced word blindness of this kind came to do so as an acquired condition, and these became fairly familiar. A Scottish ophthalmologist named James Hinshelwood, for example, observed one such patient for nine years, after the fifty-eight-year-old language teacher suddenly found that he could no longer read his pupils' exercises. He could see every character clearly, Hinshelwood found, and could read numerals unproblematically. He could even write fluently. But he had completely lost his "visual memory" for letters and words, to the extent that he could not construe what he himself had written a moment before. This man attempted for a year to relearn the skill of reading by using a child's primer, but in the end he gave up in despair. When he died, Hinshelwood arranged for an autopsy and announced that the cause of his problem had been discovered in a brain lesion. Others undertook similar brain analyses, and it became conventional to blame lesions for such problems.[46] But these cases, being acquired as a result of diseases or accidents, were relatively untroubling. Much more worrisome were instances in which a child or adult could not read from the outset, with no sign of a disorder or event that might cause such an inability. Hinshelwood became fascinated by such cases, and he christened the phenomenon *congenital* word blindness.

By the early 1920s, a number of case studies of such puzzling individuals had been published. But it was hard to diagnose, and there remained considerable doubt about the reality of congenital word blindness per se. The controversial Sir Cyril Burt, writing in England, suggested that short of some disastrous brain injury it was probably nonexistent, and Arthur Gates at Columbia University added that he had never come across a clear case of it.[47] As well as congenital word blindness, moreover, the condition also accrued a multitude of names, each of which carried a slightly different connotation. It was by now also being called *alexia*, and sometimes, as in Gray's report, *dyslexia*. In any case, researchers in the tradition of Judd, Buswell, and Gray routinely found themselves called upon to reflect on the condition and its implications. Those implications extended to the nature and acquisition of "normal" reading abilities, because if they could not explain why some people *did not* learn to read then their account of how most children *did* might be deemed wanting.

The most sustained attempt by the Chicagoans to tackle this question occurred at the hands of Clara Schmitt, a psychologist formally attached to the Chicago Public Schools' Department of Child Study. Schmitt watched a number of pupils of oddly poor reading abilities, some of whom were placed in a "special room" with a teacher dedicated to the teaching of reading. She found that students given such attention could, in fact, learn to read, at least adequately, but that it required dedication and a special method. The specialist used a variety of engaging techniques to help pupils associate sounded-out words with their printed representations. She thereby helped correct what was otherwise a failure in readers to forge "meaning" by combining the experience of perceiving written letters with previously experienced sensations—"memories of smell, taste, color, tactual qualities, kinaesthetic experiences, relations to other objects of perception, etc." In other words, fluent reading involved the habitual conjoining of experiences and memories from all the senses, and people with "word blindness" needed extra training to make those conjunctions routine. That extra training must centrally involve a strategy of active engagement, to attract the interest of the student and fix habitual links in the mind. Moreover, just as a good typist would often have to supplant early and clumsy skills in order to achieve the requisite fluency, it was also a matter of overcoming bad habits created by prior, poor teaching. In Schmitt's terms, the acquirement of early reading skill evidenced a kind of "mental plasticity." As a result, she reasoned that "congenital word-blindness" was not a result of a lesion in the brain at all, let alone, as some liked to claim, evidence of "atavism"—a counter-evolutionary return to precivilizational behaviors that might be brought about by such a lesion. It was instead a matter of tuning the teaching to the child. Her fortunate students had simply benefited from exposure to "a new method" of teaching, central to which was a focus on "phonics"—that is, the sounds to be associated with certain letters and their combinations. Schmitt also suggested that although "word-blindness" itself was an extreme manifestation, far more children experienced milder versions of this kind of problem than anyone had yet recognized. She particularly singled out those whose first language was not English, and "the colored children now coming to the city from the rural districts of the South, where little educational opportunity exists." Such children would naturally find that conventional reading lessons did not work well, because of course they were designed with different kinds of children in mind. Chicago's standard lessons did not engage with "the meaning complexes of their lives." These students were not "mentally defective," then, but ill-served. In the end, Schmitt even averred that in actuality *all* learning to read took place by "the phonic method," even though teachers themselves by now generally believed otherwise.[48] In that light, the question offered a key to understanding successful reading itself. It was a powerful point, not least because it was so much at odds with what the scientists of reading liked to maintain about silent reading and the importance of larger units of meaning. And it hit home. Judd effectively signaled as much when he essentially reprinted Schmitt's report—but omitted her conclusions—as the relevant chapter of his own *Reading: Its Nature and Development.* In 1922 William Gray likewise relied substantially on Schmitt in listing dyslexia as one among a number of problems to be addressed in his *Remedial Cases in Reading*.

By the 1920s, then, scientists of reading were well aware of this strange and apparently rare condition in which otherwise able people could not learn to read. Dearborn, among many others to draw on Schmitt, claimed that it was at least partly an inherited condition, and he attempted to draw up "family trees" charting generations of sufferers in one family.[49] Johns Hopkins education professor Fowler Brooks likewise characterized the condition as one among a plethora of reading "deficiencies," which extended even to the retention of a taste for "trashy books and worthless magazines," which he attributed partly to schools' failing "to apply the laws of learning to the teaching of appreciation."[50] But there was little consensus on what it itself was, or on how these "laws of learning" might need to be tweaked as a result. The implications would become clear only when it ceased to be seen as a matter of isolated individuals and was recognized as a problem that was social as well. That occurred in the mid-1920s, and it happened in rural Iowa.

In 1919, just as Schmitt's work saw print for the first time, Samuel T. Orton arrived in Iowa City to take up a position as chair of the local university's department of psychiatry. What had tempted him to move west, after spending five years as a pathologist in a mental hospital in Philadelphia, was the chance to take charge of the state's new public psychiatric hospital. The son of a geologist who had served as inaugural president of Ohio State University, and a cousin of President William Howard Taft, Orton was an ambitious and sometimes opinionated clinician. Educated at Ohio State, the University of Pennsylvania, and ultimately Harvard, he had also studied for a while under Alois Alzheimer in Breslau, Germany, and he was keen to advance the claims and authority of cerebral physiology over a range of mental and other conditions.[51] But Iowa City turned out to be a very different environment from these urban centers. Orton found himself in charge of a central institution expected to address the needs of a large hinterland of dispersed and poorly served rural communities. By 1924 the demands for some kind of traveling clinic had become pressing, and he yielded to them. Orton agreed to set up an experimental "Mobile Psychiatric Unit" to address the "mental hygiene" needs of distant towns.

The location chosen for the first trial, in the freezing January of 1925, was Jefferson City, a town of 3,400 in Greene County, right in the center of the state. Orton and a small team lodged with willing locals and took over rooms in the courthouse for their temporary clinic.[52] Having circulated blank forms in search of referrals, they received more than 200 suggestions of people to appraise, and examined 173 of them. "Special reading disability" was included from the outset as a psychosocial category, along with things like alcoholism, epilepsy, and "Organic nervous disease" (the classification scheme came not from some august institution but from a semipopular book devoted to listing the various provinces of what it called *The Kingdom of Evils*). Fourteen of the eighty-eight children referred to them as "dull" or "deficient" were found to have this condition. It immediately caught their attention more than all the other problems they encountered. A Stanford-Binet test implied that in these cases the cause was not any conventional lack of intelligence; these pupils simply had difficulty in the one skill of reading. Orton rapidly came to realize that such innate difficulty, far from being rare, was an everyday reality in the schools of the county. He particularly noticed two students whose problem was so severe as to be counted as congenital word blindness, and of those one, a sixteen-year-old he denoted M. P., stood out. M. P. "seemed bright," to the extent that Orton believed his score on the IQ test to be seriously deceptive, and further tests indicated that he would have placed in the top 1 percent of World War I army draftees. He could give the names of letters and numbers placed before him. He just could not *read*. Moreover, his inability took specific forms. He showed a marked propensity to parse letters and syllables backward, for example, as if reading them in a mirror. The traits were odd enough for Orton to have him transferred to Iowa City for further study.[53]

Curious, Orton went back to the founder of the science of reading, Émile Javal. He noticed that Javal had already observed in the 1870s that left-handed writers tended to produce mirror writing. He then asked around and was told by a colleague about the prevalence of right-to-left sequences in ancient scripts like Hebrew and Phoenician. In the early scripts that became Latin, he noted, inscriptions often included letters written backward compared to their modern alphabetic descendants, and the flow of words might go alternately from left to right and right to left in successive lines—in which case the pattern was called a *boustrophedon*, connoting the turns made by an ox team plowing a field. For a moment it seemed that the entire history of civilization had converged on this unfortunate Iowa teenager, and Orton leaped to a conclusion that was correspondingly grand. In the evolution of writing itself, apparently, it had originally not mattered which way round characters or words were written, as they were readable either way. The adoption of "dextral" orientations must have been merely conventional, then, being "fixed by custom and education."[54] For an ambitious cerebral physiologist the suggestions were intriguing. Orton already knew that experiments with the manuscope—an instrument that revealed which eye was dominant—showed that ocular dominance corresponded to handedness, and that both were correlated with hemispheric dominance in the brain. The tendency of word-blind people to read "backwards" implied to him a connection to this kind of work. He decided to call the phenomenon exhibited by readers like M. P. *strephosymbolia*, a neologism connoting twisted symbols.[55]

In light of this case, Orton noticed similar behavior, first in the other fourteen students, and then in more people, including adults. The Rockefeller Foundation now gave him $60,000 to investigate further. Cases rapidly multiplied as the mobile unit moved across the roads of rural Iowa; eventually the unit examined over a thousand children with roughly similar issues. It is tempting to suggest that it was the internal combustion engine that made dyslexia visible. All too many schools, he found, reported a proportion of otherwise intelligent pupils—about 12 percent overall, but with wide variations from school to school—who were backward in reading, and whose inability to read well obstructed their general development. It was not that they were unintelligent, but that their strange reading disability held them back in other areas of learning. These people often read and wrote mirror writing as efficiently as they did conventional writing, were clumsily ambidextrous, stuttered, and showed problems of balance while standing or walking. Sometimes they had been born left-handed, only for teachers to attempt to change that.

Almost inevitably, Orton was joined in the effort by a PhD student in psychology from the University of Chicago. Marion Monroe had a background in researching perceptions of color; now she undertook tests of a larger cohort back in Iowa City, and attempted to train them to be better readers by using kinaesthetics. The method was one that had been invented by Grace Fernald in California, and described in a widely read paper that Fernald coauthored with Helen Keller. It involved coordinating physical movements with visual perceptions. Here it meant that she encouraged students to move their fingers as their eyes ranged across lines of text, Orton's proposal being that this would help correct the problematic reversal of orientation. (Among scientists of reading, Arthur Gates would take up the idea and propose inventing movie-projector-like machines to guide eye movements even more efficiently.)[56] At length he inspired decades of neurological, therapeutic, and pedagogical researches, ranging from summer schools for slow readers to animal studies of cerebral physiology. Unlike Hinshelwood and his neurologist successors—but like Schmitt—he was disinclined to attribute the problem to a physical brain lesion. Instead he came to agree with Schmitt that "the sight method of teaching reading" (that is, the "look and say" method), while it might indeed work for many children, was disastrous for a few. The consensus that dyslexia was a discrete and idiosyncratic condition was therefore false: there was in fact a spectrum of disabilities, of which outright word blindness was merely the extreme. And if teaching techniques gave rise to the problem, then teaching techniques could remedy it.

Yet Orton's perspective remained that of a cerebral physiologist, albeit one informed by the science of reading. He came up with a sophisticated theory linking the operation of the human eye to specific locations in the brain. The central part of the retina of the eye—the specific part that provided for close focus, and hence the part that mattered for reading—was connected by nerves to two distinct locations of the brain, one on each hemisphere. When one looked at a line of text, therefore, impressions from both eyes went to both hemispheres simultaneously. In normal circumstances one side dominated, giving rise to a single sensation in a reader's consciousness. He reasoned that when one learned to read, one stored up memories of the letters in corresponding locations in both hemispheres. But the orientations of the letters and words were opposite in the two hemispheric locations. The orientation did not usually matter, because prolonged experience of reading led to a strengthened association with the correct orientation, which came from the dominant hemisphere. But in certain circumstances it was possible for this hemisphere to cease to be dominant. "Engrams" from the eyes were in fact received on both sides, and in such circumstances those on the usually nondominant side came to the fore for the consciousness. What was occurring in his poor readers, Orton suggested, was that these normally ignored

engrams were no longer being ignored. Instead they were becoming confused with those from the dominant side. This would not have mattered,
but for the fact that the engrams in the nondominant hemisphere were
mirror-images of those in the dominant. This was enough to delay the
mind making its associations with sounds, and hence to make the reading of syllables, words, and phrases very difficult.[57] And this was why his
subjects showed errors of such a particular kind—errors of "direction and
orientation," as he put it. They very often displayed a tendency to reverse
sequences of letters and syllables, such as confusing *p* and *q*, *saw* and
was, or *Gray* and *Gary*. At the same time, they could read mirror writing
better than expected. Such odd phenomena had often been noticed, but
they had never been explained, and they had tended to be disregarded
as simply strange. For Orton they became "cardinal symptoms." *Here* was
where the key to reading lay.

Orton's contention about strephosymbolia, like Schmitt's about alexia,
carried a very important and obvious implication. It implied that dyslexia was a matter of inappropriate habituation—and therefore that it
could be addressed. He suggested retraining readers' eyes by using the
finger to trace a path across the page and sounding letters out loud—an
old-fashioned method of learning, he averred, "under taboo today." This,
he thought, would help sufferers habituate their brains to reading in the
right sequence. Experimental efforts by Monroe along these lines certainly seemed to lead to real improvements. And, Orton added, his insight explained why some schools had twice as many reading-disabled
students as others: those that already used this "phonetic method" produced far fewer lagging students. That suggested a dismaying conclusion,
especially in the context of Schmitt's revelations about alienated Chicago
children. Not only did the prevailing method of teaching people to read
fail to address this disability: it might well *produce* it.

Orton's identification of reading disability as a product of cerebral habituation thus had at least four major implications, each of which cut to
the heart of the science of reading as an enterprise. One was that "congenital word-blindness" was not a rare and discrete condition, fundamentally distinct from the everyday range of reading abilities exhibited
by Americans. It was instead merely the extreme end of a spectrum of
disability—one that had never been noticed before, and that had real
consequences for the lives of countless people. The second was that these

children and adults could not properly be characterized as "defective." No problem of intelligence underlay this condition. Indeed, the phenomenon of strephosymbolia cast the whole enterprise of intelligence testing, then in its heyday, into serious doubt. Orton himself recommended adopting the relatively impartial term "disability" to characterize it. But this disability nevertheless did have real and distressing consequences, because it led to lasting feelings of inadequacy, distance, apathy, and paranoia in those who experienced it.[58] Third, and relatedly, it was not (or not only) a matter of unalterable and innate incapacity, but of habits. It could therefore be addressed—and quite possibly alleviated—by means such as the retraining of one's eye movements. And the fourth implication was that while the currently accepted method of teaching children to read—a method sanctioned by figures like Judd and Gray—might work on average, for many individuals it was not merely ineffective but counterproductive. It actively prevented them from learning to read, or at the very least presented obstacles to them as they worked to make progress. Strephosymbolia was a result of ill-training so severe as to affect the "normal physiological pattern of action" of particular brain locations. Its effects could well be permanent. America's schools, then, might well be stopping unknown numbers of people, even those who seemed able to read adequately, from experiencing "wholesome personality development." And the population of those harmed in this way was likely to be far greater than had ever been apparent before. The implications of Orton's theory were therefore certainly not limited to dyslexic readers alone. They extended to the entire population of readers and would-be readers—all those millions of people whom the science of reading had promised to help.[59]

The consequences of recognizing "word-blindness" as a problem of this specific kind—socially and geographically distributed, pedagogically inflected, and indistinct from a much wider range of reading abilities— would have major ramifications for the science of reading and its applications. One of those ramifications was the generation of fundamental doubt about the central premise and promise alike of this science. Maybe it was not measuring what its champions claimed it to be measuring; maybe its recommendations were not as beneficial as they maintained. When a high-school principal in Dover, Delaware, published a warning in *Harper's Magazine* in 1946 that among educators it was "common gossip" that at least a third of high-school students were unable to read effectively enough to use textbooks properly, and that this was by no means a result of any mental disability, the revelation resonated because there was already a basis for fearing this to be the case. The august W. S. Gray himself went further still, and suggested that about half the adult population were unable to read general-interest books and magazines. Existing educational practices, the Delaware principal added, were ineffective even for the two-thirds who *were* functional readers. He suggested essentially surrendering to reality: schools should resort to showing students films instead of asking them to read books. Such a recommendation had little impact at the time, but it betrayed a profound unease at the real state of literacy and schooling across America's myriad cities, towns, and counties. For the moment largely ignored, that unease constituted the combustible fuel that a campaigner named Rudolf Flesch would soon find it all too easy to ignite.[60]

4

What Books Did to Readers

In the 1930s the science of reading moved out from the laboratory and into the wider world. It was drawn out into that broader domain by the urgent need to tackle a crisis that the Depression and the rise of authoritarianism in Europe made all too evident. If democracy and modern industrial society were to survive and prosper, and if the challenges of economic disaster and geopolitical competition were to be met, then the American populace must be full participants in the complex informational environment that this generation inhabited. That meant not merely being literate—the old ideal of the eighteenth and nineteenth centuries—but being sophisticated practitioners of reading in everyday settings. Americans—all of them, not just urban white professionals— must exercise discernment in what they read and how they read it, enriching culture, enhancing understanding, and strengthening communal ties rather than fostering prejudice and discord. The problem was that America's leaders in the 1920s and 1930s were confronted with stark proof that this was not the nation that really existed. Something must be done, and the science of reading was one possible source of answers.

The science that attacked this ambitious target, however, was not that of Judd, Buswell, and their colleagues, even though some of them— William Gray especially—were prominent advocates for it. What arose to meet the need was not a laboratory science of eye-movement cameras and tachistoscopes, but a *social* science, dedicated to understanding and ameliorating reading as a collective, sociological practice rather than an individual, psychological one. Yet it too had roots at the University of Chicago. Indeed, in some ways the roots were even stronger and the affiliation more distinctive than those of Judd and Buswell's laboratory enterprise, because the so-called Chicago school of social science, then at its peak, was a major influence on its creation and conduct. The university's pioneering roles in urban sociology and anthropology, for which it was famous at the time, provided opportunities for kinds of research into reading that took a quite distinct tack.[1] Its social scientists furnished the new social science of reading with sophisticated methods, tools, and representational techniques, as well as valuable resources of authority and credibility. In this context, then, reading research did not attend to readers' eyes, and it did not depend on complex precision instruments. Nor did it reside in laboratories. It attended to reading *communities*. Its research sites were the streets and neighborhoods of cities like Chicago, Cleveland, and New York and—momentously—the often-neglected byways of rural America, most notably in the southern states where segregation reigned. By exploring the community practices of reading in these diverse environments, the social science of reading posed serious and fundamental questions about democracy and its future. Its studies were often underwritten by foundations, but they were also supported by the Roosevelt administration's public agencies, becoming thereby part of the institutional machinery devised to respond to the Depression emergency. And they gave rise to a large number of books, widely circulated and much discussed in the years before World War II. They had quite un-Juddian titles: *People and Print*, *The Geography of Reading*, *What People Want to Read About*, and the oddly disquieting *What Reading Does to People*. Together they constituted a field of knowledge quite unlike any that had existed before.

The social science of reading also emerged in response to two more specific societal issues of the interwar years: the rise of a variegated ecology of "documentation," and a stubbornly persistent problem of "illiteracy" made manifest by a new culture of pervasive testing of intelligence and reading alike, in the wake of the creation of the Stanford-Binet IQ test and the Thorndike-McCall *Reading Tests* issued from 1921. The first of these meant that reading must become much more evidently a variegated practice in its own right, and one that—such, at least, was the promise held out by champions of these new media—could elevate every village and town in the nation to a level of equal access to information that previously had only been accessible by travel to a metropolis. The second, however, threatened that utopian dream. After pioneering tests of military recruits in World War I had revealed the discomfiting fact that a large proportion of them were functionally illiterate, the spread of educational testing in the 1920s indicated that, even though literacy in the old sense of a simple "mechanical" ability to parse characters had been achieved at levels of around 90 percent in the population as a whole, major regional variations remained. Moreover, all too many Americans even in the industrial and commercial cities of the North were not fluent or everyday readers. This new form of "illiteracy" therefore became the subject of urgent, even panicked, attention in the mid-1920s. And as the Crash and the Depression hit American society, so efforts to tackle it grew more ambitious, coordinated, and extensive. The prime focus of the new social science would be on the institution that housed and facilitated reading as a communal practice, and on which the possibility of a modern information democracy seemingly turned: the library.

Reading, Science, and the Crisis of Democracy

The Slump, as eminent sociologist William Ogburn wrote in his preface to Douglas Waples's landmark book *People and Print*, "was like the explosion of a bomb dropped in the midst of society."[2] It was utterly unexpected, and its effects were socially as well as economically devastating. As it threw all existing economic, political, and social-scientific orthodoxies into doubt, efforts to engineer recovery would demand new knowledge of the culture and economy that actually existed in the United States. At the same time, while democratic institutions seemed powerless to solve the problem, Nazism and Stalinism offered what their defenders touted as modern, efficient, and effective strategies, which they claimed were in line with the latest science. Some of those defenders were American. The United States did not lack for charismatic would-be leaders—Huey Long, Father Coughlin, and, later, the pilot Charles Lindbergh—promising swift and sweeping remedies that the existing system seemed too sclerotic and compromised to deliver. Articulating what was worthy about a democratic order was consequently one of the most serious tasks for social science in general—as witness the work done to relate science itself to democratic social orders by sociologist Robert K. Merton, for example. The methods, data, questions, and infrastructure of the new field of reading science were another response to the same set of needs.

American society in the 1920s and 1930s was not only a culture of print, but a culture of manufactured information in a myriad of technological forms—hanging files, mimeographs, microfilms and microfiches, card indexes, punch-card systems, and more.[3] Since around 1900, America had been living in an age, not just of published works like books, newspapers, and posters, but of what the visionary bibliographer and information-science pioneer Paul Otlet christened *documentation*. The complexities of classifying, organizing, and investigating this mass of information were profound. In both Europe and the United States, the period from around 1890 had seen a plethora of projects, ranging from the mundane to the utopian, to take on the task. The most elaborate occurred in Europe, with Otlet's own *Traité de Documentation* (1934) their massive, prophetic, and minutely detailed manifesto. They also included Nobel Laureate chemist Wilhelm Ostwald's pre–World War I plan for an institution called "the Bridge" (the manifesto for which appeared in Esperanto),[4] and enterprises in particular disciplines such as Henri Poincaré's universal bibliography of mathematics based on card indexes, to which distant readers could subscribe by mail. Within corporations and institutions, a similar wave of schemes moved across the United States and other developed countries. Hospitals, for example, inaugurated ambitious plans for medical information management, of which John Shaw Billings was a renowned champion. It was already quite apparent by the close of the nineteenth century that some new form of classificatory science was needed to handle this domain, and in 1887 Melvil Dewey inaugurated his own proposal to meet that need in the form of "library economy," a field he created at Columbia University. The Dewey system proved widely influential, of course—although Otlet and his collaborators disliked it on epistemological and cultural grounds. This was made all the more apparent by the spread of public libraries through America's towns, sponsored lavishly by the Carnegie Corporation. By the time of World War I, it was nonetheless widely acknowledged that the proliferation of information and media alike was certainly a blessing, but it was one the benefits of which could not be reaped without a science to manage it. What was needed, it seemed, was something like a science of information.

In the period immediately before World War I, the Carnegie Corporation spent more than fifty million dollars founding public libraries. With the armistice it turned to considering what to do next. The corporation commissioned its officer William Learned to carry out a study, which it published in 1924 as *The American Public Library and the Diffusion of Knowledge.* Learned's report was expansive and optimistic, as befitted the technological aspirations of the moment. It inspired a new commitment to consolidate the profession of librarianship for a modern world in which libraries would be the lynchpin of a "Community Intelligence Service" operating in towns across the country and extending across media. But that aspiration implied an endeavor to investigate how best to provide such a service, and an institution to train the people who would bring it to life. There was only one obvious place to establish such an enterprise—the home of the science of reading itself. So the plan included an allocation to the University of Chicago of one million dollars (plus annual grants totaling another $385,000) to found "a graduate library school of a new type." The new school would occupy a position vis-à-vis these officers of the new informational world that would be analogous to those of Harvard Law School and the Johns Hopkins Medical School for their existing professions.

Such a proposal was not to be treated lightly, and a university committee was established to consider it. The committee, on which Judd served, soon approved the plan unanimously, and the new Graduate Library School (GLS) duly came into being in 1928. Its purpose was quite specific. It was not to train librarians in everyday administration—many other library programs already existed to do that, and in fact Chicago would require such training as a prerequisite for admission. Rather, the new GLS would undertake fundamental research on the nature and consequences of what one might call bibliographical culture, and it would circulate the findings in order to undergird continuing social progress. A new social science of reading would emerge here, then, underwritten by the Carnegie Corporation and centered on questions of (to use Learned's own term) public "intelligence."[5]

The intellectual leader of this second Chicago school of reading science was no stranger to the approaches of Judd, Gray, and Buswell. Douglas Waples was a member of the Education Department himself until he transferred to become one of the founding faculty of the new Graduate Library School. Unlike the nominal first leader of the GLS, George Works—

who was an efficient and convivial administrator but not a major intellectual figure—Waples had been interested in reading for years, dating back to his days of graduate research at the University of Pennsylvania, where he had investigated the role of "interest" in learning practices. Much more than Works, Waples had a vision for the new social science of information. So it was Waples who established himself as the new school's most dynamic member. Throughout the 1930s, he would use his position to steer a number of initiatives that posed big, urgent questions about the public sphere, democracy, intelligence, and propaganda. His influence was critically important for aligning the study of reading less with psychology, the laboratory, and the elementary school, and more with sociology, history, the military, and the state. It was under his effective leadership that the GLS became the nation's hub for a second science of reading, and created a new discipline of research in the social science of information.[6]

Like any ambitious new cultural venture, the social science of reading was immediately controversial. Waples and his colleagues interpreted the phrase "library science" in an unprecedentedly broad sense, while at the same time insisting that the "science" part of that phrase had real implications for the methodology and authority of their enterprise. A major problem at the outset was that their interpretation was a minority opinion within the existing profession of library staffers and administrators. Even among those who got their point—indeed, perhaps *especially* among those who got their point—it was by no means consensual that the initiative was a good idea. Waples's own notion, announced in the first issue of the school's house journal, *The Library Quarterly*, was lifted directly from John Dewey. The practice of library operation on this account was a pragmatic "social enterprise," which involved discovering "what human needs are responsive to reading" and then furnishing "the best reading in the best way." The essential nature of the endeavor was to provide "whatever helps to reveal and satisfy the needs that reading can meet." Reading was clearly at its heart—the term cropped up three times in that definition alone. But identifying how best to put this vocational enterprise into practice, and then training people to do it, would inevitably have to be a collaborative, boundaryless venture. So the GLS program of study would be designedly interdisciplinary—there was to be no "building of high fences about intensive professional interests." Establishing the necessary social networks would take time, however, and

Waples proposed that in the meantime library science should remain "in the laboratory stage" (a phrase he meant metaphorically) rather than being taken to suggest practical policies. But the clear implication was that such recommendations would indeed be on the agenda soon enough.[7]

Waples presented his vision as a consensual one, but in fact there was little agreement on either the overall ambition or the concept of science that supposedly defined it. He certainly had supporters, most notably the dean of Case Western University's School of Library Service—the term *service* being a deliberately modest one, chosen to avoid the kind of confrontation that Waples now ignited. Charles Williamson's speech in favor of the new venture preceded Waples's as the first article printed in *Library Quarterly*. But Waples also found himself assailed, and on two opposite fronts. On the one side, the venerable American Library Association—another beneficiary of Carnegie largesse—disdained the new vision of library science as at once pedantic, amoral, and authoritarian. For these custodians, librarianship was not a science at all, but a humanistic art, steeped in cultural history. University of Pennsylvania librarian C. Seymour Thompson, for example, caused a kerfuffle in mid-1931 by delivering a speech that reiterated the ALA's nineteenth-century ideals and asking whether a library science were desirable at all. The terms in which he did so were striking, because they implied that behind Waples's science lay Judd's. Quoting some of Williamson's less diplomatic terms of boosterism for the new project, Thompson invoked a specter of "psycho-sociological laboratory methods" being forced on readers in town libraries across the country. The very notion of library science implied that patrons would be subjected to spuriously scientific tests, he claimed, on the basis of which soulless apparatchiks would then "prescribe" them a "program of reading" suited to their condition. Their range of reading choices would be constrained in the name of efficiency and modernity. Spurious objectivity would generate real oppression. Cartoonish as it was, the complaint touched a nerve. Even within the University of Chicago itself Thompson's rhetoric resonated, not least because some faculty believed that Judd had actually engineered the creation of the new school in order to enlarge his personal empire. Judd had won a cost-free "appendix" to his School of Education, whisperers implied; there were rumors that he was angling to become the next president of the university. Perhaps this Svengali of the academy had intruded Waples as his placeman.[8]

This attack provoked Waples's colleague Pierce Butler into action. A renowned figure in his rather desiccated field of bibliographical history, Butler had come to the university from the Newberry Library, where he oversaw the John M. Wing Foundation devoted to the history of printing. Before that he had led a colorful life, including earning a doctorate in divinity, failing as a religious minister, and toiling as a casual laborer; he was working in a railroad freight office when a lucky meeting led to his Newberry appointment. Butler now composed a distinctly opinionated *Introduction* to the new discipline, in direct response to the ALA polemic. It became the first volume to appear in the GLS's new book series. In truth, his short but engaging tract was less an introduction than a manifesto. It contained almost nothing that aspiring library scientists could put to use, but much that evangelized for the cause itself. But Butler did not quite agree with Waples in this professed introductory work. He reckoned that library science should be *more* scientific.

Butler set out to answer the question of what made the new enterprise a science by expatiating on the meaning and nature of science itself. His account was historical, and in some ways uncannily prescient of how later generations of historians would portray the subject. He began by noting that the concept of science had been introduced only a few decades earlier, as a result of "revolutionary changes" in the earlier domain of "natural philosophy." It had struggled to win acceptance for years, he explained, but by the early 1930s its meaning was broad and largely uncontroversial. It referred to "all knowledge that is distinctly of a modern character." Quantitative measurement was one distinctive attribute. A second was a dedication to causal explanations of generally observed patterns, rather than individual instances. And a third was that science made use of instruments. Interestingly, Butler's example of a science experienced mundanely in the modern world was that of life insurance, and he proposed that library science be thought of as a similar kind of practice: quantitative, social, predictive, professionalized, and theoretically sophisticated. (He might also have pointed out that insurance used instruments, such as adding machines and Hollerith cards, that bore a family resemblance to the latest bibliographical tools.) Library science would be dedicated, however, not to actuarial rationality, but to "the transmission of the accumulated experience of society to its individual members through the instrumentality of the book."[9] And so a library, on

Butler's account, was the "social apparatus" by which this scientific enterprise worked, and hence that enabled modernity itself to exist. It did
so by allowing for the accumulation of culture over time. It must therefore be disinterested, objective, and quantitative in its operations. It must
see ideas and books as products of society as a whole, regardless of the
convention of attributing them to single authors. And it must apply this
"essentially social quality" to reading, too. Butler insisted that reading
was an intricate and variegated activity, with any particular experience
of it being shaped by current social realities and past history. The reason
it was social was that what he called "the reading chain"—the sequence
of "graphic symbols, words, grammatical combination, ideas" that linked
traces to *knowledge*—could never have acquired any meaning except by
collective, historically evolved "conventions." That explained why readers experienced a text differently from each other, and why an individual
experienced it differently every time he or she reread it. Broad "types"
of reading did exist, however, and these needed to be identified, investigated, classified, and integrated into the education of librarians, whose
métier it would be to attend to them and suggest the most appropriate
books for their clienteles in light of that knowledge. It was quite possible,
for example, that a reader's practice might become "morbid" and even
"vicious"—the kind of thing that befell people who took "engrossing delight" in "pornography, crime, and anti-social unconventionality"—and
these pathologies had to be watched for. Writ large, books in general were
"a neural apparatus for social memory," and reading was the social body's
act of remembering. So "learning by reading," Butler concluded, was a
kind of societal "metabolism"—it had to continue ceaselessly "if the normal social state of well-being is to be maintained." In that sense, library
science might even be thought of, not just as social science, but as social
medicine.[10]

"Effective librarianship," then, was for Butler "largely a matter of accurate psychological diagnosis." An important aspect of such diagnosis was
that the science must be historical, in the sense that it could not avoid
taking a stance on the development of knowledge itself: whether it were
cumulative or evolutionary, and whether what mattered about a book
was its timeless excellence or its having played a part in an important
historical process. This applied not only to formal science, moreover, but
to all of what Butler idiosyncratically called "scholarship," by which he

meant "the total residual effect of an individual's intellectual experience." As an example, he pointed to how the conviction that "mental activity is related to the physiology of the nervous system" had become as universal as it was by the 1930s. What distinguished library scientists was that they cared not about the ultimate truth of such beliefs but about their uses in a particular time and community. The "primary concern" of librarians must be libraries' "social effect." For any given book, their province was objective knowledge of "the quality and the amount of its influence on the intellectual life of the people who use it." In effect, they practiced applied pragmatism. At length, Butler envisaged (as, in fact, had Learned) pooling the various libraries in every region into a single "organized system" rather like a coordinated network. It was a frankly utopian idea, which would only become viable if librarians doffed their current dedication to particular collections in favor of an impartial, objective, delocalized ethos—a scientific ethos, in other words. It would also need substantial investment to create a mature, networked system of documentation in Otlet's sense, involving microfilm, catalogs, interlibrary loans, mobile distribution services, and the like. But it was vital if library professionals were to become America's scientists of cultural influence.[11]

The social science of reading thus emerged amid a nexus of resolve and controversy. It did so, moreover, under the unexpected but acute pressure of the Slump. The Depression would inevitably permeate everything it then did. As the Roosevelt administration ramped up its recovery efforts, it set great store on the gathering of social-scientific knowledge about the country it was governing. The Works Progress Administration, and in particular the arts component of the WPA known as Federal One, sponsored projects to foster representations of American culture, such as murals, travel guides to every state, and documentary movies.[12] Knowledge about reading habits became one of its most important and consistent desiderata. Library science would now be a major resource for that effort. One reason was that usage of libraries nationwide increased dramatically when people could no longer afford to buy books, yet the funding of such libraries plummeted. The combination made it all the more important to arrive at an authoritative understanding of what libraries could and could not do for their communities. Another reason was that library-science researchers rapidly developed unequaled experience in conducting surveys of readers' habits at a variety of scales, from the ultralocal (a particular school, halfway house, or prison) to the national. Surveys employed people, so they provided work. And they generated potentially useful information, which made them that much better as vehicles for federal and philanthropic investment. So the federal government's programs coalesced with ongoing grants issued by the Carnegie Corporation, the Rosenwald Foundation, and other groups. The Social Science Research Council (founded in 1923) and the American Association of Adult Education (founded in 1926 in Chicago by the Carnegie Corporation) joined forces too. The kind of work that this cluster of institutions sponsored was largely that championed by Chicago's GLS, and by Waples and his allies in particular.

With this kind of support the social science of reading thrived through the 1930s. It gave rise to a long series of publications, all underwritten by acronym-laden foundations. Waples's own *People and Print: Social Aspects of Reading in the Depression* (1938) was supported by the Carnegie Corporation and the American Association for Adult Education, and by the Social Science Research Council too. William Gray and Ruth Munroe's *The Reading Interests and Habits of Adults* (1930) was one of eight volumes of Studies in Adult Education sponsored by the Carnegie Corporation, the ALA, and the American Association for Adult Education. The ALA and the AAAE jointly established a Committee on the Reading Interests and Habits of Adults, which itself recommended books like Waples and Ralph W. Tyler's *What People Want to Read About* (1931) to the Carnegie; in this case the American Historical Association also furnished support. And the GLS itself launched a project in 1933 to explore the relationship between libraries and American governance, which generated three books published in the Studies in Library Science series: Carleton B. Joeckel's *The Government of the American Public Library* (1935), Waples's *People and Print*, and Arnold Miles's *Public Administration and the Library* (1941). All three carried forward the vision of libraries as a lynchpin of American social democracy and public intelligence. By 1940, Waples could survey what was a substantial and wide-ranging shelf of literature and see in the long sequence of books, papers, and semipublished mimeographed reports that had flowed forth in these years the emergence of an interdisciplinary social science.

The social science of reading was explicitly and repeatedly identified in endpapers, prefaces, and forewords with other initiatives in Depression-era social science, such as those dedicated to crime, the family, migration, minorities, recreation, religion, rural life, consumption, and health. One particularly important aspect of this work, moreover, was its attention to how issues of inequality in access to books—and to information in general—related to America's racial politics, especially regarding recent immigrants and African Americans. For example, Eliza Gleason's doctoral research, published as *The Southern Negro and the Public Library* (1941), referred to ten other major projects on similar themes to have appeared in a handful of years, under various funding and institutional auspices. A notoriously under-studied subject was finally receiving due attention, and, as we shall see, to immensely consequential effect.[13]

The Depression-era context helps explain the otherwise peculiar character of the reading studies that emerged from library science. "Inasmuch as the 'way out' may lie in the command and interpretation of facts which explain progressive changes and evolving social life in a dynamic world," as Gray put it, "the place and purpose of reading cannot be disregarded." The studies of this initiative should therefore be approached as akin to the period's groundbreaking studies of poverty, race, and other social issues, and even to movies like *The Plow that Broke the Plains*. At the apex of them all stood the Roosevelt administration, which quickly learned to use surveys, correspondence, and public-opinion research to inform itself of the state of citizens' views and try to shape a national "democratic mind." In that campaign, library science was by no means a pettifogging enterprise—in a sense, it was one of a handful of ventures on which the fate of the nation might hang. It was telling, for example, that when Roosevelt appointed Archibald MacLeish to be the new Librarian of Congress in 1939, he did so precisely because, as Felix Frankfurter attested, MacLeish was *not* a librarian in the traditional, humanistic, erudite mold. He was a reformer of the Waples and Gray camp: someone prepared not only to articulate the centrality of the book to democratic modernity but to define print's place alongside the new media of radio, cinema, and, imminently, television.[14]

The Geography of Reading

The outstanding achievement of the social science school of reading was the demonstration that reading in America had not only a psychology, but a history and a geography too. That meant that it had a politics, and a politics that demanded urgent attention. It had always been clear that reading was a constitutive element in the modern American polity. But the work done by the social scientists of the GLS proved that the converse was true too: politics had played a signal role in constituting a nation of readers—and, more to the point, of nonreaders, or at best of halting and awkward ones. The story that these investigators—social scientists, historians, statisticians, and mapmakers—told was explicitly a tragedy. The politics of reading in the 1930s, they showed, were harmful enough to endanger the nation's very future as an industrial democracy.

The most important protagonist in this effort was Louis R. Wilson. A North Carolinian by birth, by the time he arrived in Chicago Wilson had had a long stint in charge of the library of the University of North Carolina, dating back to the beginning of the century. He took over as dean in 1932 and ran the Graduate Library School for a decade, overseeing much of its most successful and influential work. These were the years of the New Deal, when federal and state agencies were seeking answers to the great social problems of the Depression and sought to recruit workers to provide them. Wilson's GLS seized the opportunity to muster resources and deploy them in extensive and often statistical studies of the social extent, place, and impact of reading across the country. It was under his leadership that many of the most influential works of Gray, Bernard Berelson, Ralph Tyler, and even Waples himself appeared. Wilson's own interests, however, centered on public libraries and the services they offered. He was fascinated by the questions of how, why, and to what effect library "service"—that is, the provision of public libraries, with their educational and cultural activities and their circulation of books—varied across the United States. And he knew full well that anyone asking those questions would have to be attentive to the stark racial disparities of the country.

Wilson's endeavor was spurred by an initiative of the Julius Rosenwald Fund, a philanthropic organization launched by a Sears, Roebuck and Co. executive who had been inspired by Booker T. Washington. The fund supported efforts to improve educational and cultural opportunities for minorities, particularly African Americans in the southern states, and, as we shall see, it also played a key part in supporting the research of Horace Mann Bond on Alabama's segregated education system. In 1929 it had undertaken a program to improve library service in the South "to all residents, urban and rural, black and white, in school and out," by providing a total of $500,000 in grants to eleven counties in seven states—North and South Carolina, Tennessee, Alabama, Mississippi, Louisiana, and Texas—in hopes that they would persuade local communities to support libraries on a continuing basis. The program lasted five years. To appraise its effects, the fund commissioned Wilson and the Library School to undertake a systematic study of these counties, comparing them to counterparts in other parts of the country. Wilson leaped at the opportunity. He recruited a Georgia-born graduate student in the Department of Education, Edward Allen Wight, and the two of them launched the first systematic attempt, as they put it, to "*place* the library"—to explain this institution in terms of the history, economy, social structure, and politics of its setting. It would have a lasting impact, not least because of the role it played in shaping Wight's future career. After serving in various library and university positions, he would join the School of Librarianship at UC Berkeley and play a major part in shaping California's postwar public library system.

Wilson and Wight approached the library issues of the South as consequences of the region's historical geography, and in particular its climate and ecology. Their focus was on the southeastern states that were predominantly devoted to cotton cultivation. "Cotton," they declared, was in this region "the arbiter of work, play, and most that enters into the making of the society which it largely supports." That applied to reading as much as anything else. The very supply of books to readers was substantially limited by the hard facts of a society built atop submarginal land that was substantially degraded by a system of exploitative farming. Sharecropping and underprovision resulted in illiteracy rates higher than anywhere else in the country. Moreover, the region was losing adult taxpaying citizens because of migration to the North, at the same time as maintaining a higher birthrate than any other in the United States. That meant that an emaciated tax base was being called upon to educate more and more children (with family sizes being typically greater for whites). And segregation only made things much worse: prior to the Rosenwald project, most African Americans in the South had had no access to public libraries at all. Wilson and Wight trod carefully here, but they made it clear that they regarded the division of schools and libraries between the races as at best grossly wasteful, and in that sense harmful to whites as well as African Americans. Moreover, the region ranked just as poorly when it came to "other media for the dissemination of ideas"— bookstores, magazines, newspapers, radios, telephones, and motion-picture theaters. The only exception was the provision of churches. The success of the Rosenwald's grand "social experiment" must therefore be appraised in the context of a stark and enduring ecological politics.[15]

Wilson and Wight's findings were dismaying. Public libraries were provided by democratic societies, they remarked, so that "every man, woman, and child may have the means of self-education and recreational reading"; they diffused "information and ideas necessary to the present welfare and future advancement of a community."[16] In other words, reading was essential to a modern society and libraries were the means by which it could take effect. But those means were widely lacking in the South, even in areas that on the face of it did have libraries, and even in the few that admitted both races. Librarians often hesitated to en-trust books to rural African American schoolhouses, for example, on the grounds that their rickety and insecure buildings were vulnerable to "prowlers and thieves."[17] The two groups making most use of libraries were "students and housewives" (women users outnumbered men by two to one), but neither group was well served; and among employed men, 6 percent of whites borrowed books, but only 0.1 percent of Black min-ers.[18] These were dreadful numbers. And the implications extended to the quality of reading, as well as its quantity. Wilson and Wight found that more than a third of books read by "urban Negroes in the fifth and sixth grades" were of "the cartoon or similar type." And this was not a manifestation of some kind of racially intrinsic tendency to bad taste. Where the Rosenwald Fund had provided "superior" kinds of fiction, the books were indeed used, which meant that the appetite was there. The library could clearly exert an influence, then, in fostering "the appreci-ation of good literature," and this should be all the more important in a region with such a large proportion of children. Well-prepared librarians could actively further this effect, seizing an opportunity to create a future generation of adult readers. But the scale of the need was immense and the investment paltry.[19]

The fund dramatized Wilson and Wight's dispiriting results in a short graphic primer entitled *School Money in Black & White*. African Ameri-can artist William Edouard Scott furnished images capturing the drastic inequalities in educational provision between northern populations in general, southern whites, and southern Blacks (fig. 4.1). In the Deep South, Scott reported, "colored children have less than one fifteenth the opportunity of education of the average American child," with Mississippi spending $5.45 per African American child per year compared to a national average of $99. No county in the region had sufficient books—let alone books in good repair—to provide an adequate service to African Americans. In principle, abandoning the region's traditional reliance on property taxes could help enormously. But outside money would be critically important too, even if it could only elevate services temporarily. Without it a library might never reach a standard high enough for the local community to recognize it as worth supporting. Moreover, the need was for more than mere repositories of books. A region's libraries, Wilson and Wight claimed, should be its prime providers of textual "materials" of all kinds appropriate to the locale's "agricultural, industrial, and social needs." Crucial to realizing their promise would be the embrace of the already growing technological infrastructure of bibliographical aids—microfiche, card indexes, interlibrary loans—to furnish materials that otherwise would remain out of reach. They also recommended a system of "book truck[s]" devoted to outlying rural areas, which could provide frequent circulations of stock from such an infrastructure and well-trained custodians dedicated to local groups; African Americans in particular should be trained to meet this need. A comprehensive national effort was required, then, and only the state could take on the task of upholding a service so central to the general good. Overall, Wilson and Wight called for an urgent policy initiative to "equalize" the distribution of resources for all the informational institutions of the age. That call resonated in what would become momentous ways.[20]

Meanwhile, Wilson's work on the Rosenwald project led him directly to a much more ambitious venture—one to chart the "geography of reading" across the entire country. The Depression made such an endeavor vital, he believed. The economic Slump had been "a stern schoolmaster" to America, and "every thoughtful citizen" needed to learn its lessons. Like other institutions dedicated to the public good (schools and colleges in particular), libraries had seen a combination of a cut in resources and a jump in demand. As unemployment increased, millions had turned to them for solace and recreation, for help in adjusting to new jobs and a

4.1 "The Black and White of It." William Edouard Scott's graphic portrayal of inequities in education funding. *School Money in Black and White* (Chicago: Julius Rosenwald Fund, 1934), 13.

new future, and "for providing equalization of educational, cultural, and recreational opportunities in a democracy." It was time to establish scientifically the "value" of libraries across the nation.[21] He attracted the support of the Carnegie Corporation to begin this effort, and called on a number of New Deal organizations, including the WPA, to help with the substantial statistical and cartographic work it required. In scale alone it would be daunting. The US library system—or rather, systems, because public, subscription, and academic institutions were largely discrete— was in quantitative terms massive. In 1934–35 alone America spent over forty-five million dollars on public libraries, which contained—Wilson recorded the total with preternatural precision—100,470,215 books.[22] But such national figures masked drastic inequalities by region, state, and county. For example, the ALA's recommended amount of expenditure on libraries was one dollar per person, but in fact the national average was about thirty-seven cents, and only one state, Massachusetts, achieved the ALA target. In Arkansas and Mississippi spending was as low as two cents. Similarly, while no *state* had zero public libraries, some 40 percent of *counties* did—and in those counties the expenditure per person was likewise zero. For about forty to forty-five million Americans—eight million of them African Americans—libraries effectively did not exist at all. And even when African Americans did live in areas where some library service existed, in the vast majority of instances (416 out of 491 counties) Jim Crow policies meant that the library itself was closed to them.[23]

The central question that Wilson addressed in his gargantuan geographical project was therefore the question of "inequality of access." Clearly, extremes persisted across the country. They might even be worsening. But how great was inequality in the different counties, states, and regions of the nation? Why did such inequality persist, and how could it be reduced? How did access to libraries relate to comparable problems of access to other forms of communication and information, such as radio, telephones, and movies, which might perhaps mitigate such inequities? And, perhaps above all, what were the consequences of this geography? To articulate these issues, Wilson deployed a technique that the Chicago sociologists of the time had pioneered—a technique at once rhetorical, statistical, and graphic. It centered on the production of dozens of tables, graphs, and, above all, maps (fig. 4.2). These maps conveyed the spatial distribution not only of libraries, but also of bookstores, movie theaters,

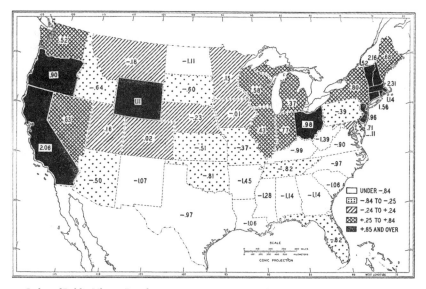

4.2 Index of Public Library Development in 1934. Louis R. Wilson, *The Geography of Reading: A Study of the Distribution and Status of Libraries in the United States* (Chicago: University of Chicago Press/ALA, 1938), 185.

and many more vehicles of culture. Conforming to a handful of templates and appearing every few pages in a volume almost five hundred pages long, they were meant to be viewed in rapid succession. A reader could not quite riffle through them in the manner of a flip-book to produce a composite image, but the intended effect was not far short of this. They should produce the equivalent in the mind's eye of an afterimage. That enduring impression would be of a gradient of shades extending from black in the Northeast and Far West to white in the Southeast, the white signifying absence and barrenness. "The fact," Wilson insisted, "begins to emerge in the similarity of the white, shaded, and solid black areas" across dozens of maps viewed sequentially: "there is a library map, or geography of reading, of the United States."

Seen in this way, Wilson's maps demonstrated the same stark truths over and over again. The greatest cultural provisions existed in the Northeast and the far West, with local strengths in the North and Midwest. The South and Southeast, by contrast, constituted a region of poverty, backwardness, and despair. Overall, Wilson arrived at what he called an "index" of library service, recording its variations across the country on a common scale. California (which already had substantial publicly mandated provision) ranked top, Mississippi bottom. Of the forty-five million Americans with no real access to public libraries, about half resided in the Deep South—which meant, notably, that about half did not. Library resources were largely inaccessible to about a third of the US population and obtainable only in "distressingly inadequate" ways to another third. Moreover, he noted, this gross inequality was not lessening over time; on the contrary, it tended to entrench itself. States with small investments in existing libraries also tended to have lower acquisitions budgets, so they fell even further behind. And by trawling through statistics on every other cultural institution and form he could think of— bookstores, publication subscriptions, radios, telephones, automobiles, movie theaters, and more—Wilson found that the inequality of access revealed by his "library map" was a general phenomenon. A rural area without a library would generally also lack all these other assets. Collectively, his maps therefore revealed a story of dire want in all too many areas. And the tale they told was not accidental, but structural; it had emerged over time as the result of a geographical and historical logic. It was truly a "tragedy" in the proper sense of the word. (Wilson's style was generally undemonstrative, so declarations like this conveyed real emotion.) That was why the library index was so important. It constituted a "thermometer," which revealed variations in "cultural temperature" at quite fine scales across the entire nation. If reading was the most important act of cultural inclusion in a modern industrial democracy—the act that, more than any other, shaped "the cultural pattern of American life"—then the climate that it revealed undermined the very existence of that society.[24]

Wilson also accentuated the scalar character of the problem. That is, the national divide was recapitulated at state and county levels. In the North as in the South, lack was far more evident and pressing in rural counties. Some 37 percent of Americans overall lived in the 40 percent of counties with no library service at all, but that average masked even greater extremes: in Massachusetts, 100 percent of the population had access to some public library, whereas in Arkansas 15 percent did, and in West Virginia only 12 percent. And at a finer resolution, it got even worse. Accessibility was concentrated in towns and, above all, cities. In Illinois, 75 percent of people overall could reach a library, but in the state's less populated counties that fell to 43 percent, and some had no access at all. Eighty-eight percent of the Americans who had no access to a library lived in rural areas like those of southern Illinois, and those areas were by no means restricted to the South. They were scattered nationwide. The issue of rurality, Wilson therefore concluded, was "America's greatest library problem."[25] But even the cities were in crisis. As the Great Migration progressed, so library investment in Chicago and other destination metropolises was falling behind. People arriving from regions without library facilities would lack the "permanent habits of reading and reflective thinking" that a modern city needed in its citizens. Unfamiliar as they must therefore be with the life of reading, how could the newcomers be expected to adapt and assimilate to a modern, fast-moving, and complex urban environment? They were hardly the adaptable and skilled workers a modern industry sought. Here too was a crisis, then. It was different in tone and implications, but it was absolutely inseparable from the crisis facing African Americans in the rural counties of Mississippi, Georgia, and Louisiana.[26]

To explain this tragedy, Wilson returned again to elementary geographical, climatic, and demographic facts. In particular, he blamed the domination of the cotton economy in the Deep South, with its associated problems of exhausted and submarginal land, soil erosion, and poverty. The cotton economy, based as it was on relentless exploitation of the landscape and reliant on a sharecropping system that kept the population rural, poor, and disproportionately illiterate, was insufficient to support an adequate information infrastructure for modernity, and was often actively hostile to the prospect. Sharecropping in particular Wilson condemned as a "blight," lamenting the proportion of tenanted farms in the South (56 percent) as "one of the most tragic figures in American life." Segregation and the "historical lag" in provision inherited from slavery made things much worse for everyone, whites included. As for race itself, Wilson again avoided taking an explicit side in the debates over inherited intelligence: but the whole thrust of his work implied that environment was the dominant issue. As he put it, if the United States was a melting pot, then the public library was the instrument that made possible the "melting process," and without this institution the national promise could never be met.[27]

But that posed the greatest question of all. What difference did it make? *Why* did information inequality matter? Wilson was acutely aware of the difficulty of answering that question precisely. On the one hand, he emphatically insisted that the viability of a free industrial nation was at stake. The training in cultural techniques made possible by libraries enabled countless humdrum citizens to "use print effectively in gaining an intelligent understanding of the present changing social order." If eighty to a hundred thousand schools nationwide did not have access to libraries, for example, then "the oncoming generation" would face a severe obstacle in its bid to become capable of navigating "a highly complex, democratic society." That was as clear to him as it was to many others. And he conceded that some progress had been made, not least in getting library and teaching staff to see the importance of training students to cultivate "permanent reading habits" of the kind that would enable them to acquire such mastery. (As Wilson pointed out, his Chicago colleagues had developed tests to measure how well-trained students were in the practical skill of library use.) New Deal institutions like the Works Progress Administration had devoted major resources to helping hundreds of libraries become established and successful. Technologies like microfiche, and bibliographical systems to organize and search for them, would make it possible to conduct a species of reading at a distance.[28] These were all genuine elements of progress. But on the other hand, they did not answer the fundamental question. What was the *significance* of reading, for individuals and thereby for society? What did reading *do*?

In arriving at that question, Wilson ended up returning to the central problem that had motivated the science of reading since the start. And here he was much more circumspect. It was dauntingly difficult to measure what effects the practice had on readers themselves, he remarked; statistics and maps could not provide answers. Circulation numbers were available, but they were recorded inconsistently, and they usually gave no indication of the kinds of books borrowed or their uses. (As Wilson and Wight had put it, "*Fun in Bed* and *A Short History of the English People* give equal weight to the circulation figures.") Besides, there were no scientific standards for measuring cultural quality.[29] But the real problem was that no mere collection of social data could address "the fundamental difficulty of measuring the effect which reading produces upon individuals." Perhaps a science could be devised to cope with that difficulty, he mused—he was surely aware that people in his own institution believed themselves to be pursuing it—but it was also possible that such effects were simply and irreducibly individual. In the end the geography of reading reinforced the urgent and profound need for such a scientific discipline. The great nineteenth-century vision of a literate populace had generally been achieved, but this achievement had only clarified the nature of the real problem: nobody really knew "what the social significance of reading actually is." There having been no "united attack" on the question, nobody quite knew how to produce "a convincing statement, based on well-attested facts, concerning the measurable values resulting from reading." Everyone accepted that reading affected "the conduct and behavior of people generally in socially significant ways," but nobody could say in what ways it did so.[30] The whole issue also made apparent the urgent need for research on "the effect of reading on conduct, attitudes, and emotional patterns." So Wilson concluded by calling upon the entire array of social disciplines, from political science to sociology, to join forces in trying to create the super-science of reading that modern society needed.[31]

The Illiteracy Panic and the Science of Readability

The social scientists of reading confirmed at least one immensely worrying truth that many had suspected to be the case for a decade or more. Even though literacy rates overall across the United States were in the region of 90 percent, meaning that the great aim of the nineteenth century had largely been achieved, all too many Americans were not actually very *good* at reading. William Gray in particular showed that a large proportion of citizens, even if they were technically literate, could read comfortably only at the level expected of an early teenager. As Wilson and Wight put it, Gray had "determined scientifically" that about 50 percent of adults could read only at the level of "pupils in the seventh grade"—and in the South, of course, the proportion was even lower. The implication was distinctly worrying. Literacy alone was not enough. *Books* were not enough. Simply making print available to the populace was all very well, but it was hopelessly insufficient by itself. Those responsible for managing the culture of print—above all, librarians at work every day in counties across the country—had to think hard about "the needs and reading abilities" of the people in their localities: they needed to work to address those needs and improve those abilities. They had to gauge what *kinds* of books to offer. Works that were too difficult for local people to read would never sustain such readers' attention. In the worst case, readers would become discouraged and permanently alienated. Contemporaries still called this issue by the name of "illiteracy," and major efforts were launched to eradicate it as such. But in fact the major part of their concern was for what might more properly have been called (to use a term that existed at the time but became common only later) *aliteracy*. Americans might be able to read, in a mechanical sense, but they were not equipped to be engaged, active, and constant readers. They were not prepared to participate fully in an informational republic.

The visibility of this kind of "illiteracy" came about by virtue of the embrace of tests that purportedly measured reading ability—and, more broadly, the rise of psychological testing in general. This was in turn an indirect outcome of the kinds of psychological approaches to reading that had been pioneered by Cattell and were then pursued by Edmund Burke Huey and his peers. In particular, it reflected work by Huey's contemporary at Clark University, the Indiana native Lewis Terman. Terman had started out working alongside Huey in G. Stanley Hall's class, and the two clearly learned from each other. Initially an advocate of Cattell's brand of anthropometry, he had seen its limitations at an early stage and adopted instead a mixture of Alfred Binet's more informational tests and Huey's science of reading. While still at Clark, Terman applied this combination of tests to a group of schoolboys in Worcester identified as "bright" and "stupid," in hopes of identifying common features of "genius." Reading ability was one of the evident indicators of such genius, he believed, along with qualities such as "invention."[32] Terman then made his way eventually to Stanford, where he became a professor of educational psychology and a staunch advocate of eugenics. In 1916 he issued a revised version of Binet's tests, with the aim of measuring "intelligence" objectively. Terman's "Stanford-Binet" tests became the accepted measure of this property, which they expressed as a number denoting an individual's "Intelligence Quotient," or IQ. For Terman himself—and in the 1920s his view was widely respected, for all that it was never beyond contest—intelligence was substantially inherited and unalterable, and such tests could therefore be used to identify innately superior individuals who could then be provided with enhanced schooling to enable them to make the most of their talents. Terman further insisted that members of African American and other nonwhite races were typically inferior in intelligence: his tests' statistics therefore supposedly warranted not only educational segregation itself but segregation into schools with explicitly divergent objectives, tending to channel Black people into menial occupational lives.

Terman's Stanford-Binet test was given its first major application when Robert Yerkes, then president of the American Psychological Association, proposed using it to appraise conscripted soldiers. Yerkes was grasping an opportunity to advance the reputation of psychological testing itself, by proposing to use it to help the war effort by guiding the allocation of men to ranks and roles.[33] Terman was recruited into Yerkes's effort, as too was his fellow eugenicist, Thorndike. As Thorndike himself put it, their aim was to "prophesy the mental ability which a man will display in the Army." But the tests that Yerkes's group initially proposed had a serious flaw: they were presented in writing. That meant that they were ill-suited to recruits who were either non-Anglophone or illiterate, as it transpired 18 percent were. To address that issue, Thorndike attempted to repurpose a standard measurement of reading ability that he had devised with fellow scientist of reading William Gray for use in schools.[34] But in practice this proved too cumbersome for a mass-testing regime that had to process vast numbers of people very quickly, so staff resorted to simply seeing whether a recruit could read a newspaper and write a simple letter. In any event, the consequence was the creation of two separate protocols, the so-called Alpha and Beta Tests, only the first of which required a mastery of reading in English. (To score highly, one also had to know quite specific bits of contemporary American culture, such as which kinds of engine were supplied in Packard cars.) The Beta Test, on the other hand—intended, Thorndike said, for men who might fail a literacy test but were still able to use a machine-gun—largely involved the manipulation of objects, images, and numerals. Some two million people took the resulting tests by the end of the war. And more than 29 percent either ended up being assigned the Beta Test or scoring the lowest possible score in the Alpha.

That statistic was among the many reported by Yerkes after the armistice in a massive summation of the testing experience.[35] Vast numbers of citizens were apparently able to read only at the level supposedly appropriate for a twelve-year-old—a level that, if extended to intelligence in general, would rank them only slightly above the technical category of "moron."[36] The report caused something of a sensation, because key numbers such as this lent themselves to exploitation by motivated writers keen to diagnose a national crisis. A veritable moral panic about "illiteracy" ensued. For example, the overtly white-supremacist author Lothrop Stoddard—in the book that articulated the concept of a civilization-threatening "under man" that would later be adopted by the Nazis—advanced particularly extravagant claims for such tests. They proved, Stoddard declared, that intelligence was hereditary and unalterable by education, that the "mental age" of any individual could be determined precisely, and that "whole nations and races" could be graded according to their "mental assets and liabilities." He maintained that Yerkes's investigations showed that the average mental age of Americans was about fourteen, and, worse still, that the trajectory was one of degeneration. Stoddard naturally insisted on crude racial correlations between low intelligence and Black or southern European origins too.[37] It was this kind of claim that prompted Walter Lippmann to write a renowned series of articles for the *New Republic* criticizing the claims made for intelligence tests in general and those by Terman and Yerkes in particular. The result was that intelligence tests both enjoyed a surge in use in the 1920s and remained intensely controversial.[38]

The main beneficiary of the army's testing experiment was the practice of testing itself. In the 1920s, intelligence testing became routine in American schools. The rise of IQ measurement has often been noticed by historians of the period, but in fact the practice of testing ranged more widely, and tests of reading abilities became ubiquitous alongside those of intelligence per se. Together they constituted, as the testing and mechanical-teaching pioneer Sidney Pressey put it, a veritable "measurement movement." Thorndike himself contributed to the development of a National Intelligence Test marketed to millions, alongside the tests for reading that he and other scientists of the subject promoted. The Psychological Corporation—the business that Thorndike launched with Cattell in these years—was intended to capitalize on the trend. In this context, as Yerkes's reported findings about reading echoed around the US commentariat, they not only sparked intense debate about the implications of illiteracy for democracy and the economy but also fed directly into controversies over race and inheritance. Stoddard was an extremist, to be sure; but for Terman, Thorndike, and Yerkes themselves, the Stanford-Binet tests did indeed measure innate intelligence, and Yerkes's report indicated that African Americans had performed notably worse than whites. (In fact, the results were more ambiguous, as Horace Mann Bond would point out, because northern African Americans had done better than southern whites.) Thorndike responded to critics who pointed to the relevance for test scores of vastly unequal educational experiences by remarking that intelligent students clearly gravitated toward places with better schools—a contention that came to be known as the "selective migration" thesis. He added that the most important single task facing the discipline of psychology was now to provide an "inventory" of the "human assets and liabilities" of the nation by comparing the abilities of different immigrant races.[39]

By the mid-1920s, then, it was widely believed that testing had shown that many Americans were not fluent or habitual readers and that African Americans and immigrants generally had lower reading ability than whites. At the very least, it seemed alarmingly clear that a large proportion of the national population could not participate fully in an information culture. One response was to ramp up the use of reading and intelligence tests in America's schools. But beyond that, two further strategies suggested themselves. If America's adult readers lacked the skills to make the most of print, then society must either improve those readers' skills or reduce the demands placed on them by books. In practice, Americans tried to do both. As a leading practitioner, Jeanne Chall, put it in 1947, "attacking the problem from both angles" would be ideal. Only in this way could the nation hope to achieve "true democracy so far as the knowledge of our culture is concerned."[40]

The panic over illiteracy first spurred a new and huge national effort at adult education. By 1924 a National Illiteracy Conference had already met in Washington, DC, to coordinate a campaign; Commissioner of Education John Tigert then appointed a further group, generally known as the Committee of Nine (although in fact it had eleven members), to be chaired by Charles Herlihy, the state supervisor of Adult Alien Education for Massachusetts. This group began meeting almost immediately in Chicago, and it submitted a report in 1925. It was then replaced by a National Advisory Committee on Illiteracy, with an initial grant from the Rosenwald Foundation and financing from John D. Rockefeller. The NACI would in turn be supplanted in the mid-1930s by the National Illiteracy Crusade. Meanwhile, the Carnegie Corporation also launched its own initiative to attack adult "illiteracy," which evolved in alliance with the AAAE and gave rise to a new *Journal of Adult Education* dedicated to the field. The Bureau of Education in 1927 issued a printed guide to *Methods of Teaching Adult Aliens and Native Illiterates*. And in the early 1930s the NACI engaged Chicago's William Gray to write a manual conveying the science of reading to teachers in adult-education programs; Gray's *Manual for Teachers of Adult Illiterates* was written quickly and came to be widely adopted. It went through several editions in the 1930s, being repeatedly revised in light of the experiences of practicing teachers and administrators.

The effort involved coordinating volunteers across the country. Many of the most active participants were middle-aged women; they frequently worked under the aegis of the General Federation of Women's Clubs, which coordinated a general survey of illiteracy by sampling one county in every state in the Union. Elizabeth Cleveland Morriss was one particularly active participant. Morriss made home visits to adults in North Carolina whom welfare organizations suspected to be illiterate; she helped set up night schools to teach them to read and, as a local newspaper put it, "thereby encourage them to better citizenship." Her manual for such efforts, revealingly entitled *Citizen's Reference Book*, first appeared in 1922 and soon became the standard reference for similarly dedicated workers elsewhere. Howard Odum (a North Carolina sociologist, and yet another student of Hall at Clark University, who had remained closely interested in work on reading in the southern states) introduced one reprint by declaring it a major element in the national campaign. Morriss's guide was also recommended by Thorndike as the ideal vehicle for applying the latest theories about the learning of reading in order to guide poor readers on to "future paths of education and citizenship."[41] Morriss herself then went to New York to work with Thorndike and Tyler at Columbia, earning Bachelor's and Master's degrees in the process. It was while she was there that she investigated the reading practices of minorities, in a project that we shall encounter in the next chapter. Like others, Morriss's experiences suggested the tricky question of the suitability of books to readers. It was all very well to say, as she and her peers did, that many books were beyond the capacity of average readers to enjoy, and to argue that this must change. How could such change be implemented safely and beneficially?

The studies of Waples and his followers relied on enormously labor- and ink-intensive statistics, often presented rather drily, and surveys of readers themselves, yet they did not attend closely to particular acts of reading. And in pointed contrast to contemporary self-improvement guides like Mortimer Adler's hugely successful *How to Read a Book*, they were not normative at the level of the individual reader.[42] (On a larger scale, of course, they were: Waples hailed Wilson's *Geography of Reading* as a model precisely because its richness allowed for reliable judgments of value to be made.)[43] The point was to use social science to make America visible to itself—its problems as well as its potential. It would foster a national conversation that would be self-critical as well as proud, and indeed proud *because* it was self-critical, but unlike the laboratory science of reading it was not itself meant to be therapeutic. In that light an apparently simple but fundamental question became central: that of "getting the right book into the hands of the right reader." In practical terms, it was not a matter of improving a reader's skills, but of providing a book that suited their *existing* skills. True, this might eventually lead to "extending and improving reading habits," but for Gray and his ilk this was a remote and vague ambition.[44] What emerged instead as a response to the illiteracy furor was a new field of research into what came to be called *readability*.

Readability was, in a sense, the inverse of reading ability. It was the degree to which a book could comfortably be read by an average literate American. A measure of readability was urgently needed, because without it there was no simple way to decide how to match print to people. Given that adult readers' capacities would be difficult to change, moreover, the standards of new printed works themselves would have to be recalibrated to fit those capacities. The current culture of publishing—a culture that was centered in cities like New York, which enjoyed high reading proficiencies—was evidently failing the mass of the American population that did not possess such skills. Changing the publishing industry's standards was thus the only way to secure in the short term a "wider, better informed, total reading public"—the kind of public on which the stability of a democracy must rest.[45] And the fortunes of the industry itself might hang on its being able to embrace radical change in the kinds of books it churned out.

So it was that organizations like the ALA and the AAAE threw themselves into the effort to replace the traditional critical prudence of publishers and editors with standards established on a secure social-scientific footing. Thorndike's series of Teacher's Word Books, starting in 1921, provided a basis for readability standards by listing thousands of commonly used words that he felt were appropriate for use in elementary schoolbooks and the like. What began as an effort to gauge the suitability of school textbooks to particular grades by appeal to these lists—the first systematic study was done by Bertha Lively and Sidney L. Pressey (the inventor of the learning machine) at Ohio State in 1923—soon extended into adult literature. The point was to recalibrate the publishing industry so that it would be geared to furnishing "books which are not beyond the reading ability of a large part (approximately 50%) of the adult population." A national Committee on Readable Books came into being in 1926–27, and the committee promptly commissioned studies by Waples, Gray, and others to constitute the foundations of its endeavor. It also endorsed a series of American Primers to be issued by the University of Chicago Press. Gray and Leary's *What Makes a Book Readable* became by far the best-known contribution to the effort, and lists like the ALA's elaborately formal 350-page behemoth going by the name *The Right Book for the Right Child* systematized the approach for elementary education.[46]

Empirical investigations of readability in adult books took off in earnest in the early 1930s. Such research, added to that of Gray, Waples, and their colleagues, made readability into something quantifiable. It could be defined and researched, and the research could then be refined and improved over time by disciplined investigators. There could be formulas and standards of readability, to which books could be compared and at which writers could aim. Gray and Leary, for example, sorted the 350 books they examined in *What Makes a Book Readable* into five difficulty levels denoted A (very easy) to E (very difficult).

The prospect of refining such standards scientifically was what led to the creation at Columbia University of a full-scale "Readability Laboratory." Its director, the Columbia professor of education Lyman Bryson, told the *New York Times* that "a group of psychologists and literary scientists, most of them at the University of Chicago," had already performed "analytic experiments" sufficient to found this laboratory. Waples and Gray had led the way. Locally, Thorndike was a major presence behind the new laboratory, which was founded "for the purpose of pushing these experiments one step further." The aim was to see whether the science of reading could be put to use to produce objectively readable books. Led by Bryson, historian Charles A. Beard, and psychologist W. C. Hallenbeck, the laboratory focused not on fiction (novels tended to be readable anyway, they thought) but on works of "current economics and social and political questions." This was where the problem of the hour lay. At the laboratory, researchers created standards to guide the reduction of numbers of hard words and the increased use of personal pronouns (which experiments had proved to make books easier to read). They also fostered "a system of 'idea' analysis" devised by Morriss in light of her experiences in North Carolina. The overarching purpose of the laboratory, Bryson attested, was to train a select group of workers in "the techniques of 'readability,'" so that they could then "test the work of others" by the same standards. Moreover, they would also go on to "rewrite or edit manuscripts into clear and comprehensible versions." The result, it was hoped, would be a cohort of readability virtuosos, who would orchestrate the production of new works that would be "trustworthy, clear and pleasant to read." In fact, Bryson claimed, "we have already succeeded in rewriting books for some of the publishers."[47] In prospect, then, was a "very salutary" change in "the whole publishing and bookselling enterprise."[48]

Very soon, the readability lab began issuing quantitative and qualitative standards by which to pursue these ambitious aims, as well as actual books that instantiated them. By the late 1940s some twenty-nine different quantitative standards had been published. These standards were mutually incompatible to varying degrees, yet they were very widely applied. They were even used to appraise university books of philosophy—in which context the standards seemed to fail, because, as was soon realized, philosophers used what were usually quite simple words like *good* and *value* in extraordinarily technical and involved senses. And that gave the champions of readability pause, especially when it came to recommending that new writers should always try to cleave to the latest readability rule. Skepticism was never hard to find, and it even grew as the hype around the principle increased. But the principle itself spread rapidly nonetheless. By the mid-1940s, it was being put to use in government institutions, and the Department of Agriculture had created its own readability lab. The National Tuberculosis Association used Columbia's latest formula to appraise its pamphlet publications, and then rewrote many of them in light of the finding that most were too difficult for 65 percent of Americans. At Ohio State University, Edgar Dale oversaw the writing of whole series of books dealing with "human relations, personal business, physical and mental health, citizenship, vocational guidance, and recreation," which were then marketed via the Government Printing Office. He also wrote a pamphlet telling healthcare authors how to write more readably. The newspaper industry was doing systematic appraisals of its own productions according to Columbia's Flesch standard by 1944, and in 1948 the Associated Press hired Rudolf Flesch himself as a consultant. "Today," as Chall declared, "everyone should be informed about hygiene, peacetime uses of atomic energy, international affairs, race relations, and means of furthering peace"—and the standards promulgated from Columbia and the other labs were tools to achieve that state.

By the time Chall wrote this, in the late 1940s, readability had become a conventional buzzword in education, government, and, not least, corporate culture. Much of the work did not give rise to publications as such, but was instantiated in ephemeral things like classes, pamphlets, and instruction manuals. Before long, for example, you could buy—or more likely acquire gratis—"a $2.00 computer" that promised to furnish "a direct reading of readability score" for any candidate text whatsoever. It was probably a kind of circular slide rule, and it embodied a standard devised by Flesch, who was a Columbia Readability Lab graduate. These kinds of materials were not commonly reviewed or even noticed in prestigious venues such as the Library of Congress or the *New York Times*, and they were themselves transient and rarely preserved. So it was already difficult to estimate their extent and effects even at the time, and it would become even harder later. But both were clearly broad and deep. And as we shall see, in Flesch's hands the questions posed by the champions of readability would soon inspire a devastating attack on everything that the science of reading stood for.[49]

From Reading to Communications

Works like Waples's and Gray's drove home repeatedly the point that books, reading, and libraries—and the social science that attended to them—were the very conditions of possibility for the informed, tolerant, critical, and sustainable public on which a democratic order had to rely. In this light, their enterprise fed into arguments about democracy, the public domain, and public opinion that had been sparked by John Dewey and Walter Lippmann in the 1920s and took on new urgency in the 1930s. (Harold Lasswell's *Democracy Through Public Opinion* was a characteristically mordant appraisal.) Particularly compelling in this regard was a 1936 ALA pamphlet entitled *The Equal Chance: Books Help To Make It*, which made use of Otto Neurath's powerful Isotype system of graphical representation to get across its message about reading with almost tachistoscopic immediacy. Reading books was crucial to both freedom and economic recovery, viewers of *The Equal Chance* learned, and the provision of such books was "not an individual matter, but a community matter"[50] (fig. 4.3).

The social science of reading originated at the point of intersection of a number of critical issues, but the scope and scale of these issues demanded that the enterprise treat print not in isolation—as had been possible before World War I—but as existing amid a swirl of different media competing for attention. The proliferation of "communication" systems vastly complicated the task facing Waples, Gray, and their allies in the Roosevelt administration. In Europe, Nazism and Stalinism were notoriously effective at propaganda, especially that using radio, cinema, and the mass press. And in the United States, Long, Coughlin, Lindbergh, and others—and, his critics complained, Roosevelt too—were all highly effective users of radio in particular. Whereas print had been the only mass medium prior to about 1920, by the 1930s this was no longer the case, and conventional understandings of culture had to be recalibrated to take account of the new reality. Waples and his counterparts were keenly aware of this need. Already by 1930 Gray and Munroe were trying to measure their Chicago subjects' relative devotion to radio, cinema, and newsprint.[51] One intriguing aspect of this was made explicit in 1940's *What Reading Does to People*, where consideration of the multimedia environment in which reading occurred led Waples to ponder the various meanings of the term *communication* itself. What made reading unique, compared to listening to the radio or watching a movie, Waples said, was that the reader controlled its pace. One could try to read at the speed of thought—that old aspiration of Huey, and of Buswell and Judd too—but one could also pause over things that needed reflection. And yet, to posit this was at the same time to suggest that it was appropriate and neces-

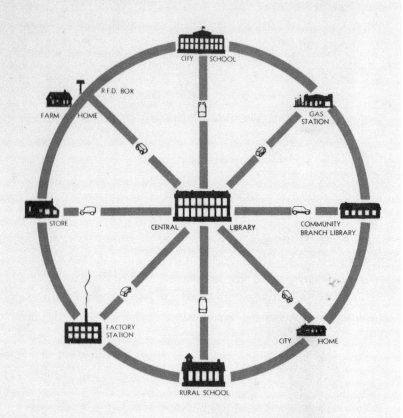

How Regional Book Service Operates

CITY SCHOOL

R.F.D. BOX

FARM HOME

GAS STATION

STORE

CENTRAL LIBRARY

COMMUNITY BRANCH LIBRARY

FACTORY STATION

CITY HOME

RURAL SCHOOL

A coordinated library system reaches every local community and every citizen. Constant interchange of books between the central library and the branches and stations make any book available to any reader in the entire region.

4.3 Isotype representation of the operations of a regional book service in the mid-1930s. *The Equal Chance: Books Help to Make It* (Chicago: American Library Association, 1936), 18.

sary to place reading alongside those other modes of communication and consider them as a whole.[52] And in that suggestion lay the possibility for a new kind of inquiry.

A fragmentation in concepts of communication was already apparent by the mid-1920s. It could be seen in the works of authorities as dispersed as John Dewey, I. A. Richards, C. K. Ogden, and, much farther removed, Ludwig Wittgenstein and Martin Heidegger.[53] In the years between the Crash of 1929 and the end of World War II, one place where the gallimaufry of meanings, dreams, ambitions, and fears attached to the concept came together was in the social science of reading. Waples again was a key figure, along now with Harold Lasswell. The prime reason was that they were both personally engaged in studies of propaganda. Lasswell's 1927 *Propaganda Technique in the World War*—dealing with World War I—began as his doctoral thesis in political science at the University of Chicago. The original article by Waples that had so provoked Thompson was on "Propaganda and Leisure Reading" in American's colleges. And the two had worked together on *National Libraries and Foreign Scholarship*, a comparative account of different countries' policies in which those of Nazi Germany were the major issue (albeit one treated, as they themselves acknowledged, with a degree of analytical coolness that was easy to mistake for sympathy).[54] The extent to which reading was not only the agent of public opinion but could easily be the means for controlling populations was a theme that therefore ran through the social science works of the 1930s.

In mid-1941, Waples summed all this work up by holding a high-profile meeting at the Graduate Library School on the role of social scientists in administering mass communications for a nation-state facing war. Published as *Print, Radio, and Film in a Democracy*, the discussions that took place at this event marked the integration of the social science of reading into a more general domain of what came to be called "communications research." The new domain would be guided by the Chicago school, as was strongly implied by the prominent role of sociologists as session chairs in Waples's conference. The cast list was an array of speakers who would be instrumental in the next few years in creating a new discipline of communication study, including Ernst Kris, Bernard Berelson, Paul Lazarsfeld, and Harold Lasswell.[55]

The applications of this perspective were not necessarily benign. In studying propaganda, Lasswell and Waples meant not merely to counter its effects but also, in some circumstances, to make its deployment more efficient. Meanwhile, behind the scenes Waples conjured up ways of putting his ideas about the management of communication to practical use. He secretly proposed that the US government establish a national network of "local collaborators" to act as information-seekers, for example. They would be tasked with staffing "observation posts" from which they would look out for subversive agents. It sounds a bizarre idea—as if little concrete bunkers would be placed at random crossroads in hopes of overhearing dangerous talk. But within the University of Chicago itself, he advocated the real establishment of such a post. In that case he wanted to call it by the deceptively anodyne term "Communications Station." It would be dedicated to watching the impact of communications on local public opinion, and, again, to identifying "subversive propaganda." Neither plan was taken up, but as the war effort accelerated both Waples and Berelson were recruited as consultants to the Office of War Information. There they worked alongside others who became pioneers of communications study: Wilbur Schramm, Paul Lazarsfeld (again), Margaret Mead, and Carl Hovland. Their charge was to research the persuasive effects of mass media in the international crisis. The Rockefeller Foundation was the funding source, and once again Waples saw an opportunity. He proposed that his university create, alongside the Library School and the Education School, a third initiative: an "Institute for Training and Research in the Field of Public Opinion." He had in mind a permanent body, dedicated to researching the "so-called 'field' of communication." It would exist mainly to serve the interests of the US Government but with a view to attracting support from marketing and media companies too. The university's president, Robert Maynard Hutchins, demurred on this, but he did greenlight something a little different.

What Hutchins approved was a new University of Chicago committee devoted to communications and public opinion. In local terminology, a committee was (and is) an interdisciplinary graduate program, established on an experimental basis for a new discipline in the making. Committees were designed to be administratively light and to eschew the inertial traditions that departments inevitably acquired. Hutchins liked them because they represented intellectual dynamism. This seemed a good candidate, so the committee duly came into existence in 1942, with the task of repurposing the kinds of work done in the social science of reading in order to create a stream of personnel trained in mass persuasion for the war effort. It was the world's first graduate degree-granting body in the prospective field of communications.

Chicago's committees were generally meant to be ephemeral, and at the end of the war the university wound up this one. But Waples's original proposal had been for a more lasting enterprise, involving a program that would endure beyond the immediate needs of the conflict. Not least, he believed, the country would need experts to manage the coming communications landscape, which was bound to be far more complex than anything anyone had experienced before. After 1942, however, he found himself on active service in Europe, working in the Psychological Warfare Division. In occupied Germany, Waples became Director of Information Control for the US Military Government, using his expertise in reading and public opinion to pursue de-Nazification.[56] Helping to rescue and rebuild the German publishing industry was one element in that mission. But in September 1945, while still based in Germany, he contacted Chicago to repeat his call for such a body. Chicago's professor of educational psychology, Stephen Corey, weighed in to agree, invoking the memory of a "communications group" that had existed "several years ago." Meanwhile, a highly publicized "Commission on the Freedom of the Press" that Hutchins had led since 1943 in an effort to define the role of the mass newsprint media in postwar society was coming to a similar conclusion about the need to understand the new media environment. One of its recommendations was that "centers of investigation, graduate study, and critical publication" be created, somewhat along the lines of the Graduate Library School, but devoted to a more general "field of communications." In 1947, then, a new committee came into existence: the Committee on Communication.

The new committee's members included Berelson, who served as chair until 1950, David Riesman, Louis Wirth, and Kenneth Adler. Waples joined when he returned from military service in 1948 to become Chicago's first ever professor of communication. The courses he taught under this label included ones on psychological warfare and strategic intelligence. He succeeded to Berelson's position when it fell vacant in 1950. The point of the program, under Waples, was to do interdisciplinary research on mass communication itself. It quietly continued to get funding from the State Department and the military for research on "intelligence and psychological warfare matters," however, and it less quietly made a point of listing its work in these areas for its backers. Waples was always proud to place ex-students from the program into US intelligence jobs. But its ancestry in the prewar social science of reading was evident too. Anchor faculty (Berelson and Waples in particular) tended to have primary appointments in the GLS, and, like the GLS, it created its own journal, *Studies in Public Communication*. In the end, the committee did not establish a distinctive identity for itself, however, partly because Waples, always its prime mover, spent much of his time traveling internationally. After he retired in 1957 it seems to have drifted. Berelson publicly declared it moribund in 1959 when he moved to Columbia, and the committee closed the following year.[57]

In short, the social science of reading that had emerged in Chicago in the late 1920s had by the end of World War II helped forge an altogether new discipline. Its major protagonists, especially Waples and Gray, often had backgrounds in education research and the experimental science of reading. But in their new role they almost entirely ignored the laboratory approaches of Judd and his predecessors, including those of Gray himself. Theirs was professedly a scientific endeavor, but their definition of science was quite different. Under the pressures and temptations of the 1930s, the social science of reading sidled up to the state with increasing avidity, becoming, in effect, a research enterprise dedicated to studying the efficacy of propaganda and the management of public opinion. It developed strong, deep, and lasting connections with the military and the intelligence community. At the same time, it increasingly embraced a reconceptualization in terms of "communication," such that reading was no longer to be treated in privileged isolation but would be studied alongside listening and viewing. The book and newspaper were now elements of "media," alongside radio, cinema, and television.[58] We shall see in the following chapters how the emergence of communications and media as fields of research affected the advent and character of the information age.

5

Readability, Intelligence, and Race

In 1935, Elizabeth Cleveland Morriss, the tireless North Carolina champion of adult reading improvement, was in New York, attending Columbia University and learning psychology with Edward Thorndike. That summer she won a Carnegie Fellowship to conduct research to explore the similarities between illiteracy among native-born whites (in this case, students at a reform school), recent immigrants, and African Americans. The study was one of several along such lines to be carried out in these years, and perhaps the most successful. She herself made its significance emphatically clear. For some twelve million Americans, Morriss declared, reading did not "function effectively," and this constituted a defining social problem of the age. The question was how to grapple with that problem and solve it. The social science of reading was set to become a key tool in the fraught American politics of race and citizenship.

The citizens Morriss engaged with had "mastered the mechanics of reading," she attested, but they did not find the materials available to them interesting, engaging, or useful. Reading was not beyond them in any sense, but it had never become an everyday part of their experience. The problem of habituating this population to routine reading was, she confessed, "baffling," and no conclusions about it had yet been "scientifically established." It was not a matter of merely measuring "ability" — existing tests could do that. The problem was that the very notion of *an* ability (singular) was mistaken. Morriss contended that readers must be recognized as possessing different reading "*abilities*" (plural), and these depended on what was to be read, when, and why. In this she and Thorn-dike agreed. Thorndike insisted even university students often lacked a sufficient range of reading skills to learn effectively from the books they encountered, and he proposed training them in a variety of different sub-categories of reading: skimming, reviewing, problem-solving, and more. (For all his flaws, Thorndike was enlightened enough to feel that most university coursework was pointless, and he liked to advise students considering taking almost any class that they would be better advised to spend the time reading a book.) To understand how those abilities arose and functioned, Morriss added, readers' backgrounds, social statuses, and experiences must all be taken into account. And because they were so various, education would not be enough to deal with the issue. Each group must be provided with appropriate printed materials—an enor-mously complex and weighty task, but also an inescapable one.[1]

Morriss's conclusion was one that others were reaching at much the same time. The problem of infrequent reading, as we have seen, inspired a new kind of focus on the books that were available for readers. It implied that public funding alone would not solve the problem of information in-equality, because adults with poor reading abilities or different acquired habits would simply stay away. So library professionals had to reconceive their role in terms of actively selecting suitable books for their communi-ties. They must engage in "the preparation and provision of simply writ-ten books relating to daily affairs." They should buy more books suitable to be read by adults of limited ability, and make greater efforts to train children and adults alike in the skill of selecting appropriate volumes. In particular, as Louis Wilson believed, they must actively seek out books "suited for service to Negroes." Only with this more active strategy in ef-fect would funding bring major benefits for future generations. That in turn posed the question of *how* to gauge the degree of fit between books and readers. And to answer that question was in part what the new li-brary science was for.

The science of readability was intricately associated with questions of race from the outset. As Morriss's experiences imply, such questions were often intrinsic to its very practice. A particularly influential early project, for example—one that played a decisive role in defining what the science of readability itself would be—was that undertaken by two graduates of Judd's School of Education, Edgar Dale and Ralph W. Tyler, in Columbus,

Ohio. Both Dale and Tyler won positions at Ohio State University in the late 1920s. Dale would stay there for the rest of his career, chasing ever more exact readability standards and managing the revision of books and pamphlets according to "the Dale formula to predict reading difficulty"; he would train Jeanne Chall, among others.[2] Tyler would return to the University of Chicago in 1938 before eventually becoming the first director of Stanford's Center for Advanced Studies in the Behavioral Sciences. At that time, as they remarked, there were "no scientific techniques by means of which to make an accurate estimate of the reading difficulty of books and pamphlets." So they decided to start their time in Columbus by remedying that. They would investigate empirically what reading materials could be understood by adults of limited reading ability. It proved a difficult task. Not only did they have to try to keep factors such as the interestingness of the topic constant across different text selections, but they also had to find a group of mediocre readers willing to collaborate for repeated sessions. To solve the first problem, they made use of an idea that Tyler had advanced in the book he had coauthored with Waples, *What People Want to Read About.* They used newspaper columns dealing with "personal health," on the grounds that these would be of roughly common interest to their subjects. They then devised tests to appraise a reader's degree of comprehension of each text, which would enable them to calculate statistically the "index of difficulty" for each piece. At that point they would be in a position to spot correlations with objective measures of the length of sentences, complexity of clauses, usage of difficult words, and similar elements. And that would allow them to extrapolate such correlations, predicting how difficult these readers would find *future* texts with similar statistical properties.

So Tyler and Dale sought out a community of mediocre readers to put their plan into operation. This turned out to be a complex task, however. Not only did adults tend to be busy and diverse in their interests, but those with limited reading abilities also tended to be reticent about admitting as much. Some were also dismissive of the whole enterprise. Night schools were not an option, because they were closed during the summer season when the project took place. A group of women at a club in a community house volunteered, but they then treated the experience as a "lark" and casually cheated on the tests: "the only value of the final results," Tyler and Dale remarked, "was to suggest very strongly that such groups were not satisfactory research subjects." They then tried a church group. It seems that they reasoned that they were likely to find a community of mediocre readers among African Americans, so they persuaded a Black pastor in Columbus to volunteer his Sunday-school class of thirty young married people. This too came to nothing, however. On the day in question the pastor slept through the appointment, and in any case only a dozen people had showed up. At length, Tyler and Dale decided that their only option was abandon their appeals to people to volunteer and instead pay for involvement. And at this point they found a group of indigents willing to participate at the Godman Guild, an education-centered community house in "a district composed almost entirely of negroes." They paid them twenty-five cents each for every session, garnering twenty-five to sixty participants at a time.

Tyler and Dale quizzed every participant to ensure that that he (they were apparently all male) was essentially literate. Each subject was given a silent reading test and then exposed to the actual test readings. With the results of their comprehension tests in hand, the two researchers then looked at twenty-five characteristic properties of the selections, looking for any that might correlate with the difficulty that their readers showed in the tests. These properties included the numbers of difficult words (about seven hundred occurring in Thorndike's *Teacher's Word Book* were assumed to be "easy" and to represent "zero difficulty"); frequencies of technical terms; and factors such as the lengths of sentences and their numbers of clauses. On the other side, they also included the number of times personal pronouns occurred, the assumption being that this implied an informal, accessible style. Dale and Tyler even recorded the number of words beginning with the letter *E*, which other researchers had suggested might be an index of difficulty. (They found that it was not.) In the end, they concluded that just three quantities really mattered: the number of technical terms; the number of hard nontechnical words; and the number of indeterminate clauses. These, then, were indicators that librarians could use to decide whether or not to provide a certain work for a certain readership. "For most purposes," they concluded, librarians charged with selecting reading materials of a given difficulty level need only look at these three numbers of any candidate work. And this process could then be scaled up to a national level. Some "central agency"—they suggested the American Library Association—could be entrusted with providing authoritative scores of the reading difficulty of all kinds of materials, which could then be used by branch libraries across the nation to decide which books to buy and circulate. Such lists would, of course, have to be checked by a large force of "field-workers" who would travel to libraries and adult-education initiatives nationwide to interview patrons and see how well the choices were faring in practice. Furthermore, the same insight could also be useful for writers charged with composing new texts. Dale and Tyler suggested "experimentation in the writing of materials appropriate for adults who have limited skill in reading." They imagined a scenario in which readers of a given group would be tested as they read sample passages and the empirically discovered values of readability then being replicated in newly written items.[3] This call became one of the principal inspirations for the science of readability pursued at Columbia. And the central place of racial difference in the pursuit of that call was no accident.

Immigration, Race, and Reading

Readability, as we have seen, was an outcropping of the moral panic over "illiteracy" and modern culture that broke out in the 1920s. The anxiety about illiteracy had several major implications for American society and political culture, all of them intensely contentious. It implied that millions of citizens were being miseducated, which was scandalous enough. The economy might be held back, and in the end democracy itself might be vulnerable if the populace could not be made into a community of readers. All those were serious enough problems. But perhaps the most pressing questions of all concerned immigration and, above all, race. Racial inequity was the greatest fault line by far in American society, and it was a rift that afflicted the development and conduct of reading as a practice too. Scientists of reading—and those of the second Chicago school in particular—were exquisitely conscious of the depth and consequence of this rift. In large part, the very point of the social geography that Louis Wilson championed was to make the gulf between the southern, segregationist states and the northern, industrial ones sharply visible and its implications undeniable. In that way, the science of reading could hardly avoid becoming an element in the struggle for equality—and later that for civil rights—for African Americans in those states and beyond.

The first issue of the illiteracy panic to make headlines, however, was not that of southern segregation but that of immigration. Indeed, it could be argued that the illiteracy panic itself was initiated through anti-immigrant agitation and only later reinforced by the publicity surrounding Robert Yerkes, Lewis Terman, and the army tests. As early as the mid-1890s, nativist campaigns had latched onto migrants' lack of English-language reading ability as a rationale for turning them away. The first federal bill proposing to establish a literacy test was vetoed by President Grover Cleveland in 1897, and similar measures repeatedly arose in the next two decades. Taft vetoed another such bill in 1912, and Wilson a third in 1915—at which point Congress very nearly overrode his veto. The following year it succeeded in doing so, and in 1917 Wilson was forced to sign into law an immigration act that banned all adult migrants from Europe who were deemed illiterate (as well as virtually all Asians, except for professionals in a small range of specified fields). The new law did not mandate that migrants had to be literate *in English*, however, and as a result the test proved far less rigorous than its advocates had anticipated. It excluded only a tiny proportion of applicants in its first few years and was soon supplanted by a system of national quotas.[4] Still, the issue of immigrants' abilities to read English was a very familiar one in political commentary of the period, and it duly became a common subject of research by social scientists interested in reading. It should hardly be surprising, then, that in the context of massively increased use of reading and intelligence tests after World War I the issue of immigrants' reading abilities would merge with that of the abilities of Americans in general.

But for Terman and many others, the army tests, along with the so-called intelligence tests carried out widely in the 1920s, also carried a message about differences within the US population. For Terman and Thorndike, the fact that such tests seemed to show consistently higher scores for (northern) whites over (southern) Blacks revealed a basic distinction between the races. "Intelligence," they argued, was an intrinsic property inherited genetically; not only was it a matter of natural capacities rather than "environment," but Terman insisted that hopes of alleviating such differences through schooling were misplaced. As a convinced eugenicist, he maintained that his intelligence tests provided good grounds on which to segregate students by ability, channeling those with higher scores into preferential schooling streams. Such arguments of course buttressed the position of defenders of what was already a drastically segregated system of schooling and library provision in the southern states of the Jim Crow era. In the old Confederacy, teaching slaves to read had been widely illegal prior to the Civil War and Emancipation. After the retreat from Reconstruction and the adoption of various explicit and implicit Jim Crow rules across the region, African Americans had been consigned to distinct and vastly under-resourced schools, and often denied access to any library facilities whatsoever. The experiences of teaching and learning themselves—and the teaching of reading in particular—became again intensely difficult and "fugitive."[5] That this deep inequity in the reading geography of the nation existed was hardly news when Wilson highlighted it so emphatically in the mid-1930s. And some practitioners of the science of reading, above all Thorndike, maintained that racial differences in intelligence were real and measurable, and that the Stanford-Binet test was useful (as he put it in a 1921 *Journal of Educational Psychology* symposium on the theme) for "prophesying how well a person will do in other intellectual tasks."[6] But for others in the field, the science had to engage critically with this stark and, as Wilson put it, tragic reality if it were to make the kind of difference they sought for it in fostering a modern democratic society.

Horace Mann Bond and the Critique of Testing

In the 1920s a number of African American social scientists publicly contested the claims of Terman, Thorndike, and their peers that intelligence and literacy tests proved the existence of inherent racial differences. The NAACP's magazine the *Crisis* regularly published stinging ripostes to such claims, as did the National Urban League's journal *Opportunity*. Among the first and most effective of these critiques were articles by an African American who produced his first when only nineteen years old. Born in Nashville in 1904, Horace Mann Bond (fig. 5.1) came from an educated family; his father was a minister and his mother a schoolteacher, and, unusually, both were college graduates.[7] But his grandparents had been slaves. Bond excelled as a student at Lincoln University and then at

5.1 Horace Mann Bond at the time of his work for the Rosenwald Fund, around 1930. Special Collections and University Archives, University of Massachusetts Amherst Libraries.

Pennsylvania State, where he realized that he could outperform almost all of his white classmates. It was then that he started publishing on the issue of race and intelligence tests, reviewing one of the most influential works to support the claim that tests measured racial differences, the *Study of American Intelligence* by the Princeton expert Carl Brigham (who later recanted his eugenicist beliefs), by focusing on the fact, which Brigham and others awkwardly explained away by the selective migration argument, that African Americans from the North often did better on the tests than those from the South. Indeed, the average scores of northern Black people were higher than those of whites from several southern states. Clearly, Bond remarked, those clever Black people who had migrated north had left their "duller" white peers behind too. The right explanation for the difference, he added, was surely the "obvious" one: that those in the northern states enjoyed "infinitely superior home, civil, and above all school conditions." What the tests were measuring, then, was not some innate property but the quality of a person's environment.[8]

Yet by the mid-1920s such tests were ubiquitous in American schools and colleges, and African Americans had to expect to be placed in positions where they would have to argue against their credibility. That applied to Bond himself: much later, he found that the University of Chicago had collected intelligence-test data on matriculants at its School of Education from his own time there, and he took grudging satisfaction in noting that the median score of Black students was above that of the student body as a whole.[9] But in the meantime Bond told the story of encountering a young Black girl who had been confronted with such "evidence" of inferiority in her Chicago school and had been frustrated at her own lack of evidence with which to counter the charge. Bond repeatedly insisted that it was now incumbent on African Americans to equip themselves to fight such battles. Notably, he was unusual in seeing the importance of a historical sensibility for that purpose. He not only likened such tests to the now-discredited anthropometric measures that had been touted in the nineteenth century as demonstrating racial differences, but devoted time to exploring the development of such measures. By 1934 he had collaborated with the renowned African American sociologist Charles S. Johnson on a serious historical study of the science of racial difference, ranging from antiquity, through Linnaeus in the eighteenth century, to the efforts of Blumenbach and others to find differences in skull sizes in the nineteenth. It was this trajectory, they charged, that had now been extended to psychological researches in the early twentieth century. Their point was to show—decades before Stephen J. Gould—that the whole history of racial science had been a long series of enterprises of mismeasurement.[10]

Bond went into details. The army's tests, he remarked, had given rise to a headline contrast that the average mental age of white recruits was 13.1 years and that of Black recruits 10.4 years. A little thought sufficed to show that the contrast was literally nonsensical, as Walter Lippmann pointed out waspishly. But it was at the heart of claims by racists like Stoddard, and it had been widely appealed to in a number of political contexts. It was allegedly on the basis of this discrepancy in IQ scores "that Nordic races have been granted the heaven sent mental superiority over South Europeans which entitles them to entry into this country; that a prominent College president and pulpit orator of the East justifies the policy of segregation in the public schools; and that one observer bewails the fact that 'There seems to be no immediate possibility of convincing the public of the necessity for preventing the reproduction of these groups.'"[11] But Bond pointed out that African Americans from Illinois and at least three other northern industrial states had in fact comfortably outscored whites from Georgia, Arkansas, Kentucky, and Mississippi. Moreover, he cited statistics to show that different states' scores correlated closely with the quality of their school systems. In effect, then, the army tests (and the intelligence tests that followed) were indeed "very accurate measures"—but *not* of inherent and unchangeable human properties. They measured, rather, "experience and education." So the "true value" of them was that they revealed precisely the gross inequities by class and region that race-determinists like Brigham systematically obscured. Intelligence tests should be deployed to produce standardization, not in people, but in the schools to which they were sent.[12]

Not only his historical sensibility set Bond apart. Unlike many critics of intelligence testing—Lippmann being the most prominent example—he was himself an experienced user of the tests he questioned, and he appreciated their value if properly used. In the mid-1920s, alongside his writing for *Crisis* and *Opportunity* against Brigham and his ilk, Bond applied Terman tests, as well as others by Sidney Pressey, to schoolchildren in a bid to identify alleged correlations between intelligence and "emotionality" in African Americans, even while acknowledging that the tests' dependence on good reading abilities had to be taken into account. (He found no such correlation, but ended up suggesting that Terman's tests might actually be better at measuring the social quality of "personality" than supposed personality tests, which were another fad of the time.)[13] By this time Bond had matriculated as a graduate student at the University of Chicago, after finding that he could not afford to go to Harvard. He initially joined Chicago's Department of Sociology, where he worked for a while with Robert Park; but his interests in the history of schooling and literacy led him to transfer to Judd's Education program, and he undertook the testing project for his master's degree. It is clear that both Judd and Buswell, like Park, thought highly of Bond, but the psychological testing researcher Frank Freeman became his principal mentor as he proceeded into doctoral research. He combined pursuing the PhD with working as an assistant professor at Fisk University. And at the same time Bond threw himself into a prolonged study of the segregated schooling system in Alabama. In 1929–31 he played a major part in a Rosenwald-sponsored study of southern Black children's education, which included administering Stanford-Binet tests himself to roughly ten thousand pupils in some five hundred schools. Judd was a member of the fund's board of trustees, which may have helped him secure the support. One explicit purpose was to demonstrate that what such tests measured was indeed "environment" rather than innate qualities. (As part of the survey Bond also queried teachers, and he found that almost half in Black schools had not reached the level expected of eighth graders.) But as his engagement in the region and its problems deepened, so he became increasingly committed to more sociological and historical approaches, and the dissertation he eventually completed in 1936 included little to none of the testing work. The piece won the university's Rosenberger Prize, and Bond had good reason to believe, as he did, that he was the finest student Judd's Department of Education had produced. It seems that he lastingly resented not being offered a teaching post.[14] But no such position was on offer, and instead he moved to a series of faculty and administrative positions at historically Black institutions. There he sought to cultivate the kinds of training that might create the generation of teachers that his envisaged future of African American literacy would need.

The Historical Ecology of Segregated Reading

Bond was always a productive writer, and his first book appeared before he had even completed the Alabama project. *The Education of the Negro in the American Social Order* (1934) supplied the broad context for such work. It argued that the problem of Black education in general had become pressing because of the sophisticated nature of a new industrial revolution. The kind of uniform, unskilled labor required by prior societies (and for which a slave system had sufficed) was no longer adequate or appropriate, and new classes of worker had already arisen across American society. Black and white people alike needed to be prepared to participate in this developing society—indeed, one of Bond's more radical contentions was that at the level of literacy the requirements of both races were in fact the same.[15] Schools alone were not sufficient any more, because no such limited formal institutions could meet the various needs of modern industrial society. The principal such need was, of course, literacy, and Bond devoted the central chapter of his book to charting the sporadic decline of illiteracy that could be charted among African Americans since the slave era. But, he concluded, literacy in this basic sense was not really the issue now, if it had ever been. As the social scientists of reading were just then pointing out, what was needed was not just the bare ability to communicate via print but a more sophisticated ability to employ fluently "the higher function of language as a means for communicating ideas" on an everyday basis. And in that sense it was likely that more than half the Black population still existed on the margins of "'mental illiteracy'"—an estimate that, given what Buswell and others were finding, might even be too sanguine. For schools to eliminate "elementary illiteracy," Bond remarked, was relatively straightforward and had for some time been achievable. Now, however, the greater mission awaited: to tackle this larger and harder task of educating students in the use of elevated readerly skills—of turning people capable of reading into fully participating *readers*. And for that, Bond agreed with Wilson. The all-important institution was not the school, but the library.[16]

The same year *The Education of the Negro* appeared, Bond and his wife Julia received the starkest possible reminder of how hard such a task might be. They went to live in a rural parish in Louisiana to help the Rosenwald Fund's "School Exploration Group" appraise the effectiveness of the fund's school-building efforts. While there, they lived as the locals did, in a wooden cabin without electricity or water supplies. What Bond found, though, was unexpected. He discovered an environment of small farm holdings, in which white and Black residents shared family names (a fact that he assiduously traced back to interracial couplings over the previous century), and where they often seemed to get along peaceably on an everyday basis. But tensions were real, deep, and serious, and violence was liable to break out at any moment. This was made brutally clear by the lynching of Jerome Wilson, a local farmer's son, which happened shortly after the Bonds arrived. The unnerved Bonds moved away to New Orleans, and Bond started to write a book about the event based on historical research and extensive interviews with the dead young man's father. He hoped to publish it and dedicate the profits to the family and the local Rosenwald school. As it transpired, however, no publisher expressed interest, the text was left unfinished, and the Wilson family disintegrated as all but the father were forced to leave for Chicago.[17] Only recently has Bond's work resurfaced.[18]

In the wake of this harrowing experience, Bond's efforts in successive offices took on even greater personal significance. Alongside them, his history of Alabamian education saw print in 1939 as what was probably his most important single work, *Negro Education in Alabama: A Study in Cotton and Steel*. Given Bond's recent experiences, the volume had a remarkably urbane, laconic, and even ironically humorous style. That it should adopt this voice was a deliberate choice on his part; he had long argued that adopting such a cool and objective tone was of a piece with a generational rise to "self-respect" on the part of African American writers.[19] The volume was nevertheless a powerful and sweepingly comprehensive historical evisceration of the Alabama system, the point of which Bond made clear at the outset. He opened with a remark to the effect that with segregation in effect in seventeen states (plus the District of Columbia), the idea of the United States as a "fairly homogeneous democracy" was fundamentally flawed, and any conclusions that might be drawn from comparisons of literacy rates between regions—of the kind, implicitly, that the intelligence mavens always wanted to make—would be unwarranted. The necessary starting point must be a recognition of the realities experienced by the African American minority "encysted" in the nation.[20] That said, Bond went on to furnish an extraordinarily acute analysis of the rise and character of Jim Crow segregated schooling in the state, explaining it in terms of the "total social complex" in which it arose. In the end, Bond's explanation was ecological: like Wilson and other Chicago-school social scientists, he began with geography. The ultimate origins of the segregated system lay, he believed, in the very nature of the environment—in soil and minerals, and in the agriculture and industry that these had fostered.

Bond traced the emergence of the plantation system of slavery out of the colonial era, arguing that convictions of Black racial inferiority emerged from that system rather than vice-versa. Under slavery, the plantation system had required "keeping the tools of civilized communication from the slaves," and had therefore enjoined "enforced illiteracy." The teaching of enslaved people to read had been made illegal for fear that the result of literacy would be insurrection. In Alabama itself an 1832 law had made it a crime to teach any African American, slave or free, to read or write; a few may have learned independently, or from masters, but generally plantation slaves were "schooled," as Booker T. Washington had put it, in muscular strength and elementary mechanics.[21] During Reconstruction there had been a brief period in which schools had been created and an effort made to kickstart popular Black literacy, especially at the hands of the American Missionary Association. The very idea had been "wholly revolutionary." And the elementary curriculum, which focused primarily on the teaching of reading, reflected a consciousness of the need to address its particular constituency: the basic reader was a special *Freedman's Book* edited by veteran antislavery campaigner Lydia Maria Child that included a eulogy to Toussaint L'Ouverture. This ambition to create an informed and self-reliant Black citizenry was soon being denounced by whites as "fanatic," and violent attacks on teachers and schools escalated. (Bond noted contemporary reports that laconically mentioned the burning-down of buildings.) They soon faced even more severe obstacles with the rise of white revanchism at statehouse level. Such initiatives, then, had not realized the radical hopes of their champions—hopes that were unrealistic given the scale of the issue. In the conservative historical views largely credited in the early decades of the twentieth century, Bond remarked, the failure to create a literate Black populace was typically explained by appeal to "theories of racial psychology," but no such "esoteric" causes need be invoked. It was a matter of the social force of racial resentment, the depressed economics of sharecropper agriculture around 1880–1900, and sheer political ruthlessness. Even African American leadership had in these crucial years been "almost entirely corrupt and venal," and it had not protected the literacy movement in the face of white claims that the ability to read would make Black people, like poor whites, unfit for share-cropping labor. The problem behind Jim Crow, then, as Bond put it, "lay precisely in the fact that education affected Negroes as it did whites."[22]

Yet in the early twentieth century, a new force had entered the scene in the form of the US Steel Corporation. USS had taken over northern Alabama's steel mills in 1907, and it came to see African Americans as more than a pool of "unschooled and illiterate masses." The new conception of the laborer saw him as useful for skills rather than sheer brawn; and besides, African Americans could be useful as strikebreakers. The result was a powerful interest in favor of Black education—Booker T. Washington appealed to it in setting up his Tuskegee Institute, which was partly meant as a "propaganda agency" to sway "public opinion" away from the idea that educating African Americans would be socially and economically harmful. But this was still to be education separate from and more elementary than that provided for whites. Literacy again was the critical concern, and even in the Constitutional Convention of 1901—an overtly racist event that set segregation policies for the new century—industry interests argued that the state must be "rid" of "the disgrace of illiteracy." But during the convention's debates, which centered on suffrage, all the attention was on *white* illiteracy, and henceforth Alabama's formal politics of education would proceed as if "Negro children did not exist."

So it was that by the time Bond and Rosenwald officer Clark Foreman (soon to join the Roosevelt administration) visited all the Black schools in Jefferson County (the area including Birmingham) and tested every child in third through sixth grades, there was complete segregation. The children did better than they would have done prior to USS's arrival, but their schooling ceased altogether at high-school level. And this, he concluded, was the reality that underlay President Warren Harding's remarks when he visited Birmingham in 1921, during which Harding invoked a "fundamental, eternal, inescapable difference" between the races, such that segregation in schooling was "natural" and, far from retreating, would inevitably spread to the northern states too.[23] The philanthropic efforts of organizations like the Rosenwald Fund—which invested substantially in Black schools in the South—had stood athwart the trend, but even those had been forced by the Depression to scale back their initiatives drastically.[24]

I have lingered over the work of Bond because it evidences the deep salience of convictions about reading—as an individual practice, as the key to self-formation, and as the essential element in an equalizing society—to midcentury understandings of race and the condition of the nation. Emerging as the prize student from the Chicago schools of Judd, Waples, and Wilson, Bond did not belabor explicitly the influence of those schools, yet it was surely essential to the strategies he adopted in arguing about what the problem of reading was, why it mattered, and how it could be tackled. To give an example of what this might mean in practice, an account of Black students in West Virginia by the young Carlton Goodlett in 1939 noted that scientific knowledge of reading abilities must precede any credible testing of intelligence, and then recommended that the local school system create a reading clinic that would merge intelligence tests with the use of eye-movement cameras and tachistoscopes.[25] It is no less significant that for all that Bond devoted extensive effort to questioning the use of intelligence and literacy tests to promote claims about racial difference, in his portrayals of southern schooling and society he remained largely silent about such arguments. Those eugenic claims were not only implausible and flimsily based but fundamentally beside the point. Reading had a geography, a history, and even an ecology; and the point was to explore and make these manifest. In an almost apocalyptic piece from 1941 confronting the possibility of a Nazi future for the world, Bond made the stakes even clearer. In the face of catastrophe, he found hope in the thought that the disciplining produced by a hostile social order tended to produce an even stronger faith in the values of democracy and human nature. It had already generated "a psychological type" for African Americans in which belief in history, democracy, and human perfectibility were far more pressing than in majority white culture. And education—into reading, particularly—could reinforce this, by using the techniques of readability. "Our scholars will take time to make themselves intelligible to the great masses of the people," Bond prophesied, "in simply-written text-books and readers." He even rhapsodized that, instead of compelling children to go to school to learn to read, families would once again, as in the days of slavery, read the one book that not even Nazis would burn—"that great history of a minority people, the Bible." With these in hand, southern pupils would be formed into adults with both messianic liberal ideals and the practical, habitual tools to put those ideals to use in redeeming democratic society.[26]

Race, Rights, and the Sociology of Reading

One could be forgiven at this point for recalling Wilson's ultimate question: what difference did it all make? What kinds of change did the reve-

lations of the social science of reading inspire? Eliza Atkins Gleason supplied part of the answer. Gleason was the first African American to earn
a doctorate in library science, which she did at Chicago during Wilson's
leadership of the GLS, with Carleton B. Joeckel, a doyen of public library
administration, as her mentor. She would later serve as the first dean
of the first institution to train Black librarians in the South, the School
of Library Service at Atlanta University. Her work proved extraordinarily
influential in shaping understandings of the role of libraries in the South,
and indeed in broader debates about race and culture, by virtue of being rehearsed in the two most important sociological studies of the subject in that era, both of which similarly arose from Carnegie Corporation
grants: *Patterns of Negro Segregation* (1943), written by Bond's coauthor
Charles S. Johnson, and Swedish sociologist Gunnar Myrdal's *An American Dilemma: The Negro Problem and Modern Democracy* (1944).[27]

Three elements of Gleason's work deserve particular note. First, in 1939
Gleason extended Wilson's geography of reading by surveying the provision of library service to African Americans across the South, showing that some six million still had no access whatsoever.[28] Others had
only nominal access; she cited the example of Petersburg, VA, where the
reading-room reserved for Black patrons was in the basement of the main
library, with a rule banning the transfer of books from the upper rooms to
the lower.[29] Moreover, in almost none of the libraries that did serve this
population could a local librarian decide which books were supplied, and
access to interlibrary loans to Black districts was routinely denied.[30] Second, she devoted a farsighted chapter to the legal aspects of segregation,
concluding that whatever the constitutional position might be—and in
fact it was unclear even then—in practice African Americans faced grave
risks in trying to make use of libraries in any southern state. The only
responsible road to follow, she suggested, would be the long one of accumulating court victories steadily in state high courts and the US Supreme Court.[31] And, third, Gleason noted that nowhere in the South was
there yet a counterpart to what Waples and Wilson had created in Chicago. There was no graduate library school at which African Americans
could even matriculate. That meant that there was precious little chance
of creating a cadre of Black experts with relevant knowledge and experience; and without such expertise, she perceptively noted, the very data
on which reform might be based could not be collected, let alone acted

upon. In particular, it was not clear what "equalization"—that term again, so familiar from Louis Wilson and the other social scientists of reading— would mean in practice.[32] Such a school in fact opened in 1941 at Atlanta University, and Gleason was its first dean.[33]

Gleason's work brought the arguments of Bond back to the problem of libraries. And over the next generation library segregation was one of the areas in which civil rights loomed large.[34] It mattered to Bond as it did to Wilson and his peers, and Gleason's efforts only made it more evident and pressing. However, change was slow, to the extent of being hard to detect at all. In 1954, on the very eve of *Brown v. Board of Education*, the Chicago-trained historian, library director, and civil rights activist Lawrence D. Reddick asked, "Where Can a Southern Negro Read a Book?" and found that the answer was no more heartening than it had been in Gleason's time.[35] The fact was that to the NAACP the issue of segregation in libraries had not been a major priority. It had seemed a secondary problem compared to that of segregation in schooling. The association's national campaign of legal actions did adopt "equalization" as its initial primary objective, but there were few resources to devote to what it felt to be a lesser issue. The struggle to desegregate libraries might well be, in the terms of Wilson, Bond, and Gleason, a struggle to provide that higher literacy that African Americans would need to be full participants in modern democratic and industrial society—and, indeed, that modern society itself would need them to have. But it was in practice a struggle carried out predominantly by local actors, often unsung at the national level. Even the ALA was conspicuously silent about it, at least until very late in the day.

Yet carried out it was. Sit-ins and "read-ins" occurred at a number of public libraries, starting as early as 1939, while Gleason was doing her survey. Samuel W. Tucker, a Black lawyer, requested a user's card at the public library in Alexandria, VA, and on being refused organized a group to enter the library, ask for cards, and sit down to read anyway. The five who were brave enough to go through with the plan were arrested and charged with disorderly conduct. Their case was never resolved, but the legal technicalities that Tucker and his friends brought to the surface seemed to imply that the library had no firm basis for refusing future requests. Unfortunately, he then fell ill and had to withdraw from his campaign; and the effort came to an end when a segregated branch opened for the Black community.[36] Two years later, a Black woman schoolteacher, Murray

Walls, who had studied at Columbia University, took local NAACP members to the main public library in Louisville to hear a radio address by Roosevelt, and they insisted on staying after the staff tried to dissuade them. It was rather a low-key confrontation as they went, because Walls had the apparent approval of the local library director; but still, she later remembered this as one of the first sit-ins of the new civil rights era. Unlike Tucker, she did not desist, and thanks to her initiative the library system in Louisville desegregated in 1948–52.[37]

Such efforts were not common, but nor were they unknown. Thanks to them, a number of public libraries in the South opened their doors to African Americans earlier than other public venues such as schools and buses. It was suggested in the early 1960s that the peaceful retreat of segregation in these spaces could occur because the African Americans who actually wanted to visit libraries were relatively "high class," whereas the whites who most vehemently opposed desegregation were infrequent readers.[38] Whether or not that was the case, however, the progress was not uniform, smooth, widespread, or inevitable. As late as 1956, the Library Services Act—the first national legislation to fund library services—effectively winked at ongoing segregation, because it delegated decisions to state library agencies that then declined to allocate funds to services for Black communities.[39]

By this time Bond held a prominent place as one of a generation of assured, distinguished African American cultural critics and social scientists. He had stopped writing on the theme of race and intelligence testing in the mid-1930s, focusing instead on developing programs in various institutions, above all Lincoln University, of which he became president in 1945. But he would return to that fray after the war. In the early 1950s, claims revived among some commentators that intelligence testing manifested real racial differences in intellectual capability. Bond, by this time a senior and widely respected educator and university leader, returned to the fight, skewering such contentions in characteristically urbane fashion. Along with work conducted by other Black social scientists in these years—especially Bond's Fisk colleague Charles S. Johnson—his work would be important in informing the cases both local and national that challenged segregation in schooling.[40] This effort came to a head in the most important of all such cases, and the one that did most to overturn the culture of segregation.

In June 1953, after a long and tortuous process, the Supreme Court deadlocked on the five cases collectively known as *Brown v. Board of Education* and referred them for reargument. The cases posed the essential question whether segregation in public schools—regardless of the allocation of resources between such schools—were inherently unconstitutional. In order to get to this core issue, the court requested that the two sides address five questions, the first two of which were historical in nature: whether the original ratifiers of the Fourteenth Amendment at the end of the Civil War had meant it to imply the abolition of public-school segregation, and whether they had understood that a future Congress or court might construe it as doing so. These two questions expressed the crux of the matter; the other three were less important and concerned the scope and scale of a possible judgment. At the Legal Defense and Education Fund of the NAACP, Thurgood Marshall, who was leading the association's campaign, recognized the historical questions as pivotal. If they could fight the other side to a draw on these two questions, which meant convincing the court that segregation did not live up to the intentions of the original (Republican) passers of the amendment, then segregation would be declared unconstitutional. For that he needed social scientists, and historians in particular.

Marshall persuaded a political scientist at Lincoln University named John A. Davis—the brother of renowned Chicago social-scientist and champion of the case against racist uses of IQ, Allison Davis—to recruit a team of scholars to answer the court's questions. After getting discouraging answers from a couple of eminent elder-statesmen of the field, Davis turned to his own institution's president, Bond. Bond accepted immediately and threw himself into the task. He moved lock, stock, and barrel to a hotel in New York City and spent the summer writing what amounted to a monograph on the early history of the Fourteenth Amendment and public education. Its central claim was that in the immediate wake of their defeat the ex-Confederate states had accepted the necessity of abandoning educational segregation in order to be readmitted to the Union. This was a radical argument, given that it had not taken long after that moment of acquiescence for Jim Crow measures to appear in the region. And over that summer Bond's allies came to find it increasingly hard to sustain, in the face of evidence from the nineteenth century that some at the time had not conceived of the amendment this way. In a sense, its credibility required the kind of purposeful, skilled, and applied reading practice that advocates had called for in the 1920s and 1930s. And it worked. Bond's contentions set the tone, and others who joined the team—including C. Vann Woodward and John Hope Franklin—then took up the story of how segregation subsequently grew and metastasized. By the end of the summer, some two hundred scholars had contributed to what became a massive, relentless research campaign. It was all tied together in September at a three-day meeting attended by about a hundred of them. The internal debates were fraught and delicately poised, but in the end worries about the more radical reading of history were overcome. It was essentially Bond's position that was made the centerpiece of the NAACP's new brief—the keystone, that is, of the argument that finally won the day against school segregation.[41]

A decade later, in the last days of 1965, Bond sat down to write an introduction to a reissue of his *Education of the Negro*. It was an opportunity to take stock, appraising the progress that had been made and the disappointments that remained. He started by declaring the most important change since 1934 to have been, not desegregation or civil rights per se, but a transformation in the "psychology of race." Recalling Brigham's work and the furor over the army tests, he noted that in the 1920s almost every leading psychologist, including Thorndike, had been on the side of the "racist cause"; now, in the era of Johnson's Great Society, not one who mattered took that position. He also recalled that in 1934 the desegregation of schools had been inconceivable to him, and that he had recommended only "equalization"—meaning, in the first instance, equalization of teachers' salaries.[42] But the NAACP's legal campaigns had in fact won that deceptively important prize as early as 1942. A trajectory had been established that had then continued to the victory of *Brown v. Board of Education*, and in the wake of that to the further struggles and achievements of the civil rights era. But as he sat writing these reflections, Bond's thoughts were notably melancholy. He noted that the McCone Commission had just reported on the Watts Riots in Los Angeles. Among its conclusions: that African American students in Watts and similarly disadvantaged areas could not "make use of reading and writing for ordinary purposes of life," and that a massive public program was needed for "educating the Negro child" in literacy. The same was surely true in Chicago, New York, and other cities. The questions that had been posed in the 1930s-1960s, and to which habits of reading had been pivotal, had not been fully answered after all. Twelve generations after slavery, America had still to escape from its consequences.[43]

6

You're Not as Smart
as You Could Be

Ever since its earliest days, one of the signature claims of the science of reading was that good readers did not proceed letter by letter but by seizing upon larger entities—words, phrases, and sentences, and occasionally even whole paragraphs. Practitioners often spoke of people recognizing "wholes," or sometimes "patterns." What exactly a whole or pattern might be was questioned, of course, and on various levels. Straightforwardly, there were differences of opinion between those emphasizing words and those championing phrases, and Raymond Dodge for one had to defend himself against claims by latter-day Wundtians that the readers he thought were perceiving word-forms were in fact latching onto "dominant" letters. But Dodge carried off his defense well, and twentieth-century experimenters took it for granted that fluent, efficient, low-fatigue reading proceeded by grasping larger units of some kind. At a more complex level, researchers differed on how to allocate responsibility for isolating wholes between the body and the mind. But in practice these distinctions were relatively inconsequential. The central task of the applied science of reading was to devise and circulate practices that would inculcate this kind of habit in Americans across the land. In effect, this basic conviction about patterns was the most important single justification for there being a science of reading at all.

How large were the units of reading, though, and how might they be made bigger? Those were key questions on which the possibility of devising strategies to improve readers' skills might depend. To answer them,

Buswell, Gray, and their peers turned to the second of the characteristic instruments of the science of reading: the tachistoscope. They used tachistoscopes to flash up words and phrases of different lengths, in the first place to ascertain what kinds and sizes of patterns particular readers could perceive in a single fixation, and in the second to try to train them to perceive longer ones. This practice was closely related to their constant emphasis on measuring and enhancing the *rate* of reading. If a reader could be trained to take in more meaning in a single "peep," they reasoned, then that reader would naturally read faster without any loss of comprehension. Eye-movement cameras would show that such a reader made fewer pauses and regressions (which implied that more was being taken in accurately during each pause). And modern society would be much the better for it. In the 1930s and beyond, this simple ambition gave rise to a panoply of machines and pedagogical tactics that fanned out across America's schools, homes, and workplaces, bringing the science of reading directly to the eyes and brains of millions of citizens.

The reason why this view of reading as pattern recognition led so directly to large-scale cultural consequences was simple: whether a group of letters constituted a pattern at all had to depend, to some extent, on the perceiver. A reader apperceived patterns, or so the theory went, by virtue of *already* possessing a set of prior experiences to which the mind would automatically compare new perceptions. In cases where a new or unusual series of characters appeared, that process of comparison might be slow—think of trying to read a word like *tachistoscope* for the very first time. In cases of extreme novelty (imagine a term such as *eleemosynary*), researchers in the wake of Dodge and Dearborn believed that a reader might operate letter by letter, a laborious process akin to the very first steps in reading of a young child. The process could be seen very clearly in eye-movement photographs that showed unusual pauses and retrogressions over the troublesome space of text. But with a familiar group, such as a word like *and* or *book*, or a stock phrase like *once upon a time*, the process became instantaneous, or almost so. What happened in fluent reading was a practical, experience-based achievement—an active seizing upon meaning, rather than a passive imbibing of it. A reader's total personal and social past was therefore involved in every moment of his or her reading. And it was precisely because reading was active and experiential that it ought to be improvable through training and practice.

In retrospect, it is tempting to associate this emphasis on reading by "wholes" with the claims of Gestalt psychology. And in fact the congruities did come to be recognized in the 1920s, after Gestalt came to the United States in the person of émigré psychologist Kurt Koffka. But the two movements in fact developed quite separately. Judd, as we have seen, adapted the science of reading to questions of image perception in the very early 1900s, in light of Stratton's campaign against physiological aesthetics. He developed an instrument that could track the movements of the eye in two dimensions, rather than just back and forth across a horizontal line. The instrument he developed then went on being refined and used in Chicago for decades, and Buswell revived its use for work on images at the end of the 1920s. He took up the belief that pattern recognition was a—perhaps *the*—central mental operation that experimental psychologists should explore. By that time, the fashion for Gestalt ideas was at its height, and the steam was going out of the laboratory science of reading as Waples's style of social science replaced it in public and political prominence. Buswell followed that trajectory too. But in his work on image perception his instincts were as acute as ever. Pattern recognition as a concept would go on to have major consequences of its own. In the end it would help shape what theories, not only of reading, but also of media, art, and information in general, could and would become. It would substantially shape everyday experiences of information to this day.

But this was not because of Buswell's own work, or at least not directly. It was far more the achievement of a quite distinct project that put the tachistoscopic science of reading to a radically new and ambitious use. This was an extraordinary venture, dedicated not merely to the enhancement of reading abilities but to the multiplication of mental powers in general. At its peak it even offered the prospect of an extension to life itself. It emerged in Ohio, and it would have profound consequences across a dizzying array of realms. Its effects started with the powers of the mind and the creativity of artists, and they reached to the boardrooms of America's corporations, the decks of its naval vessels, and the cockpits of its aircraft.

Psychological Optics

One day in 1923, the main attraction at a meeting of the Rotary Club in Kalamazoo, Michigan, happened to be a "memorizing wizard." His appearance had been much anticipated, and people made a special visit to see his feats. The most astounding of them turned out to involve a deck of cards. Handed the shuffled deck, the performer studied them for a few minutes and then returned them to his host. He then proceeded to call out the entire series of cards in sequence, without a single mistake. The audience were mystified. How had he done it? Could anyone else manage such a feat?

One member of the audience that day was less impressed. A young psychology professor at Western Michigan State Teachers College, Samuel Renshaw had been invited by a physician friend to come along to the lunch specifically to see this prodigy. But he found the performance humdrum. The friend, perhaps a bit miffed, challenged him. He bet Renshaw ten dollars that he would not be able to match the wizard's magic. Renshaw took the wager. A few weeks later, he emerged victorious. Not only could he equal the achievement: after just thirteen lessons, he had fifty students all capable of replicating the feat.[1]

How exactly Renshaw achieved his success is not clear. But he himself remembered it for decades, and he attributed it to what became one of his characteristic convictions: that "seeing," as opposed to vision, was a practical, trainable action, involving all the senses in a cooperative grasping of some whole image. From the 1930s to the 1950s he would build up a scientific career, a huge military enterprise, and a considerable commercial culture around this belief. Central to it was the instrument known as the tachistoscope.

Renshaw (1892–1981) had earned a first degree at Ohio University, and when he attended that Rotary Club event he was already spending his otherwise empty summer days studying for a second at Ohio State University in Columbus. He obtained his master's there the same year and immediately launched himself into a doctorate. For a supervisor he attached himself to Albert P. Weiss, a staunch behaviorist and advocate for exploring human sensory perceptions in terms of their sustaining "sensory-cerebro-motor system." Weiss introduced Renshaw to the craft of inventing scientific instruments, starting with a device called a pursuitmeter that assessed someone's calmness by asking them to use a stylus to follow a bead moving on a glass plate. The experience was apparently "similar to an industrial worker tending a highly complicated machine."[2] Weiss hoped that the device would find a use in criminology, and in fact Renshaw would develop a version that did indeed find such a use. The device revealed physiological changes associated with "certain mental states, such as fear, anger, and deception," and despite his insistence that it was not a lie-detector, it was soon used as precisely that on a condemned murderer.[3] He also devised a "rather elaborate" machine to classify people as more or less active while at work or rest, which was promptly seized upon by an opportunistic bed company in Pittsburgh.[4] Apparently regarded as a star in the making, in 1925 he was appointed to the Ohio State faculty as soon as his doctorate was conferred.[5]

The opportunity for Renshaw to make a real mark came in April 1934, in the shape of another apparent mental wizard. Polish accountant Salo Finkelstein was what people then called a "lightning calculator." He had become renowned in Europe for prodigious feats of memory and, by virtue of that, extraordinary abilities in mental arithmetic. It was said that for eleven years he had been employed by the Polish state, flawlessly doing Treasury work that had hitherto required thirty calculating machines and their operators. Such prodigious powers attracted the attention of a Danzig psychologist, who enabled Finkelstein to attend an international conference of insurance mathematicians in Stockholm, where he amazed the participants with his skills. Realizing the potential for gain, he then decided to go on a world tour. Finkelstein proceeded first to Berlin and then to the United States, where he arrived in 1932 and conspicuously subjected himself to testing at several institutions. His reputation spread rapidly, and by 1934 he was wowing attendees at the American Psychological Association. Einstein himself reportedly "marveled" at his speed. It was in the wake of the APA meeting that Renshaw persuaded an initially hesitant Finkelstein that it was Ohio State's turn. He came to Columbus, and for about eighteen months resided there, giving occasional public demonstrations while Renshaw put him to the test.

Renshaw sought an instrument to measure Finkelstein's abilities. He may have found it in his one-time adviser's work: Weiss's account of the nature of thinking, published in 1925, had centered on a thought experiment about two men being asked to design a tachistoscope.[6] At any rate, Renshaw now decided that a tachistoscope was exactly what he needed. The device was a staple of experimental psychology. A tachistoscope revealed a set of digits, an image, or a word—or a set of words—for a small, precisely defined period of time, generally one-tenth of a second or less.

In some versions this was done by making a printed text visible through some kind of small window; machines like the Metron-O-Scope that were intended for schools used this technique. Laboratory tachistoscopes, however, were generally projection devices. They cast their images onto a screen in a darkened room for a fleeting moment, to be viewed there by an individual or a whole class of students. So this was the kind that Renshaw seized upon. Working with experts at the University of Chicago, he soon claimed to have perfected the instrument to a degree unmatched by any other practitioner.

Like the other scientists of reading, Renshaw focused intensely on speed, seeking to measure and increase reading rates before anything else. As with athletics, he believed, velocity was the key. His idea was to see how short a time Finkelstein needed to register a sequence of digits—which had to be taken in as a whole, in a moment that was too brief for the impression to be subject to interference by memories, preconceptions, and the like. But Renshaw was not content with investigation. He then wanted to use the machine to *train* Finkelstein to need shorter and shorter times. The tachistoscope, he believed, could make even this mental athlete into a faster, and implicitly better, reader. And if it could do that for someone who was already a marvel, what could it do for average Americans? Turning measurement into intervention, Renshaw would soon make the tachistoscope into the keystone of an extraordinary enterprise in public science.[7]

Renshaw not only analyzed Finkelstein's powers, then, but improved them. Using a projecting tachistoscope, he found that after practice Finkelstein could remember a nine digit number in a three-hundredth of a second; he could do a fifteen digit number in 1.47 seconds, and a twenty-one digit number in three seconds. These were remarkable feats—world records, indeed. Furthermore, it seemed that Finkelstein's much-vaunted arithmetical ability rested entirely on this speed of perception and memorization.[8]

Renshaw saw an opportunity in this achievement. He decided to try to improve readers who did not share Finkelstein's original talent. So he began a program to accelerate students' speed in reading letters and numbers. Before long, he was seeing success, and in a few years his charges were exceeding even Finkelstein's performances. Howard Wright, for example, "a Negro student," could read a fifteen digit number in 1.45 seconds; another student managed 1.08 seconds. Eventually Renshaw had two graduate students, Robert Mauer and Chester Pheiffer, who could read nine digits in a two-thousandth of a second. Such figures were "almost incredible," Renshaw noted, and yet he saw no sign that his charges were approaching any natural limit—a fact, he added, "of great importance."[9] He even mooted trying a flash of one-millionth of a second, which would be at the very edge of contemporary electrical technology. By such means, he suggested, he could train "virtually anyone" to become a "mental wizard."[10]

Renshaw's approach to revolutionizing the reading abilities of his students began relatively easily. He would start his subject with a two or three digit number flashed by the tachistoscope for one-tenth of a second. Most people could read such a group readily, at least after a couple of tries. He then increased the length of the "word" and reduced the exposure time, gauging the pace of change by the performance of the reader. A normal reader's top speed, he found, was about one-hundredth of a second for a nine digit number; but experience with the device generally led people to increase their abilities quite quickly, and many had not reached a maximum by the time they hit the limits of the machine. Skilled readers, Renshaw inferred, did not proceed by taking in individual characters, but by recognizing patterns—precisely as Gestalt psychology suggested, and the Chicago researchers on reading too. A tachistoscope simply trained a reader to grasp such units faster, and to read larger units at one go.

Having used the tachistoscope to increase the size of units that a reader "saw," Renshaw would turn to the eye-movement camera. He would record the motions of the reader's eyes during the reading of a text, exactly as the Chicago-school laboratory scientists did. But what Renshaw now found was, he attested, radically different from the orthodoxy. Judd, Buswell, and their colleagues knew that all reading practices involved saccades and fixations, and they focused their efforts on reducing and regulating those leaps and pauses. Renshaw's readers, however, approached an ideal state in which fixations and saccades disappeared altogether. Their eyes, Renshaw said, glided over the page "like graceful skaters on a frozen pond," taking in "whole paragraphs at a glance."[11] In other words, his training process was achieving much more than met the eye. He was not merely improving an action but transforming it into a different action altogether. It was as if one compared the swing of a champion golfer to the clumsy flounderings of someone who barely knew what a club was—the one was not merely a better version of the other but something different in kind. The new action was characterized by smoothness, efficiency, harmony, and muscular economy, where before there had been an ugly hotchpotch of jerky and wasteful thrashing about. Far more than a matter of mere muscular training, or so Renshaw came to believe, his practice led to a really distinct kind of mind-body coordination.

In Renshaw's understanding, the reason why his tachistoscope achieved such apparently spectacular effects was that it was tapping into the fundamental nature of reading—indeed, of seeing itself. And that nature involved the perception of "forms," or of what he called, taking his cue from the Gestalt movement, "structure." Remembering things like faces and names—and lines of numbers, and for that matter sequences of playing cards—depended on *not* seeing them as chains of discrete objects, but on grasping their structure. Occasionally Renshaw would call on evolutionary principles to help explain what he meant. Telling a story that might have been grabbed wholesale from Edmund Huey, he explained that the earliest communication between prehistoric humans had been by means of gestures and postures. It had been a matter of the "motor" powers of bodies, so it had involved all the senses, not just vision. Renshaw believed—and as the 1930s and 1940s wore on, he bolstered this belief with an ever-increasing range of psychological and scientific authorities from Bishop Berkeley to the latest Gestalt theorists—that sensory communication retained this character. That was how *seeing* was different from *vision*. Vision was the mere optical behavior of the eye—a matter of lenses, refractive indices, and the like. To *see* properly required the active involvement of the mind to select and interpret what this optical system conveyed, and it also needed the smooth coordination of vision with inputs from other organs than the eye. Just as it was generally easier to understand someone else's speech if one could see his or her face, so seeing was generally more successful if it involved multiple senses. Moreover, Renshaw was convinced that children saw things in this holistic manner naturally. They were drilled out of it at school, where their chances of acquiring a true reading skill were destroyed by the drudgery of proceeding letter by letter. His favorite proof of that point involved a particular photograph (fig. 6.1): the average schooled American, "looking

6.1 The cow image used by Samuel Renshaw to convey the importance of grasp-ing whole structures at once. The original was rotated 90 degrees counterclock-wise, making it all the more difficult to interpret. K. M. D[allenbach], "A Puzzle-Picture with a New Principle of Concealment," *American Journal of Psychology* 64, no. 3 (June 1951): 432. Copyright 1951 by the Board of Trustees of the University of Illinois. Used with permission of the University of Illinois Press.

at things piecemeal," found it baffling, but "every small child and every ta-chistoscopically trained adult" immediately saw it as the cow it was. The tachistoscope thus undid a profound miseducation.[12]

So it was that in the mid- to late 1930s Renshaw used the tachistoscope to develop an experimental and therapeutic program that was directed not only at reading, and not even only at vision, but at the practical uses of all the senses in concert (fig. 6.2). That program continued through the war years and on into the early 1960s. As it proceeded, his accounts of the venture became more extravagant. He liked to say that humans typi-cally used only 20 percent of what he called their "sense modalities"—or rather, as one journalist put it, that "most people are only 20% alive." To improve that percentage, he used psychophysics instruments to accen-tuate the senses one by one. The reading technologies he developed—he took out a patent on his tachistoscope—were therefore just one element

6.2 Samuel Renshaw with his Tachistoscope. D. G. Wittels, "You're Not as Smart as You Could Be," *Saturday Evening Post* April 24, 1948, 31. Photograph © SEPS licensed by Curtis Licensing Indianapolis, IN. All rights reserved.

in a multisensory enterprise to teach people to "see better, taste more keenly and develop prodigious memories."[13] It was all a matter of training, to the extent that Renshaw even became convinced that most cases of poor vision were not really cases of poor *vision* at all but cases of poor *sight*. That is, myopia, for example, was not ascribable to a physical problem in the eye, but was due to poor habits. It should be corrected with retraining rather than corrective lenses. Tachistoscopes could replace much of optometry. So Renshaw campaigned to have the optometry profession nationwide abandon its conceptual foundations and embrace what he trumpeted as a new science: he called that science *psychological optics*.

For an optometry profession that was already under attack by critics championing the renegade physician William Bates's fatigue theories—the most prominent of those critics being Aldous Huxley—such pronouncements were less than welcome. The optometrists of America split into rival factions, for and against Renshaw's science. But Renshaw did have experience, experiments, and technology to bolster his credibility.

He had been instrumental in setting up America's first degree-granting program in optometry.[14] And he had powerful allies in the trade. At a conference in 1937, in particular, he met Arthur Skeffington, a prominent optometrist well known for cofounding an institution called the Optometric Extension Program to further what he called "behavioral optometry."[15] Skeffington and Renshaw immediately made common cause. For more than two decades Renshaw now maintained a campaign for psychological optics by means of a semipublished periodical issued via Skeffington's program. Called *Psychological Optics*, it was a rather samizdat kind of publication, of the sort that new subdisciplines often use to establish and sustain their social networks. Renshaw was its only author, and it served solely to articulate and defend his science of psychological optics. This it did in every way from discussing optical illusions in Gestalt fashion (in the early years) to (in later times) explaining everything in terms of information theory and neurology. The journal circulated through Skeffington's OEP to optometrists across the country.[16]

Since Renshaw embraced so fully the doctrine of the "unity of the senses," what was true of vision was true of the other senses too. "It is practically impossible," he declared "to disassociate vision, hearing, kinesthesis, smell, taste and the organics." We heard with our eyes, saw with the aid of our memories. All the senses could therefore be honed in parallel ways, and the key was to enhance the learning processes to achieve this.[17] He claimed that touch, for example, could be improved 700 percent, and taste could similarly be trained to become far "keener" (fig. 6.3). That would be beneficial to public health, Renshaw pointed out, because we might then rest content with only one sugar-cube in our coffee, or a smaller dose of gin in our evening G&T.

The point about taste immediately raised a question of standards. Since the nineteenth-century political controversies over adulteration and the emergence of agencies like the FDA, the food and drinks industries had been acutely conscious of the complexities involved in maintaining the "same" taste for a given product. Attempts to sustain and police standards had proliferated.[18] Renshaw pointed out that the US government retained a "Board of Tea Experts" to taste imports of tea and set standards in that area, while the New York Coffee Exchange had an unofficial panel that performed a similar role for coffee. A little later, in World War II, Renshaw himself would do work for the army on developing

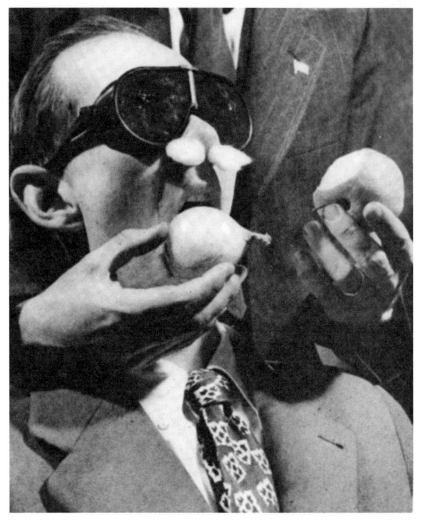

6.3 A Renshaw experiment with the senses. D. G. Wittels, "You're Not as Smart as You Could Be," *Saturday Evening Post* May 1, 1948, 30. Photograph © SEPS licensed by Curtis Licensing Indianapolis, IN. All rights reserved.

a standard taste for dried milk.[19] Breweries employed beer tasters, and whiskey companies retained people called "Tongues" in a bid to keep flavors constant year after year. This last provided Renshaw with the opportunity he needed. The whiskey industry was receiving complaints that flavors were varying, and Renshaw offered his help to discover why. He gave identical samples to three tongues, who produced divergent reports.

That meant that the problem was one of "calibrating the reporters," which Renshaw did by testing the tongues in the laboratory and generating mathematical corrections for their individual sensory reports—a kind of "personal equation" for liquor-drinkers, as it were. The grateful company hired him as a consultant. Out of the proceeds, Renshaw—who did not receive substantial support from the university for his research—was able to build an entire laboratory devoted to exploring sensory phenomena.[20]

That was not Renshaw's only entrepreneurial gambit. Since the early days, scientists of reading had voiced the hopeful idea that their claims might be taken up by corporate America, and that managers keen to enhance "efficiency" and minimize fatigue might adopt their science as a service. Renshaw moved to achieve this. By the late 1930s, he had begun selling his techniques and devices to a number of major US corporations in the electrical, aerospace, and other sectors, the largest of them being General Electric. He told these companies that he could train their managers to read faster, and that this would transform their efficiency and reduce their fatigue. This kind of venture would scale up radically in future years. But in the meantime, with global war in the offing, Renshaw saw a chance to tap into an even larger market.

The Science of Reading Goes to War

Imagine that you are a US airman, flying a P-40 Warhawk fighter on patrol over the Mediterranean one afternoon in mid-1944.[21] The air is clear, and visibility is good. You are at fifteen thousand feet and your ground speed is about four hundred miles an hour. Suddenly you notice a glowing speck in the sky ahead of you. It is getting larger quickly—something is coming toward you, and fast. If it is a German fighter, say a twin-engine Messerschmidt 110, then its likely speed will also be about four hundred miles an hour. That means the combined speed of approach is eight hundred miles an hour—almost four hundred yards per second. You have fleeting moments before the other aircraft will be upon you. And the same is true for the other pilot. The decision whether to open fire is a life-and-death one on both sides, and it may well have to be taken in a fraction of a second. If you hold your fire and it is a Messerschmidt, you may be dead in a moment. If you open fire and it is a Mustang or Spitfire then you may well kill a friendly pilot. What do you do?

Now scale up the problem. Imagine that you are piloting a P-51B Mustang with the Eighth Air Force, flying escort to a massed formation of B-17s and B-24s on one of their stupendously dangerous daylight raids over Nazi Germany. The Luftwaffe's fighters generally leave incoming bombers alone until they cross into German territory, but then they swarm up in large numbers. Suddenly you face the same critical decision as that of the P-40 pilot, only multiplied by perhaps a hundred, and repeated incessantly for hours. Moreover, the casualty rates among Eighth Air Force crewmembers are so high that it is very likely you are already suffering psychological symptoms—jitters, nausea, and disorientation. One of the common symptoms is impaired vision. Your field of view is full of swirling, diving, and climbing aircraft, large and small. Again, what do you do?

These kinds of scenarios plagued the Allied forces in World War II. Pilots, soldiers, and seamen had to recognize fast-moving, camouflaged, and evasive objects, often at night or in terrible weather conditions. A pilot's visual acuity declined above about fifteen thousand feet and in stressful contexts, and the plight of ships' lookouts in North Atlantic convoys was desperate. Inevitably, they often took the wrong decision. In battle after battle—starting with the initial German invasion of Belgium and France—the militaries of the United Kingdom and the United States found that dismaying numbers of aircraft were being shot down by their own forces or those of their allies. In the early weeks of the war, for example, almost 20 percent of the losses suffered by the RAF were attributed to their own side; and in 1942 Renshaw learned from the Ohio State University newspaper that 40 percent of planes shot down were victims of Allied action. Worse still, it was reported that in the ill-fated Dieppe raid in August that year, sixty of ninety-two Allied aircraft lost were destroyed by British fire.[22] As a champion of Renshaw's techniques lamented, "What monstrous consequences in terms of life and freight may rest on the faintest imaginable visual cues!"

For someone like Renshaw, the nature of the problem was obvious. The crucial judgment had to be taken on the basis of perceiving the whole form of a plane, ship, or tank, and that perception had to be done as fast as possible. "On the minutest differences of *visual form pattern* and *movement* rests the judgment—friend or enemy?" In effect, the situation of that fighter pilot over the Mediterranean was the same as that of the student sitting beside a tachistoscope in Renshaw's laboratory. Both were tasked with understanding some "whole" in as short a time as possible, and with integrating their vision with bodily actions. The problem was not unique to aerial warfare, then, although it was most pressing there because of the speeds involved. Renshaw's science of enhanced reading could be brought to bear.

Being able to identify distant aircraft rapidly was all-important for servicemen in all branches of the war effort. And aviators, soldiers, and naval officers were, in fact, trained to do this. But the existing approach did not accord with Renshaw's philosophy. The approach was known as WEFT ("Wing, Engine, Fuselage and Tail"), and it involved the spotter identifying such individual characteristics one by one, piecemeal as it were. The technique was an almost exact counterpart to the pedagogy of reading that Renshaw so despised. So in November 1941 he started to concoct an alternative—one based on his own, tachistoscopic, science of reading. Working fast, Renshaw found a number of glossy magazines and clipped photographs of aircraft from them. These he used to construct a series of tachistoscope slides.[23] He then tried them out to see how quickly viewers could identify the different types, and retooled his tachistoscopes so that they could project for a larger class. With the help of a couple of student assistants he started to draft course manuals. By April, five months later, he was ready. He arranged a demonstration before a one-time secretary to the Arts College, Howard Hamilton, who was now serving as a US Navy officer. The point, Renshaw told Hamilton, was first to perceive potential enemies quickly and accurately, and second to expand the visual field within which this kind of rapid perception happened. With his help, he predicted, a serviceman could take in the whole of what a report would later call "a skyscape full of aircraft" in a fifth of a second. In the chaotic airspace over Germany, that ability might well be strategically decisive. Hamilton was initially skeptical, but then impressed. He badgered the navy into allowing an experimental program to be undertaken.

An initial detachment of thirty navy officers arrived in Columbus on April 20, 1942. Renshaw was given until May 10 to show demonstrable results. When those results came in, the success seemed a "miracle." Not only did officers see more than ever before, and do so faster. They also recognized aircraft seen from unusual angles, and they could even perceive minor but potentially crucial alterations to an airframe—such as the difference between existing planes and the new Avenger torpedo-bomber, for example, which could fire backward, as Japanese pilots would imminently find to their cost at Midway. He got the approval he sought, and on September 1 a new Recognition Training School opened on the Ohio State campus to impart his technique to the navy at large. It used two thousand tachistoscope slides, representing 168 planes and ships from a range of angles. By repurposing his science of reading, Renshaw meant to render his charges able to recognize exact types of aircraft and ships in a second, at most. As the *New York Times* reported a year later, his system "trains the eye to take in a whole plane or ship at a glance, just as we take in whole words instead of individual letters when we read."[24]

The initial class of ninety-six men took over a dormitory at Ohio State that fall and renamed it USS Baker. Classes took place all day long at Derby Hall or, later, at nearby Arps Hall. Renshaw promised to train the men to recognize 125 different kinds of aircraft and ships in 1/150th of a second. The first personnel graduated in October 1942. Word soon spread, and the venture expanded fast. Enrollment for the sixty day course doubled to two hundred, and by May 1943, 550 navy officers were participating. Soon Columbus was seeing about five hundred new trainees every eight weeks. And soon after that the navy took control and made the operation a formal command post. Army officers were still skeptical of the "structure" idea, so they were shown a tachistoscopic image of Betty Grable; this convinced them that they could indeed appreciate an image flashed up for such a short time, and they approved the military's own version of the program.[25] Soon the Royal Navy and RAF were sending their own personnel too. The US Marines, the Royal Canadian Air Force, and New Zealand forces followed suit.[26] By the time the navy moved out of Ohio State's buildings to its own facility at the end of 1944, the school had trained some four thousand naval officers, who in their turn had taught Renshaw's system to 285,000 airmen. The Chicago company Three Dimension had mass-produced thousands of tachistoscopes and millions of slides, which the officers took to ships and naval air stations worldwide. A year after Renshaw's program started, almost every US Navy vessel had at least one "Renshawed" crew member aboard. And according to a contemporary estimate, the military ended up imparting the technique to "literally millions of men and women in all parts of the world." When listing the university's contributions to the war effort, Ohio State's news office listed the program alongside, and sometimes ahead of, its role in the Manhattan Project.[27]

Renshaw's military entrepreneurship attracted attention, and its effects would continue after the war. But it should be noted that he was pushing at an open door. In the years before the war, and all the more during the conflict itself, countless authorities had warned that visual acuity would be an essential contribution to the effort. For fighting men, clearly, it was vital. But it was almost equally essential on the home front. As a supporter of Renshaw noted, such acuity was as crucial for a factory hand reading micrometers in a weapons manufacturing plant as it was for a pilot in a Mustang or a midshipman on a destroyer. Such rapid and reliable reading had to be continued for hours, moreover, in spite of fatigue. War thus accentuated the visual sociology of American modernity that had inspired the science of reading in the first place. Theirs was an "inter-dependent society," Americans were cautioned, in which the survival of any one person depended on the sensory capacities, not only of themselves, but of all those fellow citizens who were exercising their eyes in domestic factories building planes, tanks, and guns.[28]

Whaam! Flash Labs, Cockpit Instruments, and the Science of Information

At almost exactly the same time that Renshaw's first pupils were arriving at Columbus to experience his tachistoscopic training in aircraft recognition, a quite distinct teaching classroom was starting work across the campus. The professor of fine arts Hoyt Sherman had encountered Gestalt principles and the tachistoscope in the late 1930s, quite possibly through Renshaw, and had become convinced of their potential for his own field. Sherman decided to embrace the instrument to train artists. In collaboration with the psychologist (and later advocate for cognitive creativity in education) Ross L. Mooney and the director of the School of Optometry Glenn Fry (whom Renshaw had helped recruit back in 1935), Sherman set up an elaborate training program. Central to it was a location that came to be called the Flash Lab (fig. 6.4). It was essentially a giant tachistoscope in which trainee artists could learn to perceive and

6.4 Students in Hoyt Sherman's "Flash Lab" at Ohio State University. OSU Archives. University Archives, Ohio State University Libraries.

create. After initial experiments in 1942, it took in its first students in fall 1943. By the late 1940s more than 250 would have passed through the program.[29]

The Flash Lab was an entirely Renshaw-ish institution, yet Sherman and Renshaw themselves seem to have barely spoken after the early 1940s. There is firm evidence that in later years Sherman felt deeply resentful that Renshaw garnered so much credit for the approach that both men advocated, feeling that he had been its true originator and that Renshaw had usurped his credit not only for the approach in general but also for the naval program in particular. Mooney explained why to an official of the Grant Foundation who was curious about the puzzling disconnection. Renshaw, he remarked, was enjoying great popular success, and his students were "receiving phenomenal offers from private industry all over the country." But his program was intellectually cramped, whereas Sherman's was open-ended. More to the point, he would have to be "a different person" for collaboration to be feasible. The two of them had come up with the recognition program jointly, Mooney said, only for Renshaw to claim it as his own invention. "He's an insecure individual, mixed up inside, hungry for status, and quite undependable. . . . Men in his own department do not trust him as a person, and many have come to the place where they find it very difficult to trust his data." Mooney added that he personally had had to "nurse one graduate student past his vindictiveness," and that "dealing with the man leaves one sick inside." We need not try to adjudicate the feud at this distance, but it is worth noting that Sherman himself was not easy to get along with (he was to be the subject of a long-running and ferocious disciplinary case at the university), and that at an early coappearance of the two at the Faculty Club to talk about the "Visual Perception of Form" in February 1941 it was already Renshaw who took center stage.[30] At any rate, it is possible that the fact of their conflict, whatever its cause, damaged the fortunes of both men's projects.

Like Renshaw's enterprise, Sherman's was based on a conviction that students must seek above all to experience a unified sensorium. Sherman called this phenomenon "perceptual unity," and he believed that the artist who had most successfully exemplified it had been Paul Cézanne. He felt that perceptual unity in the arts could best be achieved if one grasped an image, sculpture, or other object in an instant. So he had students experience objects and images flashed tachistoscopically in an otherwise pitch-dark room, after which they would make art in the dark. They worked standing up, unable to see their own drawings, to bring "the tactile sense" to the fore. A turntable played music as they did so, "to keep the body attuned to its full possibility of movement" and to "'draw off' the excess of attention which the students might otherwise give to the particular muscular movements which the manipulation of the charcoal may seem to require." In effect, the perceiving self floated at the center of "a concourse of kinesthetic, tactile, psychological, auditory, and optical sensations." The kind of "seeing" it did from that position, Sherman added, was "an aggressive act." It was not just a matter of optics, then, but a whole-body experience in which the mind actively seized upon holistic patterns. As Renshaw too always insisted, that meant that the practice required *un*learning acquired habits. The Flash Lab consequently banned all talk about prior art, including remarks about classics, perspective, symmetry, and the like—and even comments on students' own work. The immediate, actively adopted unity of the senses was all.

The Flash Lab experience certainly seems to have been a profound one for some of those who participated. Here is how one of them expressed it poetically, in the middle of his MFA thesis in 1949:

Therefore, you must use your hand
To make the felt thing seen,
Rather than your eyes
To see to say.
Nor can you feel what you have seen
Until you see what you have felt.
The truth of nature's structure
Comes to you through work,
And is then projected through your eyes.
So looking without touching would uncover for you
None of the world's structure,

. . .

You will see what you feel.

Roy Lichtenstein, who wrote these lines, served as Sherman's teaching assistant in the Flash Lab before going on to become wildly successful as America's greatest pop artist. A convinced acolyte of the Flash Lab technique, he would repeatedly try to build his own version in his later life, never quite successfully.[31]

Sherman himself was explicit and extravagant in his claims for the new technique he was pioneering. Its implications, he argued, extended to all education: to reading, certainly, but also to music, surgery, architecture, and dentistry (he inaugurated a program utilizing "an aesthetic approach to the teaching of dentistry"). They even extended to sport. As a case in point, his crew invented a "flash helmet" to train the Ohio State quarterback to spot teammates faster and more accurately during football games (fig. 6.5). After extensive statistical analyses of games carried out over several years in the 1940s (focusing, naturally, on the performance of the quarterback against Michigan), it was concluded to have helped restore the team's fortunes.[32] But the fact that Sherman launched the Flash Lab at the height of a world war meant that claims of this extravagant kind had to be made good in new settings, and with much higher stakes than before.

War produced moments in which rapid reading and viewing became a matter not only of personal betterment or aesthetic expression, but of life and death. No less than Renshaw, Sherman was acutely aware of this. To illustrate the kind of visual acuity that he claimed his laboratory

6.5 Hoyt Sherman (L) helping to train Ohio State's quarterback by using the "Flash Helmet." OSU Archives.

inculcated, he cited the example of landing on an aircraft carrier's deck at night (fig. 6.6).[33] As he surely knew, that problem was a very real one, as pressing indeed as that of aircraft recognition. During the later years of the war, American and British bombers raided German cities day and night; and in the Pacific theater American aircrews underwent long, exhausting flights over vast distances. Fatigued pilots heading back to an

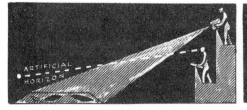

6.6 Hoyt Sherman's illustration showing how tachistoscopic devices could be used to train pilots to conduct night landings on aircraft carriers. H. L. Sherman, *Drawing By Seeing* (New York: Hinds, Hayden and Eldridge, 1947), 56.

airfield or carrier after a mission were expected to land their aircraft safely whatever the conditions—in darkness, often in poor weather, and not infrequently with a damaged airframe and wounded crewmembers. Thousands faced these white-knuckle experiences. Crashes on landing were literally an everyday—and every night—occurrence. What the air force and navy discovered, however, was more disquieting still. Hundreds of planes were crashing even when none of those conditions existed. Their crew were apparently unharmed and the aircraft themselves unscathed, and the weather conditions were acceptable. They had made it back undamaged, only to crash at the last moment. The accidents did not seem to have anything to do with enemy action at all. And losses from such mysterious events also took place far outside combat zones. In the period from December 1941 to mid-1945, according to official statistics compiled immediately after the war, the air force alone lost 13,873 aircraft—more than ten planes per day, every day—*within the continental United States*. Almost fifteen thousand people were killed in aviation accidents of one kind or another in that period, and a proportion surely died in incidents of this unexplained type. If Sherman's claim about the Flash Lab could be believed, then it might make a vital difference to the war in the air and at sea.[34]

In the event, the science of reading *was* critical to solving the problem of unexplained crashes, but it was eye-movement work that proved worthwhile, not Sherman's tachistoscopic training. (It seems that Sherman blamed Renshaw for his failure to break into the military market.) The reason was that the problem came to be treated as arising from a combination of the reading of instrument displays and the coordination of such reading with the manipulation of controls. The traditional response to crashes on landing, dating back to the end of World War I, was to assert that they must be the result of "pilot error." But in an age when new scientific techniques could reveal so much about human interactions with the world, that hardly seemed sufficient. Since 1941, a group of psychologists had been working in Washington to assist the war effort by helping the USAAF in the selection and training of pilots, and in 1943 Arthur Melton oversaw the first psychological research on aircraft equipment design under the aegis of the School of Aviation Medicine. Melton was a psychologist who had done detailed work to trace how a patron moved through a museum or art gallery viewing exhibits, and he treated the pilot in his cockpit as inhabiting a comparable scenario. Now, in the closing days of the war, the air force turned to one of Melton's protégés to investigate the mystery of the hundreds of uncaused crashes. Paul Morris Fitts (1912–65) was a Tennessean, educated initially at the University of Tennessee, where he earned a BSc in psychology in 1934. He had then gone east to Brown University and the University of Rochester, where he did doctoral research under the animal behaviorist Leonard Carmichael. He had been working back in Tennessee until he joined the air force's advisory group of psychologists. In mid-1945, he found himself suddenly catapulted at Melton's suggestion into the directorship of a new Psychology Branch at the Aero Medical Laboratory at Wright Field in Dayton, Ohio. His major task was to figure out the problem of uncaused crashes.

Fitts went about his mission energetically. Mid-1945 saw him spending three months in Germany looking into how the Third Reich had addressed similar issues. He also discussed them at length with the Cambridge authority on memory, F. C. Bartlett. Then, back in the United States, he commenced in December what became a hundred interviews with airmen who had survived the kinds of crashes he was investigating, plus the gathering of five hundred more written contributions. Carefully weeding out anything but direct testimony of experiences, Fitts found himself with an extraordinary finding: fully 50 percent of the crashes about which he heard had arisen from what he called "substitution errors" in the cockpit. That is, they had come about because the pilot had momentarily confused one control for another, and had operated the wrong control—flaps instead of wheels, for example, or the wrong engine control. Moreover, "practically all" pilots in the air force, of whatever degree of skill or experience, made such errors. No pilot, however experienced, was immune to them.[35] What that implied was that the causeless crashes were in fact resulting from a serious mismatch between the sensory capacities of human pilots and the layouts of the cockpits in which they sat.

In any "skilled motor activity," Fitts pointed out, "such as walking, driving an automobile, or piloting an aircraft," movements of the arms, hands, and legs should ideally be simultaneous and coordinated. They had to occur in sync, and along with the reading of instruments, "with a *minimum of conscious visual control over separate acts.*" The problem was that existing cockpits did not allow for this kind of automatic sensorimotor coordination. Indeed, Fitts showed that the situation was almost the worst one could have conceived: different cockpits had knobs in the same place and the same colors, but they controlled totally different elements of the aircraft. So it was quite possible that thousands of hours of experience in one design might result in an ace pilot being *more* likely to make a catastrophic mistake when flying in another. And the fundamental reason for this was that, as recent psychological research had shown, a human viewer saw an image as a "total pattern" and would "recognize a pattern even though the specific elements are varied considerably." This being a military report, Fitts did not specify which psychological research he had in mind, but it is striking that this view could almost have been transcribed word-for-word from the researches of those scientists of reading who, like Renshaw or Buswell, insisted on the primacy of pattern recognition in learning. Indeed, it was as though a pilot experienced the obverse of Renshaw's tachistoscopic experience. Whereas Renshaw wanted readers (or spotters) to pick up precise patterns immediately, doing that was fatally risky in cockpits as they then existed.[36]

That thought led directly to Fitts's next move. If the problem was partly one of controls, he realized, then it was also one of instruments. It was not just that pilots reached for the wrong handles, but that they looked at the wrong dials or "read" the wrong numbers. Indeed, this seemed the harder problem to fix. So Fitts took 270 more reported "pilot error" episodes and analyzed them in terms of experiences of "reading and interpreting" the texts displayed by cockpit instruments. He identified nine distinct varieties of what he called "errors in reading." The commonest, for example, was to misread the altimeter, which registered the aircraft's altitude by a combination of a pointer and numbers visible through a window. Because of the instrument's design, it was all too easy to mistake one's altitude by a thousand feet. But there were errors of simple legibility with some of the other instruments too, and a number of other problems that arose repeatedly for several displays. It also happened very frequently that what a pilot's visual sense told him from reading the instruments came into conflict with the rest of his bodily sense organs. And because the instruments were not standardized, a pilot must "change his mental set each time he shifts his eyes from one instrument to another," and this took precious time. Again, the consequences could not be escaped by any pilot, however skilled or experienced.

Overall, Fitts recommended a thoroughgoing program of research to determine "more satisfactory methods of display for information," including an investigation of how humans interacted with such devices. The air force needed to know "the variables influencing instrument legibility" and "the degree of reading precision possible with different styles and sizes of dials, scales, pointers, and numerals." It would also need to discover how readily human eyes could move between different instruments, and how to minimize the time of confusion when they did so. It must take into account the figure-ground relationship so central to Gestalt ideas in order to understand pilot readings of dials and needles. And it must tackle the cockpit as a "total display," too, which was being apprehended moment by moment by a pilot with multiple senses all operating inseparably. Finally, it ought to distinguish between different *kinds* of reading—"check reading" was not the same as "quantitative reading," and it might be conceivable to design different instrumental displays for each. This distinction between kinds of reading, Fitts acknowledged, was "a radical departure," but he insisted that it should be considered. The range of experiments must be broad and ambitious, then, and they must imagine radically different systems such as instruments that would operate by sound rather than sight. Potentially, he added, there might need to be changes that would produce a "major readjustment in the pilot's perceptual activities."[37]

Fitts's conclusions were emphatic. Many of the accidents conventionally ascribed to pilot error should be attributed instead to "the design characteristics of aircraft instruments." It should therefore be possible to eliminate a large proportion of accidents by redesigning those instruments. Moreover, experiments should be done to discover how to "reduce the amount of time required for instrument readings" and radically increase the "efficiency" of pilots' interactions with cockpit devices. That meant, among other things, that the number of cockpit instruments should almost certainly not be increased, because the task of perceiving them all was already "near the threshold of human ability." And the advent of jet engines meant that future aircraft would be much faster, making the need for rapid perception all the more urgent. The faster the plane, Fitts pointed out, the shorter the time in which a pilot must "see, comprehend, and act." And the worse the consequences became when he failed to do so quickly enough.[38]

In other words, Fitts concluded that the problem of mysterious aircraft crashes was a problem of a certain kind—and it was best tackled by redeploying the science of reading. So he moved to do this. In fact, inquiries in this area had already begun on a much more modest level elsewhere. As he had already learned, in the United Kingdom, Bartlett's Applied Psychology Unit at Cambridge had been researching aviation equipment design for the RAF since 1940, and the Germans too had made desultory efforts.[39] The first such investigation in America was a joint US-UK committee on aircraft standardization that met in 1943, for which Fitts had served as psychological consultant. That had led one William McGehee of the Naval Instrument Flying School in Atlanta to do an early and very provisional eye-movement trial with pilots, and in mid-1944 Melton had supervised his own study of the legibility of aircraft instrument dials. Melton had used an ersatz tachistoscope to produce short exposures of 0.75–1.5 seconds and see how well a pilot could perceive an instrumental display.[40] The air force's School of Aviation Medicine, too, had taken up such work to a small extent even before the formation of the new Aero Medical Laboratory branch that Fitts now led.[41] But it was Fitts who gave such work its rationale and who turned it into a systematic, continuing research program.

Fitts started by installing an eye-tracking camera into the cockpit of a C-45 that he persuaded the air force to let him use. The C-45 was a small, twin-engine, general-purpose aircraft, familiar to virtually all pilots, so it was ideal for this purpose. He had dozens of experienced pilots practice landings while the eye camera recorded their glances, saccades, and fixations (fig. 6.7).[42] Thanks to reference photographs taken earlier, the device showed which instruments the pilots looked at, when, and for how long. On the ground, multiple researchers then examined

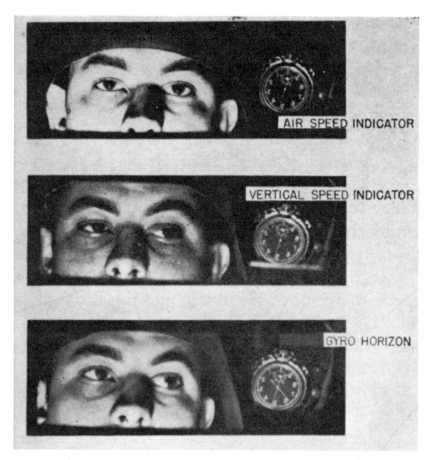

6.7 Paul Fitts's experiments on aircraft pilots as readers. P. M. Fitts, R. E. Jones, and J. L. Milton, "Eye Movements of Aircraft Pilots During Instrument-Landing Approaches," *Aeronautical Engineering Review* 9, no. 2 (January 1950): 25.

and reexamined the films frame by frame—all 76,800 of them—to determine the sequence. Fitts created a graphic style for presenting these eye-movement results on schematic images of instrument panels—a counterpart to the conventions for showing a viewer's fixations that Judd and Buswell had devised—so that a busy military officer could *see* the patterns of efficiency and hesitation that he wished to convey (fig. 6.8).

Fitts then followed up this eye-movement work by having tachistoscopic slides made up displaying the different gauges in use and exposing the same pilots to these images. The tachistoscope revealed which gauges were more or less "legible"—that is, which of them conveyed information

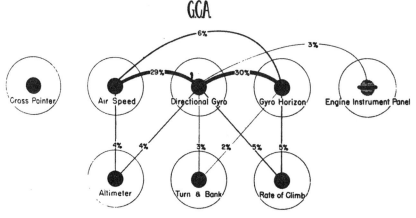

EYE MOVEMENT LINK VALUES BETWEEN AIRCRAFT INSTRUMENTS
GROUND CONTROL APPROACH
GCA

6.8 The movements of pilots' eyes between instruments on approach to landing. P. M. Fitts, R. E. Jones, and J. L. Milton, "Eye Movements of Aircraft Pilots During Instrument-Landing Approaches," *Aeronautical Engineering Review* 9, no. 2 (January 1950): 28.

that a pilot could take in accurately at a glance. It also suggested ways in which displays could be improved, for example by changing the font of numerals on the dial or altering the shape of the needle. The combination of eye-movement camera and tachistoscope thus gave rise to recommendations for both improving and standardizing instrument displays.[43] And, of course, there was also the question of the *arrangement* of instruments, and in particular of the distance that the eye had to move between them.[44] What had been hundreds of different dial formats and arrangements in use in 1940 was going to be reduced to order.

That first burst of work was just the start. Fitts's initiative set the agenda for what became a continuing, explicitly planned, research enterprise. Centered at Dayton, it involved a team of psychologists spending decades devoted to extending the approach and articulating its implications.[45] They produced a stream of reports and published papers, extending well beyond the departure of Fitts (with his longtime collaborator at Antioch College, C. W. Simon) to a faculty position at Ohio State. The investigations were still continuing in the 1960s and beyond. And their methods became common coin in a growing discipline that Fitts dubbed "engineering psychology," the practice of which extended internationally.[46] As early as 1946, R. C. Oldfield (later a major neuroscientist) was using similar science-of-reading techniques in Bartlett's Cambridge laboratory, in a project for the Royal Air Force to investigate how pilots scanned their field of view for an expected but inconspicuous body. As well as tachistoscopic work, Oldfield used an ophthalmograph borrowed from the American Optical Company, this being the commercial version of Guy Buswell's eye-movement camera marketed to make the science of reading portable. That meant, of course, that the object his putative pilot was watching for had to appear somewhere on a horizontal line—just as if it were a letter in a line of text.[47]

By repurposing the science of reading in order to create his enterprise of engineering psychology, Fitts made his work enormously consequential. In the first place, it proved fundamental to the subsequent history of aviation, and thus of the infrastructure of the modern world. It gave rise to the standard cockpit layout for all mainstream aircraft to this day, and thus permitted the generalization of training and management regimes on which airlines and air forces depend. But that may not be its most important outcome. In 1948 Fitts moved to Ohio State, where he took over the Aviation Psychology Laboratory and turned it into a venue for research on general issues of human engineering and systems research, mostly funded by the military. A decade later, tempted by his old mentor Arthur Melton, he moved on to the University of Michigan. By this point he had become fascinated by issues of the human interaction with informational devices in general, such as radar screens and air-traffic control systems (a subject in which Renshaw himself was involved, as well as a certain young researcher named J. C. R. Licklider). Fitts did experiments and simulations of human-machine interactions in information-theoretic contexts, some of which sound uncannily like early video games: he had students try to "shoot down" simulated aircraft on radar screens, or act as air-traffic controllers in a simulated airport. He did in fact provide advice to personnel handling the Berlin Airlift, during which air-traffic control was a major bottleneck.[48] Above all, he reimagined his old work on pilots' attention in terms of the exciting new discipline of information theory. The result was a logarithmic formula—soon to be known as "Fitts's Law"—that captured mathematically the trade-off between distance and accuracy when a person's eyes or hands moved between points on a surface.[49]

By the late 1950s, then, Fitts's use of the science of reading to tackle the mystery of crash-landing aircraft had given rise to a curious mathe-matical law relating human capacity to information density in interactive displays. As with Buswell's work on images, nobody then recognized this as particularly revolutionary, although it did become a commonplace of what became known as "human factor engineering." At the time, Fitts attracted rather a different sort of attention. He became central to the hopes of a small group coalescing around the philanthropist Harry Frank Guggenheim (who had a background in naval aviation and remained in-terested in air safety) that was interested in analyzing domination strat-egies taken to be intrinsic to human nature, in hopes of developing ways of mitigating global violence. As it turned out, Fitts died of a sudden heart attack in 1965 just as that project was getting off the ground, so it did not realize Guggenheim's grand aspirations—although it did give a powerful boost to efforts at accounting for social violence in terms of deep evo-lutionary trends, which became a foundational motif of sociobiology.[50] Only later would interest in his aeronautical work revive among a new generation of information technologists, and thereby become one of the foundations for our own everyday encounters with computers. That, however, is a story for a later chapter.

High-Gear Reading

The success of Samuel Renshaw's recognition program was widely hailed. As *Flying* magazine put it in August 1945, air warfare had been on the verge of "a suicidal free-for-all" until his system debuted. It had equipped America's fighting men with a speed of perception that matched the pace of "this jet-propelled war." Renshaw himself had withdrawn from the pro-gram a little earlier, as the navy took over control. With the war heading to its conclusion, he increasingly shifted his focus to creating credibility in schools and businesses. In the postwar world, he reasoned, parents would want their children to be able to read quickly and well in order to succeed; and corporations would be eager to enhance efficiency and speed by having their executives, increasingly bogged down in paper-work, liberated to think creatively. At the moment when Vannevar Bush famously toyed with the thought experiment of the Memex, a sort of desktop workstation designed to help readers cope with the informa-tional flood of contemporary research, Renshaw proposed the alternative of training Americans to read really fast. So he reemphasized the use of the tachistoscope in the teaching of reading skills, both to children and to adults. A trial got under way in the schools of Gary, Indiana, with his Chicago-built machines, and he confidently predicted that "within 10 years" his method would be "a necessity" in all schools.[51]

Schools were all very well, but it was really in the business and industrial worlds that Renshaw saw most potential. He told *Flying* that he had proved that engineers could increase their reading speed by 150 words per minute, while also increasing comprehension by 70 percent. He added that in the future "the Renshaw system" would ensure that airline pilots could check their instruments for signs of trouble "in the stroke of an eye."[52] Travel would be safer as well as more efficient. "What all this implies for post-war education is obvious," the reporter enthused. Renshaw himself traveled to corporate conferences, trying to persuade managers that his "psychological optics" could prevent costly industrial accidents by expanding the "form field" of a worker's sight. He sought to convince business leaders of the importance of the kind of fast reading his tachistoscopes could furnish. Renshaw recommended that in future a good businessman would read a report quickly, using tachistoscopic training, in order to create a "structure" into which any more sustained reading could place the elements to be remembered. At a time when mail was overwhelming corporate officers, it made a big difference if a manager could get through the paperwork in only four hours as opposed to six: think of the gain to corporate efficiency![53] He took to the radio waves to urge Americans in general to accelerate their reading.[54] And Walter Cronkite even brought an entire TV crew to Columbus to create a TV special devoted to Renshaw's enterprise.[55] Confident of a bonanza, the Chicago company Three Dimension ramped up production of his reading-improvement machines. By the early 1950s a device called the Renshaw Tachistoscopic Trainer was being widely used by the Farm Bureau and other institutions to improve "visual habits."[56]

The press was not reticent in noticing all this. As early as 1948, the *Saturday Evening Post* published a series of articles fawning over Renshaw and his techniques, under the enticing title "You're Not as Smart as You Could Be." In the wake of the reports, that cultural bellwether Robert Heinlein invoked Renshaw explicitly in at least three of his works—a novella called *Gulf,* a novel named *Citizen of the Galaxy,* and perhaps his most famous book of all, *Stranger in a Strange Land*. In each case the "Renshaw tachistoscope," as Heinlein called it, was used by characters to accelerate learning and allow a subject to use the senses—all of them, not just vision—to maximum capacity. The novelist adopted Renshaw's own terms to explain that without its aid humans were only "one-fifth efficient in using their capacities to see, hear, taste, feel, and remember." In *Stranger*, the device enables a photographic memory. In *Citizen*, the slave-boy trainee is able to develop such sharp senses through being "Renshawed" that he can shoot down a pirate spacecraft because of his intimate engagement with the ship's fire-control panel (shades of Fitts). Heinlein even set his super-agents in *Gulf* apart from the hoi polloi by having them speak a language called *Speedtalk*, which shaded into telepathy.[57]

Thanks partly to Renshaw and his followers, the early 1950s became a goldrush period for the science of reading. He did not create the boom single-handedly—others had seen the potential too, some even earlier—but he was tireless in his efforts to bolster and capitalize on it. From his bastion at GE, his protégé William Schwarzbek avidly collected dozens of manifestos for such devices and programs. The media proved both receptive to and champions of the claims of this battle-proven science. "You Can Read 2,000 Words A Minute," trumpeted the *Brooklyn Eagle*, and this would bring about a kind of modernist, domestic utopia. College students would reduce the time it took to complete assignments by three-quarters, while businessmen would finally find the time to do the pleasure reading they so missed. After vacuuming her suburban home, a housewife would get through *Gone With the Wind* in just four and a half hours. All that they needed was the "new scientific machine" that made all this possible. Thanks in part to such affirmations, speed-reading became the self-help practice of the age. It was, as one critic remarked, "a free-enterprise bonanza," with all kinds of self-described "educational practitioners" offering gizmos and techniques guaranteeing personal and professional transformation.

For American adults keen to accelerate their reading and equip themselves for the postwar age, there were actually several "scientific machines" from which to choose. The one the *Eagle* was advocating was perhaps the simplest. Called a "Reading Rate Controller," it was the device invented by Guy Buswell back in the 1930s when he was hoping to make the laboratory science of reading a viable enterprise in schools. It was supposed to separate voice from vision and accelerate the habit of reading by forcing the eye to move down the page at a prescribed rate (fig. 6.9). Now the Three Dimension Company of Chicago and its counterparts were producing them by the hundreds of thousands.[58] Often called "reading accelerators," these devices were pressed into the hands of would-be reading-wizards all over the country, along with teaching plans, hortatory pamphlets, and correspondence schemes. Another device was the Phrase-O-Scope, which was marketed by Chicago's Foundation for Better Reading. This was an ingenious spring-driven tachistoscope, rather like a domestic, one-person metronoscope. It came with a full set of slides and a meticulously produced textbook by one William D. Baker, PhD, of Michigan State College, which explained carefully all the details of fixations, saccades, habits, perceptions-by-wholes, and so on, and provided a schedule through which the reader-in-training should proceed. It was also possible to get an individual tachistoscope called a tachitron. In addition, the foundation sent out what it called its "Eye Analyzer," a "weird" (according to *Newsweek*) pair of Jastrow spectacles set in reverse to allow a tutor to watch your eye movements as you read.[59] And there were more still. A "panoply" of instruments, films, books, and techniques tempted America's families—"motor-driven, spring-operated, gravity-pulled, thumb-catch released, projected and non-projected." Extravagant sales pitches for these machines "flooded the mail of everyone who had an address."[60] Any self-respecting school reading lab should have at least twelve tachistoscopes and twelve accelerators, one commentator advised—guidance that would have been insanely extravagant in Buswell's day, but was a little less so now. After all, mail-order companies could send you them by return.

The Foundation for Better Reading tended to loom particularly large among these ventures. The foundation was a profit-making concern set up by an entrepreneur named Steven Warren, with funding from Avery Besser, a wealthy Chicago arts patron. Warren was a graduate of the University of Pennsylvania who had worked for a company marketing magazines to salesmen. He had branched out into the reading field when he discovered that his customers did not have time to read the magazines he sold them, and after hearing (surely from some media account of Renshaw) that most people used only 20 percent of their "reading capacity." His foundation sold courses at thirty to a hundred dollars each. They used

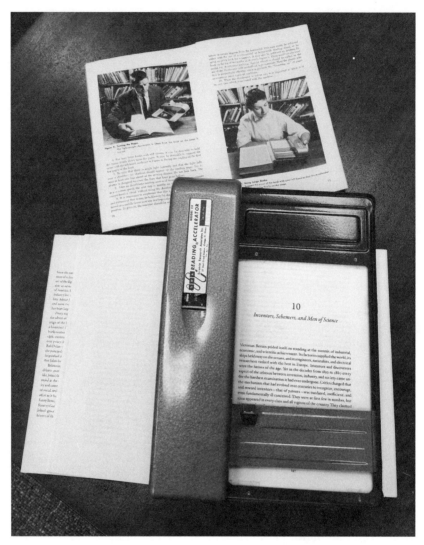

6.9 A cheap and widely-distributed reading-rate controller. This particular "Reading Accelerator" was fairly sophisticated and well-made by the standards of such devices, and it came with an instruction manual by Elizabeth A. Simpson, erstwhile clinician with Guy Buswell at the University of Chicago and at this time (1958) director of the Reading Service at the Illinois Institute of Technology.

the whole armory of eye-movement cameras, tachistoscopes, and electrically powered reading accelerators to train people to take "bigger visual bites" and improve their "mental digestion." Warren claimed that these programs could double reading rates even while increasing comprehension. Before long he had developed a client list including over a hundred companies, with major corporations like Hotpoint, Motorola, and Bissell prominent among them.[61]

As was the case with the devices embraced by Warren's foundation, none of the reading machines on offer in the 1940s and 1950s was really a self-contained entity. All were meant to be used as part of some coherent, designed program. It was the program that one paid for, after all. In pursuit of profit as well as enlightenment, commercial ventures sprang up in many of America's cities. In Boston, for example, one Samuel Joslow founded and directed the "Reading Institute" and spoke avidly of how many businesses on the East Coast were having their managers take classes to boost their speed and comprehension. Joslow was a keen user of the Renshaw technique, adopting both what he called a "flashmeter" to get people reading by phrases and a "reading accelerator" to stop them doing regressions. To prevent them verbalizing as they read, he gave them lollipops to suck. A decade later, Joslow was still enough of a believer to patent his own handheld tachistoscope.[62] In New York, meanwhile, you could attend the Reading Laboratory at 500 Fifth Avenue, right next to the New York Public Library, that was directed by Lynn Draper; he would take eye-movement films of you and then give you bespoke tachistoscope training in the Renshaw mode.[63] In Baltimore, John F. Kennedy himself attended the Foundation for Better Reading's satellite institute in 1954–55; he was said to have attained a reading rate of twelve thousand words per minute, but he left before graduating and the claim was dubious (although he did champion such programs in the White House). And in Chicago, again, DePaul students were exhorted to walk over to Warrens and Besser's foundation so that they could get their work done in half the time and have fun with the hours they freed up. In 1952 Warrens and Besser launched a home improvement course: if you signed up, they would send you an "electric pacer" (a reading accelerator), a stopwatch, and even your own projecting tachistoscope—twenty-five pounds of equipment in all.[64] They had to stay ahead of the competition, after all: as intrepid journalist Jack Cooper discovered, there were enough rivals in the field to fill a whole series of articles in the *Tribune* on the ways in which second-city dwellers could improve their reading. Cooper himself took a course at the Illinois Institute of Technology, where, by becoming habituated to "seeing in wholes" and by using the tachistoscope in another iteration of the Renshaw program to widen the "eye span," he went from reading 270 words per minute to 923.[65] Cooper also reported on yet another new type of portable tachistoscope invented at the University of Chicago's Reading Clinic, which was demonstrated both as a reading-improvement machine and as a device to hone aircraft spotters for the navy. One of the inventors had been trained on a Renshaw device in the Marianas Islands, and complaints of their failings on aircraft carriers in rough seas had helped inspire the invention.[66]

As with IIT and the University of Chicago, many colleges and universities embraced such systems of their own. To give one instance, in 1949 Cornell started a program that used tachistoscopes, reading accelerators, and ortho-raters (a device to appraise visual acuity); it attracted a thousand students in its first semester, and soon overwhelmed the faculty who had initiated it. The same report that described that experience also detailed similar programs at Purdue, Marquette, the University of Wisconsin, the University of Texas, and a variety of smaller colleges.[67] Most college reading programs that still thrive today in America's colleges were launched in these years, from 1950 to 1955. Some four hundred replied to a survey intended to provide a snapshot of the field in 1956, which revealed that they had adopted the technologies of the science of reading wholesale. That included tachistoscopes and eye-movement devices, as well as sets of 16 mm films—produced by Harvard, Iowa State, and Purdue Universities—that mimicked the behaviors of tachistoscopes, metronoscopes, and rate controllers.[68] With the sudden advent of all these "laboratories," "clinics," and "programs" came a new genre of guidebooks and textbooks for them to use, perhaps the most influential being *The Art of Efficient Reading*, by the two reading-laboratory directors George D. Spache (University of Florida) and Paul C. Berg (University of South Carolina). Spache and Berg's book was first published in 1955 and reissued in radically different forms several times in later decades. New national channels of networking helped the members of such programs see themselves as pioneers of a discipline in the making.[69]

Speed-reading programs both commercial and collegiate thus proliferated across the United States. So much so, in fact, that it is plausible to suppose that every literate American of the postwar generation must have been exposed to the hard-sell of these techniques by one institution or another, whether it was Columbia University, the Book of the Month Club, or one of the countless "reading laboratories" that sprang up. And it might just as easily have been your employer, too, who introduced you to the new science, because many corporations did embrace the opportunity to enhance efficiency. Naturally, it was the biggest companies that adopted it most wholeheartedly. Here the key figure was Renshaw's favorite PhD student, William Schwarzbek. Schwarzbek had undertaken the key tachistoscope work with Renshaw in the 1930s—his doctoral dissertation is in fact the most detailed account of that work—and after military service he joined General Electric, where he spent decades dedicated to projects in the improvement of reading alongside other industrial psychology ventures.[70] Having trained officers in the air force using the tachistoscope technique, he persuaded GE to commit to it from top to bottom: he reported that on average it increased managers' speed of reading by 30 percent and their comprehension by over 80 percent (fig. 6.10). And General Motors was quick to embrace the future too. The initial request came from the corporation's AC Spark Plug Division, where engineers had come to feel overwhelmed by the amount of reading they had to do. They requested that the GM Institute in Flint develop a reading program. In response, company officer Harry Patterson devised a pilot "General Motors Reading Improvement Program," involving eye-movement measurements, tachistoscopes, and accelerators. Launched in early 1953, it was not experimental for long. Thirteen divisions were soon demanding similar help, and in roughly three years five hundred executives were trained using such techniques, with three hundred more on a waiting list.[71]

The programs for managers and staff that Patterson and his peers devised persisted for decades, although their content and conduct did change over time. And Patterson himself became a cheerleader for the integration of the science of reading into American industry in general. The need to achieve this integration was urgent, he pointed out, because of the sheer "mass of reading" that an executive, foreman, or supervisor in the postwar corporation had to get through. Accelerating the act of reading meant eliminating one of the most frustrating bottlenecks in the way of an efficient, modern, information organization. Any manager

6.10 A worker at General Electric applies Renshaw principles to operate faster and more efficiently. William C. Schwarzbek keenly supported the development of reading-improvement and similar programs at GE for decades after World War II. D. G. Wittels, "You're Not as Smart as You Could Be," *Saturday Evening Post* April 17, 1948, 21. Photograph © SEPS licensed by Curtis Licensing Indianapolis, IN. All rights reserved.

inclined to doubt the cost-benefit ratio of breaching it, Patterson pointed out (with apparently unconscious irony), need only measure how much the flow of paperwork through an organization increased with a reading program.[72] It was, in fact, imperative for any company wishing to survive. The broader point had to do with the new discipline of management. In that context, the doctrine that "time is money" applied: if a meeting took a whole afternoon when it could have taken an hour, the wastage might be seventy-five dollars in time alone. Improving reading was part and parcel of the new science of management.

In particular, Patterson focused attention on the new professional figure that he dubbed the "engineer-executive." This was the kind of company man who was both a leader and a technical expert, and who was charged with the task of sustaining original research in corporate settings.[73] It was a delicate role, and Patterson found that the relentless torrent of reading risked making it unviable. In one division of GM, for example, so many journals arrived every day that the hundred engineers working there had to divide them up just so that they could manage the amount of paper. It was even rumored that executives sometimes delegated such reading to their secretaries. In mid-1956 Patterson surveyed 165 companies in the Fortune 500, from sectors including food, banking, energy, manufacturing, and transportation, and found that just under 40 percent had developed reading programs in the last five years; another 15 percent planned to start one soon. Almost five thousand managers had participated already, many of them being key leaders in the aerospace, automotive, and electrical industries. Virtually all these programs used tachistoscopes, at least initially, and most had reading rate controllers too.[74]

The state was not about to miss out on this great moment of progress, of course. The solution to the old problem of "efficiency" was apparently at hand, and with it the end of mental fatigue. These were vitally important matters to the military. So in 1949 the air force, ahead of the other services, created its own Reading Laboratory at the Pentagon. It also launched another at its "Air University" college in Alabama, attributing it explicitly to Renshaw's influence. The CIA built one in its headquarters at Langley, too, tracing its lineage back, through Renshaw, to Buswell. And the army embraced tachistoscopic learning for "all officer classes." It was only sensible, after all: as a contemporary report noted, "time is money" was a saying in the military as well as in corporations.[75]

And from these initiatives, the science of reading expanded through the federal government. By 1956 there were twenty-three reading laboratories operating in government agencies in the Washington, DC, area alone, all but two of which were geared to improving professional skills. (The others were for remedial training.) Their point of origin was the air force laboratory at the Pentagon, to which other government agencies sent people to see an efficient reading lab in action. They then inspired programs back at their home offices. Because the Pentagon's operation was "largely mechanical"—that is, it was based on reading rate controllers, tachistoscopes, and ophthalmographs—they followed suit. Most of them used ophthalmographs, then, virtually all bought tachistoscopes or tachitrons, and all had reading rate controllers.[76] Most originally adopted the air force lab's training manual, too. But they adapted quickly to their various civilian contexts, for example by replacing that manual with Spache and Berg's *Art of Efficient Reading.* Courses typically lasted thirty hours over six weeks, and the laboratories were all in communication with each other through a DC-wide "Reading Improvement Laboratory Group" that allowed them to share practical wisdom. Although their explicit emphasis was on "key civilian and military personnel," in fact they trained every level of employee, from typists to presidential appointees. In all cases, the point was similar to that at General Motors or General Electric: to liberate the corporate citizen by reducing the time required to deal with the paperwork of modernity. Directives, reports, the latest literature on management and marketing—all would consume half as much time.

There was great optimism in this moment, when the science of reading offered to make the dreams of modernity possible. Efficiency, leisure, domestic happiness, personal fulfillment, public culture—all seemed within the grasp of the postwar generation. The same skills that had defeated the Luftwaffe would guarantee a world that would make the costs of that victory worthwhile. What middle-managers glimpsed in the momentary flashes of their tachistoscopes were not just individual words or sequences of numbers but visions of a smoothly harmonious future state. Executives and engineers felt that this might be attainable through wise managerial policies; officers of the air force and other services sought it in the reading laboratories modeled on that at the Pentagon. Federal agencies were going to be freed to work imaginatively for the public good. And yet: there was also nervousness behind the buoyancy. How sure could

they be that these machines would really work the wonders their backers promised? Would the benefits they did bring prove lasting? And how
would improvements in reading—if they really happened—map onto existing social, racial, and other distinctions? A pervasive anxiety existed
here too—an unease about the place of the individual in an age of state
and corporate mechanisms.[77] Perhaps an age of reading technologies
might not help to ameliorate the social divisions that dogged the Eisen-
hower era but instead exacerbate them. Nobody knew for sure. But meanwhile, America was being readied at breakneck speed for a dawning age
of information.

7

Exploring Readers

On May 4, 1960, Samuel Renshaw entertained a special guest at his laboratory in Columbus. His visitor was on sabbatical from his regular teaching position at the University of Toronto and was devoting his year of leave to a mission. The president of the National Association of Educational Broadcasters, Harry Skornia, had sought him out after he had made a string of provocative and enigmatic statements at various venues in Canada and the United States about how changing media intersected with social and cultural transformations. Skornia had asked him to compile a report on that topic, which might then be used to prepare syllabi suitable for eleventh-grade courses in American schoolrooms. It seemed self-evident that in fulfilling that charge he should pay a visit to the man who claimed to be able to use media technologies to drastically speed up perception, extend human capacities, and unify the senses.

Marshall McLuhan was not yet the famous media prophet that he would become, but he was well on the way there. The first recorded instance of his using his most notorious phrase, "the medium is the message," had taken place a couple of years earlier, in May 1958.[1] McLuhan had been participating in a conference in Vancouver on the future of radio broadcasting, alongside Paul Lazarsfeld, Wilbur Schramm, and an array of experts from the Canadian, US, and British systems.[2] The event was one of an ambitious multiyear series at the University of British Columbia to rethink media in general; the following year McLuhan would return to teach a summer school. Technical questions were to be treated alongside more academic, "theoretical" themes, and administrators hoped the combination would make UBC into Canada's "major centre for the study of communications." Never one to observe stuffy intellectual conventions in any case, McLuhan took the hint and spoke expansively about media in general. As a result, when he voiced his famous motto, he was not talking at that moment about a new medium at all. He was not even talking about radio. He was talking about print—and the significance of print, he announced, lay in its enabling of people to read in a particular way. By allowing people to read silently and "at high speed," print had given rise to "a totally new set of mental operations." That was why "the medium is the message."[3]

To read silently and rapidly—in coining his motto, McLuhan picked up the two major and perennial themes of the science of reading. He liked the phrase that this inspired from the start. At the end of the Vancouver meeting, he flew to Washington, DC, and repeated the same point to an audience interested in education and television. "We should long ago have discovered that the medium is the message," he told them: "The effect of reading is far more decisive than anything that gets said from moment to moment on the page." Indeed, he went on, the page itself was not "a conveyor belt for pots of message" but "a producer of unique habits of mind." The very "nature of thought itself" emerged from our encounters with it.[4] And in October he voiced similar thoughts yet again, both in person before the annual convention of the National Association of Educational Broadcasters (NAEB) in Omaha, Nebraska, and in a brief memo he sent to Skornia.[5] Soon he was saying much the same once more at a conference in Cambridge, Massachusetts, on "Myth and Mythmaking," initiated by the Harvard cognitive scientist Jerome Bruner as part of a project on the Unity of Learning. Print, McLuhan now added, made reading into an experience like viewing a film at the cinema: "in reading print, the reader acts as a kind of projector of the still shots or printed words, which he can read fast enough to have the feeling of re-creating the movements of another mind."[6] And by March 1959 he was in Chicago, talking now before a thousand-strong audience of members of the American Association for Higher Education. The "electronic revolution," he intoned, would mean radical change, and yet educators still insisted that "Johnny acquire the art of reading." Of course they did, McLuhan continued. Like a child in a bygone age, the reader "must roll his eyes in lineal, sequential fashion."

But that was the attitude of people for whom a car would always be a horseless carriage. Electronic media were *different*, and they brought with them different styles of thinking. The mode of understanding now would be "configuration" rather than causation, simultaneity and pattern rather than sequentiality. Fighter pilots had to learn to read their instruments instantaneously. In the meantime the movement of information had become America's largest industry, with corporate education programs—of the kind we encountered in the last chapter—already costing "at least three times" the annual budget of formal education in North America. The statistic may have been implausible, but there was no mistaking McLuhan's intent in citing it.[7]

There is something remarkable about this sequence of presentations that McLuhan gave during the period before he and Renshaw met in May 1960. His "medium is the message" slogan would become the key mantra of 1960s media theory, but it turns out that it emerged from articulations of the great themes of the contemporary science of reading. McLuhan's reader was not so much cinematic as serially tachistoscopic. By early 1960, he was starting to put these idiosyncratic points together to form a quasi-psychological argument about the human senses. If media were extensions of the senses, McLuhan reasoned, then the impressions reaching the brain through the sensorium must be dependent on them, so major changes in media environments would inevitably give rise to profound transformations in human nature. He was thrilled by this idea and threw himself into exploring it. McLuhan charged ahead even in defiance of his own deteriorating health. By mid-1959 he was sickly and starting to experience blackouts, and at some point in 1960—we do not know exactly when—he would suffer a stroke. It was serious enough that a priest was summoned to perform the last rites.

Somehow, during this time McLuhan managed to visit half a dozen experts in the United States to tap their brains on the subject of media and culture. They included management guru Peter Drucker, William Allen of RAND, Harvard sociologist David Riesman (a veteran of Chicago's communication committee), and Gordon Brown, dean of electrical engineering at MIT. McLuhan also attended a presentation by the systems theorist Richard Meier that persuaded him of deep analogies between sensory information and media, on the one hand, and the economic performances of developing-world nations on the other—an encounter that clearly fascinated him and the significance of which has not been adequately noticed.[8] It was in the middle of this odyssey that he arrived at Columbus. There, he reported, he enjoyed "a most exciting and profitable session" with Renshaw, discussing all his "media hypotheses" and exploring the psychology of vision. Renshaw recalled his wartime tachistoscopic program, and mentioned that in order to tune his subjects' "visual virtuosity" he had found it necessary to train their ears as well as their eyes—something that had not been reported at the time. McLuhan seized upon the observation as important confirmation of his rapidly evolving concept of the mediated sensorium. Renshaw handed him a stack of papers to read, detailing how training one of the senses allowed it to use less and less stimulus because, as McLuhan put it, "inner effectors take up an expression role." It is tempting to suppose that these may have been copies of the samizdat journal *Psychological Optics*, where Renshaw had addressed these "effectors" at some length. He also told McLuhan to visit Bell Labs, telling him that researchers there would be "happy to tackle my hypothesis," and that he must meet one J. M. Allen of AT&T (presumably Jonathan Allen, an expert in human engineering later renowned for programming computers to read and speak). Before he left, Hoyt Sherman's collaborator Ross Mooney showed McLuhan round the "perception laboratory" that they operated for their art students. Mooney was by now interested, like so many in those years, in "creativity," writing rhapsodically about how elemental forces of creation in the natural world could be harnessed to foster creativity among students.[9] What he showed McLuhan was not the Flash Lab itself but a venue for exploring optical illusions, in the manner of Adelbert Ames's famous "room" at Dartmouth. (Sherman and Ames were friends and collaborators, and the Flash Lab had evolved in this fashion.) In any case, given the centrality of Cézanne to the Sherman program, it seems plausible that the experience helped inspire McLuhan's subsequent conviction about the pivotal importance of that painter's visual theories. At the end of the day, McLuhan departed confident that he was on the right track.[10]

By this point McLuhan was in fact quite far advanced in his project. He delivered a draft of his report on "Understanding New Media" at the end of June, and the NAEB issued it publicly some months later. It ended up serving as a kind of prototype for both of his most famous books, *The Gutenberg Galaxy* (1962) and *Understanding Media* (1964). The report is sometimes described as a first edition of the latter. I think that is stretching a point, but it is fair to say that all of what became McLuhan's distinctive "take" on media, the self, and culture was present in some form, if to varying degrees, in the text that he submitted to the NAEB. As an educational schema it was utterly impracticable—or, at least, nobody ever tried to put it into practice—but he had treated the project as a way of thinking through his own ideas, and in that light the paper was pivotally important. By the time he completed it, he was already working hell-for-leather on his book about the end of the typographic era, *The Gutenberg Galaxy*.

This is not the place for a full-bore exegesis of McLuhan's various claims and arguments. He was a notoriously elusive writer and thinker, and he himself insisted that his claims be treated as "probes" rather than elements in a grand, coherent thesis. But one component did loom large for many years: a distinctly psychophysical concept of the human sensorium. The sensorium was an ancient concept, associated by McLuhan and his interpreters with Aquinas. For the distinctly Thomist McLuhan, it was the locus in which impressions from the senses were combined into perceptions. The interdependence of perception and media took shape in this location. The idea is also reminiscent of Renshaw's program, of course, which relied entirely on such a notion. It was by virtue of his concept of the sensorium, moreover, that McLuhan could so persistently claim that the senses bore a "ratio" to each other, and that exposure to a new array of media had the effect of accentuating one of them at the expense of the others. In short, a media revolution was a moment when the ratio of the senses was thrown out of kilter in the sensorium, such that humans felt disorientated until they had had the time to recalibrate their perceptions of the world. It was this kind of process that McLuhan believed lay at the heart of the historical roles of writing and (especially) print, and that would be equally central to the roles of cinema, radio, television, and electronic media. In the case of print, for example, he maintained that typography had elevated the visual sense at the expense of the others, and that this had led to all the effects cataloged in *The Gutenberg Galaxy*: perspective, linearity, calculus, Newtonian physics, cultural homogenization, individualism, nationalism, and, ultimately, the nostalgic Romanticization of oral authenticity.[11] It was, on the face of it, a peculiar idea, both excessively literal in its invocation of sense ratios and—like much in McLuhan—exasperatingly hard to render into a specific, useful, or testable claim. What numerical values could one assign to vision, for instance, as opposed to smell, so that the notion of a ratio could make more than the most vaguely metaphorical sense?[12] But on the other hand—and this was also typical of McLuhan—it was also somehow intuitively plausible. In an odd way, it seemed to "make sense." The combination explains why critics in the 1960s found the line of argument as frustrating as they did.

The point is that McLuhan's theory was psychophysiological through and through. That wasn't *all* that it was, of course, but psychophysiology was a core element. Where did he get this from? He did in fact refer to some scientific authorities for the contention. Notable among them was the biologist J. Z. Young, whose work on the nervous system of the octopus would be foundational to neuroscience.[13] As with many instances of "citation" in McLuhan's work, however, his use of Young was as obfuscatory as it was helpful. But he did mention another inspiration, and here he was more helpful—for the paradoxical reason that he did not name a specific author at all. Instead he made what was for him an unusual move and referred to a broad cultural phenomenon. He invoked the many commercial programs that sought to improve Americans' reading. Many of those programs were based loosely on Renshaw's ideas and instruments, as we have seen—and Renshaw himself had used them programmatically to rebalance the senses in ways that he had conceived of in at least vaguely quantitative terms. Why, after all, did McLuhan take the time to visit Columbus at the moment when he was struggling to finish his report and even facing the prospect of physical collapse?

It is more likely that his visit to Renshaw reinforced McLuhan's commitments than that it generated them ab initio. In a broad sense, beliefs about reading lay behind many of McLuhan's claims, both about print and about other media. In the periodical *Explorations* that he and anthropologist Edmund Carpenter had edited in the 1950s, McLuhan had already written about the unity of the senses and about how print accompanied speeded-up reading; "the speeding eye of the new reader" was to blame for cultural effects such as the downfall of scholasticism.[14] *The Gutenberg Galaxy*, which he wrote at great speed in late 1960 and 1961, was full of pronouncements on the psychophysiology of reading, from the more than one hundred pages on the medieval practice of reading aloud to the claim that "the reading of print puts the reader in the role of the movie projector."[15] Indeed, McLuhan drove home his central claim about how new media led to "changing sense ratios" by invoking this kind of change in particular. We could grasp the "specific technology of writing," he added, thanks to what he called the "new institutes for teaching speeded-up reading habits." McLuhan surely had in mind enterprises like the Foundation for Better Reading, and he claimed, rightly, that these ventures sought to separate eye movements from "inner verbalization." And in fact some early reading-reform initiatives, such as Walter Pitkin's *The Art of Rapid Reading* (1929) and James Mursell's *Streamline Your Mind* (1936), *had* likened the mind to a cinema projector. Dearborn too had recommended using a film projector to mimic a tachistoscope so as to control and accelerate reading rates.[16] By the 1950s so-called foundations like Warren's were using machines like reading accelerators and tachistoscopes to train clients, in McLuhan's words, "how to use the eye on the page so as to avoid all verbalization." As reading became truly silent, so "our cinematic chase from left to right" would accelerate, and we would perfect "the mental sound movie which we call reading." McLuhan himself, of course, claimed that ancient reading had been reading aloud and that print had caused the voice to quieten and the visual practice to accelerate. So the institutes existed to replicate what was in his view a deep historical process. What the science-of-reading businesses were doing, in short, was exploiting—and hence confirming—the process that lay at the very heart of McLuhan's media theory.[17]

As we have seen, McLuhan's familiarity with these institutes was nothing unusual in the late 1950s and early 1960s. In fact, he may well have been less acquainted with their details than many of his students. Businesses like the Foundation for Better Reading advertised widely to college populations, and at McLuhan's own Toronto campus a commercial operation calling itself the Career Development Institute offered to increase students' reading rate by 1000 percent, reassuring them that it offered the "same scientific course as taught by McGill, Yale and Cornell." Toronto's students were so enthused that they even created a tachistoscopic program of their own.[18] In general terms, the applied, commercialized science of reading was a substantial enough enterprise in these years that its conduct and applications were familiar to a huge swath of the citizenry in North America. As we have seen, they encountered it in many everyday settings, and they believed, perhaps rightly, that their lives and fortunes could be reshaped by the properties that this science made manifest and the skills that it allowed them to acquire. Moreover, this commercialized science of reading did often involve the manipulation of the relative attention that the mind gave to impressions from the various senses. Some of the Renshaw processes did depend on the conceit of a balance between the senses, and did routinely involve tweaking that balance. The instrumentation used in the science of reading thus constituted a set of technologies to extend the senses and rebalance the sensorium. Although it would be an exaggeration to say that McLuhan's media theory was nothing more than a transposition of the science of reading into the domain of new media, such a transposition was certainly present.

McLuhan was, of course, something of a sui generis character, and he very much liked to market himself as such. Tom Wolfe's famous challenge—"what if he is right?"—flummoxed everyone who read his work or encountered him in person. Few at the time knew what to make of him, to McLuhan's own evident delight. But it was widely accepted that he was a hugely important cultural figure, and that his work had to be grappled with. As George Elliott replied to Wolfe, even if McLuhan was wrong, he mattered.[19] As "media studies" came into existence—partly out of the earlier ventures of Lazarsfeld, Waples, and Lasswell, partly out of communication and information sciences, and partly out of the efforts of McLuhan's followers—McLuhan took on the status of a charismatic leader. That reputation would eventually fade, yet interest never entirely disappeared, and in the media it would revive quite markedly in the age of the Internet. Boosters of digital culture like those at *Wired* magazine found it congenial to identify him as a neglected prophet.

Controversy about McLuhan's ideas was rife at the time, continued through his years of eclipse, and will surely never cease. Yet I want to suggest something that has perhaps never been said about him in all those decades of debate—that is, that McLuhan's philosophy was, at its core, *ordinary*. His ideas and approaches were in many ways absolutely representative of important intellectual, cultural, and practical currents of the time—and, perhaps more significantly, of everyday middle-class experiences. But that does not make it uninteresting. What should interest us, in fact, are precisely the ways in which it was so ordinary. To see that—to recognize his viewpoint as ordinary—we need to refamiliarize ourselves with those currents.

It is clear that a moment of profound intellectual and disciplinary re-calibration took place in North America in the postwar generation. A slew of novel manners of inquiry came into being in the years from 1945 to 1965, many of them at least as ambitious and initially odd as McLuhan's.[20] Some succeeded and became mainstream disciplines (or "interdisciplines"), such as cognitive science. Others either melted away or were subsumed into different fields, as happened to cybernetics. The science of reading had a foundational role—or at least a preparatory role—in all of them. As behaviorism dissolved in favor of an array of cybernetic, information-centric conceptions of creativity, so scientists of reading sought to stake their claim to a central part in an emerging disciplinary ecology. Although it might seem that the midcentury science of reading was a behaviorist endeavor par excellence—serious charges were leveled against it on that basis, as the next chapter will show—in fact its practitioners from Huey onwards had made a point of insisting that one could *not* simply reduce reading to a series of responses to stimuli. And their contentions had dispersed throughout American culture. As a result, ex-periences, machines, and theories from the science of reading now came to inflect not only media theory, but also information science, human engineering, cybernetics, cognitive science, psycholinguistics, computer science, and even the prospective field of artificial intelligence. What these new enterprises tended to share was an overall orientation against doctrinaire reductionism, in contrast to which they emphasized the au-tonomous, creative, exploratory properties of human nature. One reason why the science of reading connected to them was simply that reading itself was *already* seen as essential to the realization of those properties.

Teaching Machines and the Law of Effect

By far the most prominent movement promising to transform American education in around 1960, however, was of quite different character. It called itself *programmed learning*. It aspired to revolutionize pedagogy by using machines to supplement, or in some visions even replace, teach-ers.[21] The aspiration was in fact decades old, and it originated in the 1920s boom in mental and reading tests. Two figures in particular played ma-jor parts in that era: Edward Thorndike and Sidney Pressey. Thorndike's doctoral work on animals (which he initially undertook after Harvard authorities stopped him from doing reading-focused parapsychology experiments on children) led him to spend decades at Teachers College trying to come up with a theory of learning in general, and his key concept of a "law of effect"—according to which an organism's responses to a given prompt, if they produced a satisfying effect, would tend to repeat in future cases of similar prompts—became fundamental to the behaviorist convictions of B. F. Skinner. Believing that reading abilities correlated strongly with intelligence, Thorndike continued to work on reading as a proxy for investigating intelligence directly. Yet his view of reading was in fact rather complex, as befitted someone who spent so much time researching the practice. He held that reading was a form of active reasoning, not of mere responsive sounding-out or letter-by-letter parsing. And he came to believe that certain kinds of books could and should be redesigned to exploit that principle. As he remarked in 1929 while describing the differences between learning by reading and learning by listening, schoolbooks all too often rested content with merely declaring what habits a reader should form: they failed to lead that reader through the process of actually forming the habits they recommended. That flaw, he felt, was not intrinsic to print itself, but merely "conventional" to the textbook genre as it had developed historically. It would be perfectly possible to create a book that would guide a reader stepwise through the acquisition of skills, and such a tool might even work better than in-person instruction. At this point Thorndike allowed himself to indulge in a moment of fantasy. If, he speculated, "by a miracle of mechanical ingenuity," a book could be devised so that "only to him who had done what was directed on page one would page two become visible," then "much that now requires personal instruction could be managed by print." It need not even require a leap of engineering. Even with an existing book, he pointed out, it was perfectly possible to pull its pages apart manually and reveal them to a student at the appropriate pace. And in fact Thorndike did have his students pull books apart and rearrange the pages to see if they became easier or harder to use.[22]

In the mid-twentieth century there were various efforts to realize Thorndike's vision of what came to be called a "programmed book." None of them got much traction. But at Ohio State, Sidney Pressey, then a young professor of education, took note and took the idea in a slightly different direction. Pressey invented a "teaching machine" that he hoped could relieve schoolteachers of the most tedious part of their work. The machine was a simple device that allowed a user to choose one of four options in response to multiple-choice questions; it would record the answer to each question and then move on to the next. In terms of Thorndike's law of effect, it therefore both trained a student and tested that student at the same time. Pressey publicized his machine widely, but it was not taken up beyond his own circle. In future years he would continue to think about such possibilities, while at the same time exploring the psychology of learning in general. He kept himself informed of Buswell's work on the science of reading, for example, using his data on eye movements to chart what he called the "learning curves" of children from first grade to college—a concept that he adopted from earlier efforts to measure the acquisition of skills in the reading and writing of shorthand and telegraphy messages.[23]

Teaching machines came back into favor in the wake of World War II. By the early 1950s, Pressey's original invention had largely been forgotten except by cognoscenti, when Skinner began experimenting with a device that performed in a very similar manner. Like Pressey's, Skinner's teaching machine would put Thorndike's law of effect into action to train human subjects. The machine displayed questions to the student, who would answer them by moving sliders and then trying to turn a knob. If an answer were correct, the knob would indeed turn freely, and a bell would ring to signal success. As the bell made crystal clear, the assumption behind it was starkly behaviorist, and in fact Skinner devised the machine as a mechanized tool to replicate in humans his experiments in getting pigeons and rats to "learn." In effect, then, it epitomized everything that Rudolf Flesch, who was in the midst of writing his shocking *Why Johnny Can't Read* when Skinner published his first description of this teaching machine, suspected to lie behind the recommendations of contemporary educational psychology. For Skinner, however, its benefits were self-evident. If mass-produced, such a machine should cost about as much as a domestic record player, he reckoned: so it would offer a cheap way to "mechanize our schools." And for a nation that had already filled its kitchens with fridges and garbage disposals, that seemed to him a perfectly reasonable ambition.[24]

Skinner was not alone in seeing such possibilities. The US Air Force was a particularly keen patron of experimenters, using video, microfilm, and various kinds of feedback mechanism in a number of machines designed to both impart information and test the user. At a time of widespread concern about the perceived inadequacies of American education—especially in technical fields—the notion of an automated solution proved immensely attractive. In 1957 the star ballistic-missile designer Simon Ramo warned that the rate of technological advance in America was so fast that conventional education could not possibly keep pace, and that a "crisis" must soon ensue. He suggested that teaching machines would provide the way out of the crisis. High-schoolers in general would learn by engaging with automatic machines that would show them films, invite them into learning sessions on the Socratic model, and keep permanent records of their performances. Spurred by such dreams, a veritable movement arose very quickly in the late 1950s to invent, legitimize, and market these devices. At GE, Renshaw's old student William Schwarzbek collected large amounts of material on teaching machines and programmed learning of all kinds, as his earlier interest in reading-improvement inventions transformed into a fascination with these new devices.[25] Some early versions were put on display at 1959's annual meeting of the American Psychological Association.[26] And barely a year later, a printed collection of canonical papers on teaching machines would already fill over seven hundred closely printed pages.[27] Teaching machines were clearly a major trend, as was the ideal of "programmed learning" that they promised to realize. The combination was hailed as "the greatest educational advance since the invention of printing." As the 1960s dawned, it seemed that a fundamental revolution in the practices of reading and learning was at hand.

In 1962 the US Office of Education sponsored the first national traveling exhibit of teaching machines and programmed learning. By this time some 123 different machines were either in development or already on the market, some costing as much as five thousand dollars. Shortly after, Bruner and his allies recruited none other than Sidney Pressey to contribute his views on "autoinstruction" to a volume on theories of learning and instruction. Pressey was by now in retirement, but he still avidly researched and published, and in the intervening decades he had kept himself up-to-date on the science of reading, invoking such experts as Buswell and Renshaw in his own psychological works.[28] He complied eagerly with the request, providing both a retrospective of the history of teaching machines—"educational toys with feedback," as he called them—and a sobering account of their real prospects. For the former, he went all the way back to his own initial invention in the 1920s. His idea had initially arisen, he recalled, partly out of the rise of mass testing in those years, which had led to desires to perform the scoring of tests more efficiently by mechanization. That early machine had been his response, and it had been so thoroughly behaviorist that it had even included a capacity to reward correct answers with candy. It had worked very well. He attributed its failure to be widely adopted to the impact of the Depression: with thousands of teachers already unemployed, there was little appetite for a machine that threatened to put still more out of work. Still, it had attracted substantial attention at the time, and Pressey betrayed a certain irritation that Skinner blithely attributed the inspiration for teaching machines to his own research on pigeons. While Skinner's basic ideas were plausible, he added, his extravagant claims to be creating a revolution were overblown. Instead of the kind of gimcrack teaching machines Skinner and the contemporary technology companies were by now promoting so keenly, Pressey reverted to Thorndike's original idea and proposed the concept of a "programmed book," insisting that it would be both cheaper and more effective.

But even programmed books had major flaws. For one thing, Pressey noted, they were prodigiously inefficient. It took a lot longer to work one's way through them, and they tended to be ungainly because they had to incorporate so much repetitious material. He had himself recently been given a free "teaching machine" after spending five dollars at a supermarket, and the device had turned out to be the size of a five-hundred-page book but to contain about as much matter as might fill six normal pages. So much for the dream of efficiency. And what would happen, he asked, if such a programmed book were allowed to replace both textbook and teacher? Nobody really had any idea. Extravagant evaluations had been widely touted, but all of them had been questionable. "Small groups of students (much smaller than a usual class) are paid well to go to a room which is not a classroom, in a building which is not a school, to do a special bit of learning in a strange way for undisclosed purposes, under the direction of a stranger so well subsidized that neither equipment nor time costs are important." No findings from such trials were worth very much. In all likelihood, Pressey suggested, in reality a machine would make little to no difference: a programmed book would be about as effective as a conventional book read silently to oneself. And a conventional book would be more flexible, more informative, and more useful. "In short, there are values in a book." Programmed learning might sometimes be useful, but it should at most be employed in small sections at the end of conventional chapters. Pressey's major point was that if teaching machines were to have a future, it would have to be one in which they broke free from behaviorism and enabled "cognitive integratings and clarifyings" in a far more open-ended way. The whole fashion for Skinnerian machines, he concluded, was merely a "bizarre and unfortunate" climax to the long and disheartening tradition of likening children to lab animals.[29] It was time for a different model of learning itself—one almost antithetical to the very idea of programming human beings. It would require a quite new kind of teaching machine.

Freedom and the American Reader: Reading as Cognition

"What McLuhan has done," explained the appropriately polymathic neurologist, theater-director, and humorist Jonathan Miller, "is to bring the nervous system right into the center of the discussion of ordinary communications and of human knowledge in general." Wolfe agreed: "McLuhan is talking straight physiology here," he declared: "science."[30] To the extent that they were right, McLuhan's psychophysical themes were nothing unusual for the time. In the late 1950s and early 1960s, the study of the self was the most exciting area of the social and behavioral sciences. Whereas in the 1930s sociology had been the social science of the moment, in the central years of the Cold War it was psychology that came to be seen increasingly as both the lynchpin of the social sciences and the most promising source for insights into the problems and potential of modern society. Expert psychologists were called upon both to explain the global calamities of midcentury and to provide ways of thinking that could underpin a viable American ascendancy. But the kind of science that could serve these high purposes was fundamentally distinct from the psychology of the prewar era. Its practitioners themselves insisted on this: they merited such confidence, they said, because they had broken once and for all with the behaviorism that had dominated for decades (in America, at least; it had never been as supreme overseas). Behaviorists had wanted to explain everything in terms of habits developed in response to outside stimuli, and they tended to dismiss as unscientific any appeal to the mind as an agent. The new generation retorted that a modern science could not avoid dealing with the mind as such, and they sought to develop scientific tools for doing just that. They treated the brain as a dynamic actor, constantly interpreting perceptions, storing and reconstituting memories, making "plans," developing "strategies," and creatively seeking ways to "solve problems." As in many cases of so-called revolution in the sciences, the binary opposition signaled by those buzzwords was to some extent a matter of perspective. It certainly served the purposes of the new generation to accentuate the degree and sharpness of the break, and even as they did so they laid claim to a handful of John the Baptist figures, like Karl Lashley and the British psychologist F. C. Bartlett, whose work before the war had presaged their own. But the departure was recognized on all sides as major and momentous. Its ramifications extended across the sciences and far beyond.

Insofar as this self-described "revolution" contributed to attempts to come to terms with what had happened in the 1930s, it stood alongside a number of intellectual projects of the time: Robert Merton's sociology of science; Ruth Benedict's ethical anthropology of Japan; studies of media and communications by Paul Lazarsfeld, Harold Lasswell, and Merton; and Theodor Adorno and his partners' work on "the authoritarian personality."[31] It also bore comparison with the emergence of areas of study that had previously been interstitial to disciplines in the humanities and humanistic social sciences, such as communications, media studies, and—with a much more evident geopolitical motivation—the various "area studies" (Southeast Asian, Latin American, and so on). The tendency to assume in retrospect that disciplinary innovations in the sciences and the humanities—and the arts too—took place in isolation from each other is especially unfortunate when applied to this era, when protagonists like McLuhan and, as we shall see, Jerome Bruner not only noticed the commonalities but took inspiration from them. More will be said about this in following chapters, because it is a hugely significant historical process, and one that bears directly on the existence of this book itself. But for now what matters is that similar initiatives were advancing across the disciplines, and that their champions were well aware of that at the time.

What set these new fields of inquiry apart from prior ventures, to some extent, was their recursive character. Cognitive disciplines sought knowledge about the making of knowledge, and their conduct supposedly exemplified the creativity they explored. Cognitive psychology, for example, would be a psychology that both ratified and exemplified pluralism and tolerance, in explicit contrast to what its supporters characterized as the inflexible, conformist, and tamped-down discipline they were supplanting. Historian of science Jamie Cohen-Cole, whose account I largely follow here, has traced the fortunes of this initiative in detail. In the postwar context, as he shows, being "open-minded" came to mean embracing a set of "cognitive virtues"—individuality, freedom of thought, and above all what the new generation called "creativity"—that were of general social and political significance as well as scientific consequence. By contrast, the characteristics these protagonists identified with behaviorism were those of the closed-minded: intolerance, conformity to discipline, a blinkered insensitivity to new ideas, stagnancy. Those were the very traits that America's leaders sought to identify with un-American habits of mind, and in particular to project onto Soviet personhood. Autonomous creativity, then, as both the subject and the characteristic stance of the new discipline, emerged as a scientific riposte not only to Communist uniformity but also to domestic McCarthyism and racial prejudice.[32]

If the antibehaviorist "revolution" coalesced around any one enterprise, then, it was this new congeries of disciplines called cognitive science. The field had its origins in wartime projects that had addressed problems of the practical engagement of humans with informational machinery, such as air-defense radar screens and aircraft cockpit displays—the kind of mission undertaken by Paul Fitts. These problems had turned out to be insoluble by any one existing discipline.[33] After the war, cognitive science emerged initially at Harvard, MIT, and Columbia, all of which had housed multidisciplinary research efforts during the conflict. Cognitive science also found ready support among the prestigious grants agencies that sustained much of America's social-scientific research. Prominent among them was the Carnegie Corporation, which had so emphatically backed Chicago's initiatives in library science and the sociology of reading for similar reasons in the late 1920s and 1930s. With the support of the Corporation and other agencies—including the Atomic Energy Commission—the emphasis became the cultivation of a distinct subjectivity, tolerant of ambiguity, a variety of perspectives, and what Waples's protégé Bernard Berelson, by now director of the Ford Foundation and a major shaper of social-scientific fields, called "lack of closure." Berelson himself would sponsor programming at Voice of America that portrayed the new sciences as exemplifying American values. An intellectual community thus came into being, the articulated values of which included conversational civility, the integrity of research without dogma, and a certain freedom to traverse disciplines. In fact, this community was largely responsible for creating a discrete genre of "interdisciplinary" work in general. Even more, it was responsible for instilling that term with the presumptively positive values that it generally still holds today. To be interdisciplinary, for the cognitive scientists, was to be freed from programmatic shackles and enabled to think with autonomy and intellectual freedom. It meant the introduction of fresh air into the dusty hallways of scholarly convention.[34] For them cognitive science itself exemplified these virtues, of course, and not just as matters of overt civility but in its very epistemology. Focusing squarely on what the previous generation had disdained as "mentalist" entities, this science investigated explicitly how the mind came to know things, to remember them, and to act on its knowledge. In contrast to behaviorism, which they cast as a dour commitment to Pavlovian stimulus-response mechanisms, the cognitive scientists started from the basis that human nature was creative, open-ended, multifarious, and emancipatory.

This had major public consequences because it occurred amid a fierce debate already taking place about American education. The midcentury confrontation with totalitarianism was only one of the debate's causes; the kind of inefficacy identified by Waples, Buswell, and their peers was another. A third was the fear that industrial modernity had a tendency to give rise to conformity, dullness, fatigue, vacuous consumerism, and bureaucratic mediocrity. America wanted to see itself as a land of individuals, and if the distinction between democratic society and totalitarianism lay in that characteristic then these anxieties about modern culture took on new force. So the social sciences were called into play to provide insight into the stability and cultivation of the individual American. This was the context in which "creativity" came to be invoked habitually—almost as a mantra, as the president of the Carnegie Corporation confessed. The term stood for autonomous, free, nonconformist, tolerant originality. The creative self was to be the antithesis to the kind of conformist mentality treated at exhaustive length in *The Authoritarian Personality*, the famous report issued by Adorno and others in 1950. If "the" authoritarian personality existed, and if it could be identified and measured by psychological tests, then measures of autonomy and the other psychological virtues of the democratic self should be possible too. And it should also be possible to apply such knowledge to inculcate those virtues more thoroughly in future generations. Hence the focus on education.

Toward the end of the war, a prestigious Harvard-based panel had already met to address these nascent questions. It concluded that the objective of educational institutions should not be to impart specific knowledge but to cultivate in every student a mind possessed of valuable skills. High-school English classes, for example, should tackle fewer texts and focus more on analytical reading practices. But the import was more pointed than that. Everyone understood that in modern society specialization was an inescapable reality. People were doomed to know more and more about less and less, producing an atomized culture in which criteria of judgment were wanting. That meant that the most important pedagogical aim of all was to cultivate the skill of skills—that of discerning "competence in any field." One could never be an expert in all the domains needed for modern life, or so the claim went, but one *could* learn to detect charlatanry. And that mattered not only for everyday life but also for the viability of representative democracy itself. Elections, on this account, were processes for rejecting incompetent candidates. That meant that it was critically important to foster an electorate qualified to make such judgments of competence.[35]

Among educational institutions, the one most immediately affected was Harvard itself. The new cognitive emphasis helped shape two major institutional ventures there, neither of which exists now but both of which were highly influential at the time. The first was the Department of Social Relations, launched in 1946; the second was the Center for Cognitive Studies, inaugurated in 1960. Both were intended to be interdisciplinary and conversational, and both fizzled at much the same time, in around 1972. The first originated in a confidential recommendation of 1943 to reorganize the social sciences completely by creating a department devoted to "Basic Social Science" that would traverse the more specialized disciplines. The backers of the bid coupled it with an effort to hire Robert Merton, on the grounds that he would exemplify the kind of "good citizen" that their new institutional arrangement would both need and foster. The effort to recruit Merton failed amid acrimony, but the call for a new unit did not. The new Department of Social Relations then brought under its ambit the programs of sociology, social anthropology, social psychology, and clinical psychology, and it had its own laboratory (even though experimental psychology remained to one side). The intent to facilitate conversations across old disciplines was clear, and although Merton was not there to personify the ideal, figures like psychologist Jerome Bruner were. Bruner would be perhaps the key figure—with George Miller—in first the cognitive revolution and then its extension, through the science of reading, to the shaping of an information society in general.

Having worked in "psychological warfare" during the war, Bruner knew from experience that matters of propaganda and the shaping of public opinion were altogether too complex to be tackled single-mindedly. His first book, *Mandate from the People*, used public-opinion surveys to tackle the kind of broad questions of social knowledge that Waples and his colleagues had addressed in the 1930s and 1940s; Bruner paid prominent tribute to Hadley Cantril and Gordon Allport for informing its approach.[36] This was someone steeped in the types of multivalency seen in the social science of reading and the new communications field, then. And yet the first attempt to institutionalize them at Harvard was not a great success. The department never really cohered, and its central endeavor—a "Project on Theory" intended to provide a coherent foundation for its activities—proved centrifugal. In large part that was attributable to Talcott Parsons, the senior figure involved, who ran the project in an explicitly elitist manner and insisted on the centrality of his own concept of "function" to everything it did. Junior participants, Bruner to the fore, balked at being required to subscribe to the functionalism expressed in Parsons and Edward Shils's intended party-platform for the department, *Toward A General Theory of Action* (1951), and the ambition was quietly dropped.[37]

Bruner got his own chance a few years later, in the wake of the publication in 1956 of his *Study of Thinking*. The book stood as a manifesto for the cognitive approach, which Bruner then sought to institutionalize in 1960 in a new Center for Cognitive Studies founded under his and

George Miller's leadership.[38] The center was another attempt to establish a particular, conversational locus of interdisciplinarity. Its mission was to research how humans thought, and, reflexively, its design reflected how Bruner and Miller believed that humans really did think. Like the DSR, its appeal would fade after a period of exciting but ultimately baffling conversations, as specialists in different areas talked past each other and its vaunted interdisciplinarity dissolved.[39] But in the meantime it was by all accounts an exhilarating social and intellectual space.

If cognitive science in these decades had an outstanding area of focus, it was the theory of what its champions called "learning." Its manner of addressing the question of how humans acquired knowledge and skills became its key distinguishing mark. Bruner and his counterparts defined themselves against an old approach that they identified with figures like Edward Thorndike, according to which humans, like animals, learned to respond to elements of their environments in habitual ways by associating behaviors with rewards. The "law of effect" went by the wayside. Their own interdisciplinary science devoted itself to the quite different task of understanding the creative engagement of individuals with what were termed "problems." The underlying epistemology involved was antithetical to the old positivism, as Bruner made clear when he argued in 1951 that one needed to start with a "heuristic" theory of perception. Stimuli (and perceptions in general) were not givens, on this account. Whenever one perceived something, one was always "tun[ed]" by some hypothesis. What mattered were not the "energy characteristics of stimulation," as in the old psychophysics, but the "cue characteristics provided by stimulation." Cognitive scientists should therefore attend to this "signaling," not to crude motor activity.

In short, perception—and everything that followed from it—now had to be appreciated in terms of *information.* As such, cognitive science drew extensively from the information theory advanced by Claude Shannon and Warren Weaver, and from the emergent interdisciplinary science of cybernetics. This was the context, for example, in which contemporary computing became an inspiration for protagonists like Herbert Simon, who likened the problem-solving "strategies" that cognitive humans devised to the programs followed by computers. Such antibehaviorists conceived of human problem-solving in algorithmic terms, and even tried to produce actual computer systems—or human proxies for them—that emulated these processes, on the principle that if an algorithm could be shown to work then it counted as a reasonable model. But at the same time they also sought to clarify the specific ways in which humans were *not* like computers. The best-known example of that was George Miller's "The Magical Number Seven"—a paper that made a great impression on Samuel Renshaw. Miller argued that although it seemed the human mind could only hold about seven items in immediate memory, those items could themselves be internally heterogeneous. They could encode information within themselves, and in various ways—something that was quite beyond any contemporary computer.[40]

But the science of cognition, aligned as it was with information theory and the algorithmic practices of 1950s and 1960s computing, also had strong roots in the science of reading. These roots were visible at the time, but they have attracted almost no attention. One reason for this may be that the connections were not avowed, and they were often not explicit at a theoretical level. They were matters of experimental practice and instrumentation. But that did not make them any the less important. On the contrary: convinced that the flaws of behaviorism had endured largely because of its use of a standard repertoire of scientific instruments, Bruner and Miller believed that it was vital that the Center for Cognitive Studies both co-opt such instruments and develop new ones. It was a central part of the center's mission. Moreover, Bruner himself had spent the late 1940s and early 1950s pursuing a sequence of experiments on perception that became fundamental to cognitive science, and that sequence also represented, *pace* Renshaw, the most sophisticated exploration of tachistoscopic reading since the days of Dodge. Much of what Bruner and his peers articulated on the basis of such work could be described as a transcription of the science of reading, via information theory, linguistics, and cybernetics, into a new idiom of cognition.

Starting at the end of the war, Bruner and colleagues collaborated on a series of experimental studies arguing for the active engagement of the mind in perception. The series as a whole was sometimes called the "New Look." They argued that perception should no longer be the domain of narrowly experimental psychologists, who treated it, they pointedly remarked, as if the perceiver were a passive recording instrument obtainable from the American Optical Company—the very corporation that had appropriated Buswell's reading machines and made all those ophthalmographs. Psychophysics had fostered a "myth of punctiformity," they claimed, according to which "sensations" were particular, extremely localized inputs that could be treated in isolation.[41] Leave the laboratory and go out into "the market place," they insisted, and everything that such work left out became very clear. In the real environment, any person—any organism in general—lived in "a world of more or less ambiguously organized sensory stimulation," such that the flow of unmanaged stimuli would surely be overwhelming. The very fact that prior experiments were laboratory-based was thus epistemically corrosive. In actuality, perception was a "goal-directed behavior," the purpose of which was to construct a meaningful environment. What someone routinely perceived—what was "*actually there* perceptually"—had therefore to be a result of a process combining external input and internal selection. To achieve it everyone constantly formed hypotheses by which to seek out, reject, or accept candidate perceptions.[42] The acknowledged fact of "sensory conditioning," such that humans could be conditioned to see and hear things, confirmed this. It was therefore time for a scientific investigation of *how* perception varied when a person was, say, hungry, anxious, or engaged in solving some problem. The very distinction between perception, motivation, and learning had to be put into question.

So Bruner and his colleagues recruited students from Harvard and Radcliffe and put them to work, running a regular Seminar on Perception at the Laboratory of Social Relations. The task, as he put it, was "to delineate the creativity of the organism in perception," and eventually to lead the various accounts of perception current in different disciplines to converge on one coherent "theory of behavior" appropriate for both laboratory and clinic. Bruner's sequence of experiments led him believe that the common thread to that end would be a "hypothesis" theory. On this account, any act of perception required hypotheses, which people generated constantly and which became "fixated" by experience.[43] Such a hypothesis, as Bruner defined it, was "a highly generalized state of readiness to respond selectively to classes of events in the environment." In other words, it was a kind of "tuning" of the organism. Without it, the sheer number of new environmental inputs would swamp our ability to receive them. In effect, perception therefore involved a three-step cycle. It began with a tuned organism. This received an "input of information," and subjected that information to a check, or a "confirmation procedure."

If the information were incongruous, then the organism would respond in one of a variety of ways, depending partly on immediate feedback, partly on personality, and partly on experience. It could then generate the third step, a new hypothesis—a new tuning. A perceiver was thus constantly forming and reforming hypotheses, by means of which the selection, organization, and accentuation of perceptions occurred. Borrowing a term from the science of reading, Bruner admitted that his model was "saccadic," or "jumpy," and that generally perception did not feel like that. But then, that was true of reading too. He suggested that humans *did* experience perception as more jagged when conditions were unfamiliar. One might liken it, perhaps, to having to read a difficult text letter by letter.[44]

This composite process of perception was very fast, but it was not instantaneous. It took time. To investigate it, one therefore needed either to slow time down or to divide it up. And that was precisely what a tachistoscope did. So Bruner, Leo Postman, and their colleagues developed an experimental practice based on the instrument. In effect, they reversed Samuel Renshaw's procedure. They would show tachistoscopic slides repetitively, starting at very short durations and increasing step-by-step, to see how a perceiver worked by stages toward an organized perception.[45] The instrument could also be used to infer the character of a hypothesis by referring to prior instruction and relating that to the attribute reported on most frequently. In general, the device helped reveal the relationship between hypothesis and perception—which meant, in this context, reading.[46] So, for example, they used a Dodge tachistoscope to measure the times it took students to identify emotionally charged words. They discovered both what seemed to be a "defense process" qualifying recognition when it might lead to anxiety (subjects reported nonsense perceptions) and, for especially threatening terms, a "selective vigilance" process that led them to respond with greater speed.[47] Others then added a galvanic skin response meter to the mix, which showed that taboo words like *penis* produced greater effects than unoffensive ones even if flashed up for such short durations as to be unrecognized by the viewer.[48]

Another experiment involved using a tachistoscope to investigate the consequences of "frustrating" a perception, by showing an image or word for too short a time for anyone to grasp it. In this case Bruner and Postman indulged in some acting, telling students "in a serious tone of voice" that they were going to be shown "something," and repeatedly insinuating that their reported perceptions were unsatisfactory. They projected for a hundredth of a second a complex picture shown in black and white and at low illumination. When subjects said that they could not see anything clearly, Bruner and Postman told them "in rather icy tones" that this was "substandard." They followed this with eighteen three-word sentences, projected one at a time for durations below and above the threshold of recognition for one word. As the two experimenters continued to press the viewers, they became tense and even swore out loud. Skipping the step of securing two words, such viewers often began to guess prematurely and wildly about all three—and they often guessed words that were themselves aggressive, such as *hurt* for the correct *rust*. Having made a premature hypothesis, a reader had to "unperceive" it to get to the correct reading. The upshot was that frustration seemed to "interfere with a nicely balanced hypothesis flow."[49]

One aspect of these experiments that deserves note is that Bruner was always interested in what his subjects said *during the process* of arriving at a correct perception of a tachistoscopic slide. This was quite unlike most uses of tachistoscopes in the science of reading, where it was assumed that no such process existed: a perception was either accurate or an error. Bruner disdained the notion that perceivers made errors.[50] For instance, he used his tachistoscope to expose sequences of capital letters one of which—an *F*, say—was reversed. He found that when they came to this character, subjects would report seeing some other letter, or an *F* in a different font; some declared that it was not a letter at all. Others reported what Bruner called an "irradiation" effect, whereby *all* the letters suddenly looked reversed. It would then take a series of elongated exposures, and progress through various different responses, before the reader would arrive at the correct perception. All this, he believed, must be addressed not as a matter of error, but as a matter of information. It confirmed his notion that perception was "a cycle of attempted hypothesis confirmations." According to a viewer's hypothesis—a hypothesis developed out of repeated exposure to normal characters—each letter in the tachistoscopic slide must be *some* conventional letter. So subjects resisted perceiving unconventional ones. They even showed some entirely unconscious creativity in perceiving hypothesis-saving alternatives.[51]

By far the most renowned experiment in this series—one of the most famous experiments in the whole history of cognitive science, in fact—

employed a Dodge tachistoscope once again, this time in conjunction with a small set of playing cards. It is worth pausing over this experiment to see in detail how the science of reading merged into cognitive science. Bruner premised the trial on what he called a "'minimax' axiom." His suggestion was that "directive processes in the organism" (that is, hypotheses) would "organize the perceptual field" such that percepts relevant to current needs and expectations would be maximized. Those inimical to current needs and expectations would be minimized. Problems would arise, he realized, when "incongruity" arose—that is, a violation of expectation of some kind. His contention was that perceivers would tend to ward off perceptions that posed such problems for as long as they could, because incongruity would eventually require a disruptive task of "perceptual reorganization." So he sought to examine how incongruity was "dealt with."

Bruner asked twenty-eight students to look at tachistoscopic projections of five different playing cards. Of the five cards, between one and four would be "incongruous"—their color and suit would be reversed. For example, Bruner used a three of hearts where the hearts were printed in black, and a six of clubs with the clubs in red. These cards were not easy to make—no playing-card company could do it, so Bruner ended up painting both the incongruous-red cards and the normal-red ones with red paint so as to get a uniform color. That done, the tachistoscopic presentations started at ten milliseconds and then increased in duration through a series of steps; when either the card was correctly identified (that is, when there were two successive correct responses) or a duration of one second was reached, that sequence ended and Bruner started on the next card. He found that the "recognition threshold" for incongruous cards was much higher than for normal ones. Normal cards tended to be recognized at 28 milliseconds, incongruous at 114 milliseconds; about 10 percent of incongruous cards remained altogether unremarked. The next obvious question was whether experience of incongruity reduced this gap (with experience, after all, incongruity ceased to be incongruous). And Bruner found that this was very clearly the case. Of course, practice in viewing tachistoscope projections in general was a contributing factor in speeding up such recognitions—as Renshaw would have pointed out—and Bruner conceded that this effect was difficult to tease out. But tachistoscopic experience with normal cards actually *increased* the threshold time for incongruous ones, and this seemed to confirm that sheer tachistoscopic practice could not be "solely or even largely responsible" for the effect.

In the event, Bruner parsed out four kinds of response to tachistoscopic incongruities. He called them *dominance, compromise, disruption,* and *recognition.* In the first instance, dominance, someone faced with an incongruous-red six of spades might simply perceive a normal six of spades or a normal six of hearts. This preserved the familiar experience of a conventional card. Almost all Bruner's subjects did this, with individual tendencies to "correct" either suit or color. In fact, several students even showed a dominance response to a *normal* card, when they misperceived the color and reported a different suit to conform to that color. In the second case, compromise, the subject would perceive something halfway between the expectation and the incongruous card. They might see a purple six of spades (or hearts), say, or a six of spades lit by a red light. Again, many students did this, reporting colors such as brown, red-black mixes, "rusty black," "grayish red," and the like. In part, Bruner realized, this could be ascribable to the briefness of tachistoscopic exposures—color took time to perceive, after all, which was presumably why compromise responses were rarer for black cards than for red. The third reaction, disruption, then happened when someone could not perceive anything coherent at all. They might not even be sure it was a playing card. More than half the students experienced disruption at some point, always with an incongruous card and typically after trying a compromise. It produced rather stressed responses. After a series of them, for example, one student blurted out, "I'm not even sure now what a spade looks like!" Occasionally someone in this situation would extend the disruption to the *number* on the card in question, having originally identified that number correctly. (Bruner pursued this by using cards he found at a magic store with markings for the fifteen of diamonds and the fourteen of spades.) A vague "sense of wrongness" persevered, often about the pips on the card (were they "reversed or something"?), which preceded the last change of hypothesis about the card's identity. Finally, resolution came with acquiescence in the incongruity of a card—an instant of shocking recognition that Bruner came to call the "'My God!' reaction." After this, students would perceive formerly incongruous cards correctly. But some students never reached this realization, because of the tendency for perceptual hypotheses to become confirmed if they were repeated. One student thus reported the wrong suit for an incongruous card forty-four times in a row. These "fixation tendencies," Bruner concluded, were "the chief block to perceptual learning."[52]

For anyone familiar with thinking about the scientific enterprise in the 1950s and 1960s, the sequence that Bruner charted here will surely seem familiar. In effect, as his students moved between what he called "paradigms," they underwent little individual scientific revolutions, in very much the manner portrayed by Thomas Kuhn in *The Structure of Scientific Revolutions* in 1962.[53] In fact, Kuhn, who at this point was teaching in James B. Conant's innovative series at Harvard on the history of science, was much taken with this series of experiments. He learned about them while he was first imagining his book, and in the end he devoted a central section of *Structure* to them. Kuhn suggested on the basis of Bruner's experiment that scientific discoveries emerged through a historical sequence of steps, and that the same elements of that sequence were "built into the nature of the perceptual process itself." His summary of the experiment cleaned it up a little in order to focus on the profundity of the questions it posed. When a student shown an incongruous card reported seeing a normal one, Kuhn suggested, it was immediately and serenely "fitted to one of the conceptual categories prepared by prior experience." And because seeing, as Renshaw had insisted, was not the same as vision, he "would not even like to say that the subjects had seen something different from what they identified." He noted that as the exposures increased, they did, however, begin to feel uneasy. They showed what Kuhn glossed as "awareness of anomaly." In the end, most students would indeed produce a new identification—and they would then proceed to perceive this way in future cases. But a few subjects never made that flip. Those who did not often experienced "acute personal distress." The whole experiment, Kuhn concluded, "provides a wonderfully simple and cogent schema for the process of scientific discovery."[54]

Two points deserve note about this remarkable transposition. The first is that Kuhn referred to the exposures as brief and defined, but he did not mention the tachistoscope as such. That is presumably why people have not recognized the lineage involved. The omission was all the stranger in light of his generally acute appreciation of the importance of specific instruments in scientific practice. And in context, his discussion of Bruner's experiment directly followed an extended account of the importance of the Leyden jar, the piece of equipment that had made all of nineteenth-century electromagnetic physics possible. More important, though, than this point about his reticence on this instrumental issue, is a very important distinction that may be thought of as philosophical. Kuhn implied that although the path to a decision could be tortuous, the final move to a new "paradigm" was a once-and-for-all shift in Gestalt. (His point of reference for Gestalt theory, incidentally, was the émigré psychologist Heinz Werner, who had also been the key source for Renshaw.) He elsewhere likened it to a religious conversion. For Bruner, on the other hand, the "My God!" moment was merely the marker of a process that was definitely *not* instantaneous. Bruner took pains to emphasize that there was a duration to the change, and that it could be disaggregated. The fact that it took time, in fact, was precisely why the instrument needed in order to reveal it was a tachistoscope.

All this work proved extraordinarily influential. By 1957, Bruner could remark (with somewhat posed puzzlement) that three hundred papers had been published on a subject that had been a backwater before his efforts.[55] And he continued to use the instrumentation of the science of reading after the Center for Cognitive Studies was launched. But he added now the second signature instrument of the science of reading: the eye-movement camera. Tellingly, the particular device that he employed was so temperamental that he found he could not use it successfully without the presence of its minder, British psychologist Norman Mackworth. Mackworth was certainly qualified to contribute to the new center, given that he had worked on man-machine interfacing in radar during World War II and subsequently headed the Unit for Research in Applied Psychology at Cambridge. In his British work he had twinned an eye-movement camera to the new technology of closed-circuit TV to show (and record) in real time how the eye moved over a field of vision, accruing direct testimony about how such movements were controlled by mental processes. His device could also show that eye movements became erratic when reading syntactically complex material—an important point for psycholinguistics, which was one of the new center's major concerns.[56] But the reason he was given an appointment at Harvard was that his eye-movement instrument could not be persuaded to work without him.

Bruner concluded that perception and "cognition" merged into one an-other. No difference of principle existed between deciding that a percep-tion was of an orange, say, and assigning a number to the class of primes. Neither process even had to be deliberate or conscious. That led imme-diately to a major epistemological consequence. It was "foolish and un-necessary," he declared, to maintain that some elementary sensory "stuff" existed that could justify "the classic doctrine of sense-data." Even if an entirely unique, unclassifiable percept of this kind could exist, it would be impossible to *communicate* anything about it. Such a thing would have to remain forever "locked in the silence of private experience." In short, the notion of observation statements—central to the logical-empiricist accounts of science pursued by thinkers like Rudolf Carnap—simply had no basis. But what should replace it? Perceptions were supposed to be "veridical" to varying degrees, Bruner acknowledged, but he reinter-preted this in a pragmatist vein. The "meaning" of a perception, now, was a matter of "model building." Perceiving was something we learned to do in childhood, acquiring a *practical skill* "to predict and to check what goes with what." To buttress this idea, Bruner appealed again to the ta-chistoscope. Suppose one exposed for half a second two sequences of eight letters, *YRULPZOC* and *VERNALIT*. In terms of Claude Shannon's information theory, both contained the same amount of information. But a viewer of the tachistoscopic exposures reported the first 48 percent cor-rectly and the second 93 percent, because "the individual has learned the transitional probability model of what goes with what in English writing." That is, the viewer was familiar with the probabilities of which letters followed each other in English words (V was much more likely to be fol-lowed by E than Y by R). To say that the one perception was more veridical than the other was therefore to say that the observer was working with an internal model of reading, which corresponded to actual English. More precisely, Bruner suggested that one should speak in terms of "coding sys-tems," to which observers in front of a tachistoscope referred when they "read" obscure characters. "Perceptual learning" consisted of acquiring ways of coding one's environment and then allocating new sensory inputs to the appropriate "categorial coding systems."[57]

In general, the tachistoscope and (to a lesser extent) the eye-movement camera were *the* enabling technologies for the New Look at perception, and they were thence pivotal for the study of cognition. No science of reading, in effect, no cognitive science. And that remains true if we broaden our focus. I have concentrated on Jerome Bruner here, partly because he was such a pivotal figure and partly because his use of the instrumental repertoire of the science of reading was so explicit and persistent. But there was much more to the cognitive revolution than Bruner, of course. And in fact the science of reading had a presence across the new fields of research that was robust, even if it was not prominently flagged. Another example would be that of Richard Lazarus and Robert McCleary from Johns Hopkins, who argued that tachistoscopes, if used alongside galvanic skin response instruments (loosely known then as lie detectors) and electric shocks, could reveal a propensity to distinguish between character strings at a level below conscious recognition. They called this propensity "subception," and it played a major part in the arousal of public concerns about subliminal messaging on TV.[58] Or one could cite Alphonse Chapanis, a Yale PhD who had worked with Fitts on cockpit designs. Chapanis went on to teach at Johns Hopkins, where he worked with Peter Bricker (later of Bell Labs) on Lazarus and McCleary's claims about subception. He and Bricker used a tachistoscope in an effort to identify information taken in by viewers even when their reported perceptions of a word were incorrect, suggesting that in such cases they were still perceiving "cues."[59] This tachistoscope work was summarized at length in George Miller's agenda-setting book for the cognitive turn, *Language and Communication* (1951).[60]

And then there was Leonard Carmichael, president of Tufts University. Carmichael was something of a personal conduit between the prewar science of reading and the postwar cognitive sciences. He was a longtime protégé of Walter Dearborn, he had built his own eye-movement camera before the war, and in the 1930s he had done pioneering work at Brown University on measuring readers' brainwaves while photographing their eye movements. He had also tried to develop an electrical neuromuscular instrument to measure eye movements without the need for complex optics at all.[61] During World War II, Carmichael dedicated Tufts's Laboratory of Sensory Physiology and Psychology to eye-movement studies for the training of night pilots. And then both eye-movement photographs and "electro-oculograms" enjoyed pride of place in a further series of experiments on reading and fatigue that he carried out with the now seventy-year-old Dearborn. They had students read either Adam Smith's

Wealth of Nations or R. D. Blackmore's *Lorna Doone* for six hours at a time, and produced some *fifteen miles* of oscillographic traces. The resulting 1947 volume on *Reading and Visual Fatigue* represented more than any other book the links between the old world of Dodge, Judd, and Buswell and a new science aligned with neurology.[62] The eye movements of readers had become fodder for cybernetics.

Information theory deserves particular attention in this context. A layperson might assume that it dealt with issues similar to those of a science of reading, and yet it is generally represented, accurately enough, as taking a form that sidelined those issues completely. The authoritative statement by Claude Shannon was very obviously abstracted from experiences of telegraphy and similar media rather than from print and reading. His *Mathematical Theory of Communication* appropriately appeared in two parts in the *Bell System Technical Journal* in 1948 before being republished as a slim book the following year. It pointedly set aside any issues of semantics or context, regarding them as "irrelevant to the engineering problem." Instead Shannon proposed a highly general and mathematizable model in which an information source sent a message, via a transmitter, through a "channel" of some kind, to a receiver, which in turn passed it on to the recipient. The message was liable to be affected by "noise" in transit. In this highly schematic model, "information" was defined in terms derived from thermodynamics, as a matter of entropy. The problem, as Shannon presented it, was for the recipient to select the actual message from a set of possible messages. But here the abstraction had to come back to the physical world. Shannon's own focus was on "discrete" channels—ones like telegraph lines, in which every signal had to be one of a limited number of elementary symbols, such as alphabetic characters. But any alphanumeric medium—a book, say—was similarly "discrete," so in practice Shannon's approach could be extended to all such media. And as Benoit Mandelbrot pointed out, the application of information concepts to "non-electrotechnical" problems constituted "the principal postulate of cybernetics."[63]

The cybernetics and cognitive-science advocates saw ample opportunity to make this application. Alphonse Chapanis, for example, insisted strongly on the work's importance for social science in general, despite its appearance as a theory of telephone engineering.[64] The question central to Shannon's theory—in effect, the defining problem of the emergent science of information itself—then turned out to be strikingly congruent with one habitually asked by scientists of reading: what did a recipient need to perceive in order to make true sense of a message? For Shannon, that old science-of-reading question became, not an experimental problem to be addressed with a tachistoscope, but a mathematical problem to be addressed in terms of entropy. Generating sequences of characters by rules that increasingly approximated the frequencies of letters and spaces in printed English, he displayed a series of textual passages that approached readable English. He estimated that some 50 percent of characters in actual printed English were redundant, and he soon increased that estimate to 75 percent.[65] Intrigued, the ex–radar engineer J. C. R. Licklider did his own calculation at MIT and came up with a similar verdict. But Shannon fully recognized that in the real world this figure would depend very much on the *kind* of text involved, and Licklider too insisted that the ordering of letters was in reality so complex that "the only statistical machine that can deal adequately with the problem is the human verbal mechanism."[66] The extent to which that was so was, of course, an empirical question.

And so—although this has not generally been noticed—Shannon's information theory, abstract as it was, led almost immediately to attempts to measure *real* human readers. And it turned out that there was a big difference between redundancy as defined in Shannon's engineering terms and actual humans' ability to parse "mutilated" English. Chapanis, in particular, found that if you removed more than about 25 percent of characters at random, readers often could not interpret what was left. And George Miller, Bruner's partner in the forging of cognitive science, worked with the expert cryptanalyst Elizabeth Friedman to show that different kinds of "mutilation" led to quite starkly different success rates. The differences in readers' skills mattered too: military codebreakers were rather better than average adults at such parsing.[67]

All this meant that the new information science had an affinity with the science of reading that was closer than generally supposed. The language of entropy, "bits," and "redundancy," introduced by Shannon from telecommunications, did not simply supplant and displace the terms that had been used for decades to describe actual reading practices, but merged with them. Investigators of reading could help themselves to such concepts, seeing them as refracted through empirical experiences, tests, and measurements. One way this occurred was by researchers reenacting the long tradition of experiments that had been done since around 1900 on reading rates, parsing them now as accounts of information transmission. Several information scientists took this tack. Licklider did something similar at MIT, for example, looking at whether someone could track a series of marks manually while maintaining reading speed—a form of multitasking, as it were, with clear debts not only to the science of reading but to Licklider's background in radar-system human engineering.[68] He thought one could indeed keep reading quickly; two Bell Labs researchers, John Karlin and John Pierce, tried it for themselves and disagreed. Karlin had worked in psychoacoustics during the war, before inaugurating human-factors engineering at Bell Labs—where, among other things, he designed the standard telephone keypad. His collaborator, Pierce, was an electrical engineer who is often credited with coining the word *transistor*. In 1955–56 the two investigated how fast readers could read lists of words aloud, in order to assess "the human being as an information channel." They told the National Academy of Sciences at a meeting at Caltech that they had found the maximum rate—which, incidentally, they believed was a natural limit, potentially discrediting the boasts of all those speed-reading businesses.[69] Meanwhile, Wilson Taylor—a student of one of the founders of communication studies, Wilbur Schramm, in Illinois—adapted Shannon's contentions about "redundancy" in printed English to create a procedure he called "cloze" for measuring the readability of pieces of prose by removing a certain proportion of words.[70] And, to cite a final example, at Harvard Richard Solomon used a Dodge tachistoscope to relate the recognition of words to their frequency in the language. He and Leo Postman repeated the familiar trial with invented words, which, as Mandelbrot noted, rendered the case much more information-theory-like.[71]

It was tempting, as psycholinguist John Carroll remarked in a wide-ranging review of this field in 1958,[72] to link all this talk about the reduction of uncertainty to processes of learning and education. After all, what did a teacher do, if not reduce a student's uncertainty? Teachers were certainly supposed to impart cognitive skills. A lot of the science of reading could be called into duty to flesh out the parallel. But much of the work that Solomon, Karlin, and others were doing remained on a level that Carroll called "molecular." It was unclear how such particular experiments could be scaled up, translated into recommendations for practice, and applied in classrooms. Partly for that reason, in the next decade Bruner himself moved away from his perception work and focused instead squarely on the question of children's learning.[73] He produced a novel account based on a notion of "scaffolding," according to which children could learn almost any discipline, in principle, if it were introduced at a suitably congenial level and then built up by repeated iterations of greater complexity. Just as he had been a principal player in the carving out of cognitive science in the 1950s by means of a theory of learning, so in the 1960s and 1970s he played the central role in attempting to translate that science into everyday life by turning theory into practice. The project was visionary and, for many, compelling. But it was also deeply controversial, and its results would be fought over for decades. And reading became one of the two most intense battlegrounds, alongside mathematics. As we shall see, cognitive science became the foil for a generation of bitter and highly politicized struggles over the learning of these two intellectual skills. Perhaps inevitably dubbed the "reading wars," the conflicts over reading in particular would afflict the public reputation of science in general well into the twenty-first century.

But that is the topic of the next chapter. In the meantime, cognitive science, like the science of reading, took on an aspirational and commercial character almost from its outset. As well as learning to read really fast, *Reader's Digest* told America's middle classes, you could also "Learn to Think Creatively." And at the same time as corporations were exhorting their managers to take reading classes, they also started to train their staff in creativity. If General Motors was the model for corporate reading science, DuPont became the model for creativity training, hailed as such by its president Crawford Greenewalt in lectures at Columbia's business school that were published in 1959 as *The Uncommon Man: The Individual in the Organization.* "Conformity in behavior is a human necessity," Greenewalt warned, "conformity in patterns of thought a human danger"; and he immediately referred to Communism to exemplify the peril.[74] In that context, creativity fitted well with the self-image of industrial research managers, and thence helped popularize a particular view of science itself, as inherently open, nonreductive, antipositivist, and market-like.[75] Writers like Arthur Koestler adopted that notion whole-sale, to immensely influential effect. The popular press, too, took up the idea. It was also put into practice in innovative multimedia works, of the sort described recently by Fred Turner—museum and gallery installa-tions where citizens were encouraged to construct their own meanings by following different routes past the exhibits, and performance pieces by composers like John Cage and Morton Feldman that permitted musicians and dancers to carve their own paths through the score. As has become well known, the CIA quietly supported modern art on the basis of a simi-lar conviction that it exemplified American values and would encourage people in socialist regimes to draw contrasts with their own cultures. And American pavilions at international trade shows adopted the same kind of approach to cast a sharp distinction between the individual freedom purportedly definitive of Western culture and the disciplined, regimented conformity expressed by Soviet displays.[76] At the same time the embrace of citizen creativity was helped by miniaturization, too. The transistor was invented in 1947 and solid-state circuits followed soon after; by the late 1950s "transistor radios" were becoming commonplace, allowing for a novel practice of private, autonomous listening, and forms of expression that at least *felt* liberated. A decade later the cassette tape would allow citizens to compile their own musical and audial narratives. One of the social kinds that creativity itself created, perhaps, was the teenager.[77]

Perhaps the clearest demonstration of the relationship between the science of reading and the cognitive revolution came in the chaotic year of 1968, when John B. Carroll and psychologist of pattern recogni-tion Paul A. Kolers persuaded the MIT Press to reissue Edmund Burke Huey's classic foundational work from 1908, *The Psychology and Peda-gogy of Reading*. They each contributed an introduction to the paperback, making it clear that they both regarded the book as a definitive classic of cognitive science avant la lettre. Carroll declared that it deserved to be restored to prominence for its willingness to "raise deep questions" even if they could not yet be answered. Huey had been braver in this re-spect than his successors of the sixties, he added. "His characterization of reading as an information-processing activity," Kolers agreed, "has not been surpassed." Huey's work had now been placed at the head of a line of intellectual descent that passed from the pioneers of reading science, through Lashley and Hebb at midcentury, to information theory, psycholinguistics, and cognitive science in the information age.[78]

A Renaissance Man Effect

Gloria Swanson, Hollywood goddess and self-described "mental vampire," moved from Los Angeles to New York City in 1938. She had been frozen out of MGM, allegedly because Louis B. Mayer believed her anti-Semitic, and she needed a new direction. Like several film-industry regulars of the era, Swanson had an abiding interest in technological wizardry—she had once come up with an idea for a personal communication system, for which a Russian inventor friend had tried to get her a patent.[79] Now she once again began thinking of becoming a technology entrepreneur. But this time she would buy inventions rather than making them herself. She rented office space in Rockefeller Center and created a company that she called Multiprises, Inc. to fill it. Then Swanson dispatched a representative to a place where she knew that excellent scientists could be found who would be willing to come to her venture: the German lands now ruled by Adolf Hitler.

Swanson's representative was a factotum of the international music industry named Iphigenia Engel. She turned out to be a "tornado." Engel preemptively commandeered the Paris lodging of an ex-husband of Swanson's, French nobleman Henri de la Falaise, as Multiprises' European office. ("I know," Engel retorted when Swanson pointed out that the company did not actually have a European office; "but the Gestapo doesn't.") Then she moved fast to seek out persecuted inventors. Swanson had her travel to Vienna, hoping she would find there the patentee of a luminous paint that she thought would sell in LA because her driver had often found it hard to identify street numbers on dark nights. And Engel almost succeeded. The actual inventor of the luminous paint had already been arrested and taken to a concentration camp, but she found that he had been part of a small circle of friends who had pooled their patents under the leadership of an electrical engineer named Richard Kobler. Kobler happened to be a fan of Hollywood movies, and he was at first incredulous that *the* Gloria Swanson might be behind the extraordinary proposal that Engel made to him. But the Anschluss had just happened, and with the Nazis in power he and his friends were in mortal danger. Four agreed to make the move, three of whom were Jewish: the twenty-nine-year-old Kobler himself, Leopold Karniol (a chemist, expert in plastics), and Leopold Neumann (an acoustics engineer—a vital field in the context of the coming war). The fourth member of the group was metallurgist Anton Kratky. It took six months to spirit the men from Paris, via Brussels, to America, and Swanson herself quietly traveled to Europe three times to help their escape. But in the end they did all arrive safely—to be followed by Henri de la Falaise himself, bearing a contract from Kobler and Neumann giving him the guaranteed job that allowed him to leave war-torn Europe. Swanson set them all up with a factory in Queens. Soon the inventions started to come, for a range of gizmos from recording technologies and a copying machine to a cigarette holder.[80] The most lucrative was a way of manufacturing plastic buttons, much needed for military uniforms.

In the end, Multiprises did not last very long. Kobler and the others were called upon to do war work for the US government, and Swanson's venture went bankrupt in 1943. Kobler himself moved on and went to work at the Thomas A. Edison Laboratory, a small research lab operated by the electrical manufacturer McGraw Edison. In the postwar years he became a senior engineer there, continuing to invent new machines, one of which was a dictation-recording device called a Televoice. It was while he was working at Edison in about 1959–60 that he came across a film that captured his inventor's imagination. It showed a young child learning to read, apparently at a much younger age than most American children developed the ability. The child's name was Venn Moore, and the film showed her being inspired by her use of a new electric typewriter. With his long-standing interest in such communication devices, to Kobler the potential was obvious. So he got in touch with the maker of the film, Venn's father, Omar Khayyam Moore.

Moore père turned out to be a sociologist and educational researcher at Yale. He proved highly receptive to Kobler's enthusiasm, and the two started to work together. What emerged from their collaboration was a novel venture in the use of technology to help children learn to read. It came to be called—inaccurately but perhaps inevitably—the Talking Typewriter.[81] Emerging at precisely the moment when (as we shall see) controversies over the teaching of reading were at their peak, the Talking

Typewriter offered a possible resolution to those angry and otherwise intractable debates. It was also a deliberate counterblast to teaching machines and the programmed learning that they embodied. And it promised to transform the inequities in American education that substantially underlay such debates, not least those to do with race. Even more than that, however, it also suggested ways in which computing itself might be paired more successfully than ever before with people, because built into the design was a conception of human beings as creative, autonomous practitioners of culture and cognition. In short, the Talking Typewriter twinned an optimistic culture of technology to the latest cognitive science of reading. By doing so, it took a vital important step in the sequence of developments that gave rise to the personal computer, and hence to the twenty-first century's information culture.

As was the case with so many of the postwar era's most innovative social scientists, Moore had served in military intelligence in the war. He subsequently worked in the Group Psychology Branch of the Office of Naval Research, where he played a major role in shaping training programs for radar operators in the US's advanced early-warning system, SAGE. That was a fairly typical background for a postwar cognitive scientist, and at Yale Moore pioneered the kind of "problem-solving" perspectives that cognitive researchers liked. Herbert Simon, among others, picked up on his work in developing his early artificial intelligence theories. His reputation was thus on a steep upward trajectory when Kobler contacted him.[82] His initiative with the typewriter had sprung from those experiences in cognition, but it also came from Moore's engagements with earlier generations of social theory. He had been working on a general account of social interaction inspired by G. H. Mead, and had been poring over the pre–World War I sociology of Georg Simmel at a time when hardly anybody was paying Simmel much attention. Out of this odd combination of inspirations, Moore wanted to develop a theory of social action embracing the postwar emphasis on autonomy, creativity, and problem-solving (a successful version of the work that Shils and Parsons had failed to produce, perhaps). Like Bruner, he went on to spend years researching early learning. It was when he began using an electric typewriter himself to write up his ideas that he discovered his daughter, then two and a half, happily bashing away at the machine's keys. He was struck by how quickly she learned to type and read its characters. Given the common assumption of the time that children did not develop "reading readiness" until first grade, her avidity was surprising. He filmed her at the keyboard, and it was this film that inspired Kobler to get in touch. As they collaborated, Kobler worked to invent an automated, computer-linked version of the typewriter, and he and Moore imagined a classroom installation to be built around it. With McGraw Edison becoming a corporate partner, they got ready to revolutionize the teaching of reading.

The automated computer-plus-typewriter that Kobler invented allowed a child to type characters and words, to which the computer would respond. The result was a virtuous feedback loop that enticed children to learn as if they were playing a game. Yet Moore himself always insisted that the device itself was not really the point: what mattered was the whole "environment" in which it was situated. He had become convinced that the key to childhood learning—and to adult learning too, for that matter—was what he called "autotelic responsive environments." A learner should be immersed in an informational setting that allowed her to explore at her own pace and that responded immediately to her actions. The exploration should be open-ended, allowing for a range of different kinds of discoveries, and the process should be satisfying in itself, by virtue of the inherent excitement of discovery. Moore emphasized that the learning process must not rely on exterior rewards or punishments. Fun was a key element in the system. In true cognitive fashion, it was to be tweaked for individual students and designed to foster their imaginative creativity.

Moore and Kobler were far from alone in the general character of this ambition, and they had chosen almost the ideal moment to launch their technology. Some in the scientific world had been urging that school education become more exploratory, more "discovery-based," ever since Dewey. In 1958 the reformers got their opportunity when anxiety over Soviet science provoked by Sputnik led to the passage of the National Defense Education Act.[83] The new law mandated large increases in federal funding for schools and other educational ventures. At the same time the National Science Foundation threw resources into the development of new curricula in mathematics and science, for which the cognitive virtue of "understanding" was to be a key concept. (McLuhan's NAEB project for "Understanding New Media" was in fact another beneficiary of the NDEA.) It was an enormous endeavor, initiated with great hopes; by

1975 the foundation would have spent $200 million on the new curricula, plus $350 million more on training teachers for them. The most important agenda-setter for the transition was to be none other than Jerome Bruner. Bruner's *The Process of Education* was published in 1960, and he played a major role in a number of ambitious—and before long highly controversial—educational initiatives, including the ventures known collectively as "the new math." Another prominent example was the social-studies program called "Man: A Course Of Study" (MACOS).[84] In fact, over the ensuing decade Bruner's book became the second-most-assigned work for teacher training in the country. It underpinned a national ambition that, for a scientific and democratic future, every American should and could think as a democrat and a scientist—creatively and autonomously. This was the ambition that Moore and Kobler tapped into.

Garnering support from the Office of Naval Research and the Carnegie Corporation, Moore initiated a plan for "responsive environments laboratories" to be placed into schools. The first of these laboratories was installed in 1960 at a day school in Hamden, Connecticut. The installation used two rooms of a prefab building. One room contained mimeographs, tape recorders, and thermofax machines, plus an assistant's post and a set of soundproof booths. Based somewhat on Moore's old model of radar operators' terminals, these booths contained Talking Typewriters for the children to use (fig. 7.1). Some of the booths were automated, others designed to work with the oversight of "booth assistants." The space was designed to be friendly to children, although gossiping was discouraged and the room had no windows because (as Moore put it) they were "an open invitation to digress." New entrants were to be shown around by existing students, not adults, and the helpers present were pointedly "booth assistants" rather than teachers. In fact, Moore and Kobler found that the automatic booths were useful in training these assistants almost as much as the children: left to themselves, they tended to be too interventionist.[85] The second room was a more conventional classroom. Called the "transfer room," this was the setting in which children could refine and combine their "correlative" skills of reading, writing, speaking, and listening. These came together in the publishing of a newspaper out of the laboratory—a task, we may note, very similar to that recommended by Edmund Burke Huey at the very beginning of the history of the science of reading.[86]

7.1 The autotelic responsive environment in action, with the Talking Typewriter at the heart of the system. O. K. Moore, "Autotelic Responsive Environments and Exceptional Children," in *Experience, Structure, and Adaptability*, ed. O. J. Harvey (New York: Springer, 1966), 174. Copyright © 1966, Springer Nature.

It turned out that Moore and Kobler had timed their invention perfectly. Having just experienced one burst of funding with the NDEA, in 1965 America's schools enjoyed another with the Federal Elementary and Secondary Education Act. Both pieces of legislation explicitly encouraged investment in educational technologies, so the nation's schools were likely to be receptive to a pitch for a cutting-edge system of this kind. New technology was very widely touted as offering radical solutions to the school system's long-standing problems, and indeed to the nation's problems at large. Major corporations were looking to capitalize, including IBM, GE, RCA, and Xerox. This first installation of the Edison Responsive

Environment Learning System, as it came to be called, was thus soon followed by four others in different cities, and the first large-scale adoptions took place in 1966. Meanwhile, even before Moore's ideas for his "environment" took firm shape the Talking Typewriter had become a focus of media attention. Between 1960 and 1973, hundreds of reports and articles were published about it, and it was given a central role in TV programming about education in the Great Society. In 1963 Moore formed a company, Responsive Environments Corporation (REC), to market the system. REC helped the boosterism, circulating films and making sure that corporate conventioneers could witness it in action. Countless packets of information were mailed, to anyone likely to show interest. Salesmen and trainers traveled across the country to market the machine to principals and administrators in America's cities and educate the educators in its use. So many people wanted to see the Talking Typewriter in situ that the company had to incorporate a requirement to permit visits into the agreements it signed with participating schools. Kobler, by this point living in Switzerland, wrote excitedly to Gloria Swanson telling her of his excitement at having invented this marvelous machine that was finally solving the problem of how to teach children reading.[87]

By the late 1960s, the Edison Responsive Environment (ERE) was apparently a roaring success. REC could point to about 150 EREs in use in some fifty locations. It also had a racial-politics edge, as antipoverty politicians championed its use in segregated cities to even up educational opportunity in the society of post–*Brown v. Board of Education* desegregation but enduring inequity. It was soon put into action in the largely African American Hill district of Pittsburgh, for example, while in Chicago children were bused from the poor and dangerous housing projects of Cabrini Green to the schools where it was installed. They were then filmed using it, to help boost support for this "secret weapon in the War on Poverty." A newspaper produced by children with the system was even entered into the congressional record in Washington.[88] Perhaps at long last "equalization" would finally take root thanks to the affordances of a computerized environment. And in 1967 Kobler himself told the annual meeting of the National Reading Reform Foundation (a prophonics group) about the "rather startling" effects on children's reading that had resulted in a deprived area of Brooklyn. In just one quarter, he announced, the Talking Typewriter had advanced them by about half a grade compared to children who did not use the Environment—at the same time as logging no acts of vandalism or theft, and a marked improvement in "grooming and dressing." This was the result, he declared, of recognizing the "genius" in every child and using a "multisensory precision instrument" that allowed for "different accentuations of his 'sensory tentacles'—seeing, listening, talking, touching, etc." The nominal cost of $250 per child massively outweighed the losses that would otherwise be incurred by society for even one "dropout," which Kobler reckoned to be $250,000. And REC stood ready to replicate the success in any school in the country. This, he concluded, "is real 'Headstart.'" It really seemed that a step had been taken toward "an 'intellectually classless society.'"[89]

As soon as the ERE was installed in more than a handful of schools, however, it became very clear that a principal element of Moore's concept was impracticable. He had envisaged teachers creating their own software for the device, tailored to the needs of their students. But local teachers did not have the time, energy, or (despite REC efforts) training to compose the kind of individualized programs that the cognitive theorist Moore had in mind. As a result, in 1967 REC established its own Program Department to create software. The company recruited Milton Katz, a Job Corps official and a veteran of man-machine interface design in Polaris submarines, to administer the department. Katz had experience with an earlier hybrid reading program that he now ported into the ERE system. The system had been developed in California by one Maurice Sullivan, yet another ex–World War II intelligence officer, and it was based on a combination of behaviorist psychology, linguistics, and phonics. Its appeal was partly that Sullivan was a champion of using tape instead of teachers. During the war he had inaugurated the use of tape-recording to teach idiomatic German to American spies, and he had subsequently set up the first "language laboratories" at Yale, Puerto Rico, and Madrid, basing them on German wire-tape technology and the installations used by radar observers. Under Katz, the Sullivan technique now became the somewhat unlikely centerpiece of the Talking Typewriter environment.[90] A rapid development cycle costing some $1.25 million soon generated 128 programs based on it, and from then on REC sought to churn out a continuing set of programs and textual materials. Impressive as it sounded, however, these materials were restricted to the teaching of reading—the only other software REC ever created was a drivers-education sequence—

whereas Moore had wanted the system to teach all kinds of skills. And although the company did add its own perspective, the overall character of the Sullivan technique was one of phonics and code-cracking, which was not really in tune with the ERE's ethos. By the late 1960s, in consequence, what had started as a cognitive *environment* had devolved in the public eye into a teaching *device*. Ironically, commentators around 1970 tended to remark that the system looked to them distinctly behaviorist.[91]

In the meantime, however, Moore's ambitions remained high. He imagined branching out, hoping to market spin-off devices such as the Voice Mirror, which used cassette tapes to feed back the user's own voice.[92] And at the end of 1967 REC proudly announced that it had appointed a visionary new adviser: none other than Marshall McLuhan.[93] It was a very public statement of intellectual ambition, and in May 1968 McLuhan confirmed as much when he agreed to act as speaker at the company's New York launch of its latest device, the "Talking Page." This was a $700 portable machine that used 45 rpm records to teach reading. Its most notable feature—apart from its portability—was that unlike other machines of the time it could take branching paths depending on a student's responses. Kobler, who had flown in from Switzerland, demonstrated its use, while an REC executive (perhaps Moore) hailed the machine in revealing terms. "Learning takes place best," he said, when the learner was engaged in "multisensory adhesion" to the material being presented. McLuhan then took to the stage, praising the Talking Page as a device that would further "the decentralization of knowledge." Just as portable radios on camels had revolutionized the Middle East, he claimed, "this kind of portable device can revolutionize education." Talk of revolution was queasily apposite in mid-1968, and nobody missed the reference to student demonstrations taking place at that very moment at Columbia University. "Under electric conditions," McLuhan went on, "every type of knowledge is decentralized." Americans already lived in a "programmed environment" in which levels of information were "fantastically higher" than anything a conventional university could offer. Thanks to new media, a child of six knew as much about this environment as an adult. The Talking Page, McLuhan concluded, was the future. "We should be training perception, not just cramming in information," he added, before fleeing the jabber of Kobler's machines with a cohort of quote-seeking journalists in tow.[94]

In the end, the Talking Typewriter did not live up to its early promise. One reason was the very holism of its conception—the insistence that it was not merely a machine but an environment. Such a commitment did not lend itself to miniaturization or mass-production. The Talking Typewriter was always going to be the size of a room, and while that was nothing unusual in 1960, by the early 1970s designers were starting to frame computers small enough to be "desktop" devices. They were coming to appreciate that size alone was a major factor in adoption. Partly for that reason, as a congressional investigation discovered, the Talking Typewriter was simply too expensive, especially when the prosperity of the mid-1960s yielded to austerity. In 1972 an installation cost $37,500, or $800 monthly for five years, and schools also had to pay for a maintenance contract, plus the programs and literature. They had to set aside physical space, too. Each lab needed at least one teacher, plus a trained attendant for every two booths. Investing in even one Talking Typewriter therefore meant foregoing several real, human teachers. And even though the outcomes from the system were reckoned to be good, with such high costs the expectations were naturally extravagant.[95] In the new economic world of the 1970s these were serious weaknesses, and by 1973 the Talking Typewriter was largely gone from classrooms. For a while, REC hoped to pivot to using new "desktop technology" to teach reading to the citizenry via a network of private reading clinics—the equivalent, it seems, of Warren's old Better Reading Foundation. It even opened a storefront in El Cajon, CA, just east of San Diego, presumably a promising location because of its Spanish-speaking population. This did not pan out either, however. Soon it was being reported that REC was on the verge of bankruptcy and looking to sell itself.

Yet the Talking Typewriter had existed for long enough for it to be heard by some very important listeners. Leading figures in the nascent worlds of artificial intelligence and personal computing found inspiration in its initial premises. They found the Talking Typewriter, as the anthropologists say, good to think with. As a result, although in the end the Talking Typewriter itself had mixed success and was soon superseded, the commitments it embodied—to learning as creativity, to individual autonomy, and to the central place of the practice of reading in shaping human-computer activities—went on to become essential to early designs for what became "personal computers." One rather disconcerting consequence of that was that the "person" for whom such computers were to be designed in those crucial early days was imagined, not as a radar operator, nor as a computer scientist, nor even as a businessman or military officer, but as a youngster. More specifically, a youngster who was learning to read.[96]

Reading a Dynabook

The most influential of the visionary computer pioneers to find the Talking Typewriter fascinating was probably Alan Kay. At the peak of REC's fortunes, Kay had arrived at the University of Utah after a rather wayward career as a musician and a programmer for the air force, and he was now doing a PhD in computer science. He was slogging away at a new-generation computer language, eventually to be called FLEX, when he met in quick succession Douglas Engelbart and Marvin Minsky. Engelbart's visionary ideas of human augmentation were fascinating, but it was the AI pioneer Minsky who really electrified Kay. He spoke at a meeting on the theme of education, delivering what Kay later remembered as a "terrific diatribe" against traditional classroom ideas and practices. Minsky invoked Piaget and the MIT computer scientist and educational reformer Seymour Papert to advocate a concept of learning as the acquirement of problem-solving skills, much along Bruner and Moore's lines. "It was clear that education and learning had to be rethought in the light of 20th century cognitive psychology," Kay recalled him arguing: now was the time to pay attention, finally, to "how good thinkers really think." The following year Kay took the opportunity to visit Papert himself, and saw him using his educational language Logo with schoolchildren. The experience, he testified, "hit me" with a revelation of "what the destiny of personal computing *really* was going to be." Children could use the system without specialist training; so could adults. It made the eclectically brilliant Kay think of the moment in around 1500 when the Venetian printer Aldus Manutius had realized the potential of print by making books small enough to be portable. What he saw on that day in Lexington was the realization of a distinctly McLuhanite ideal—an individualized computing machine acting as "a personal dynamic *medium*."

Returning to Utah, Kay raced through his doctoral work, while on the side building a cardboard mock-up of a computer that someone could hold in their hand. He called it the Kiddikomp. Later, after a spell at Stanford's AI lab, he would take the idea with him to Xerox's Palo Alto Research Center (PARC). And there, working with a collective called the Learning Research Group, he turned it into one of the most renowned of early personal-computer designs, and *the* most famous to envisage a truly portable device. It now went by the moniker "Dynabook."[97] But although the name may have left any overt associations with childhood learning behind—and later hagiographic memorializations would reframe it in terms of the subsequent development of laptop computers and tablets— the fact is that Kay's vision remained clearly and explicitly centered on the kinds of concerns that Moore and Kobler had tried to address. This was made extremely clear in the latterly famous text that Kay composed in 1972 outlining the scope of his proposed machine and setting out the rationale for building it. The text was a memo for others in Xerox PARC, and its title was "A Personal Computer for Children of All Ages.'"

Kay's document has a major place in histories of computing and information, because of its extraordinary projection of a handheld, networked computer and its manifold uses, long before such machines became first feasible and then commonplace. But before Kay got to his computer idea at all, he focused on children and learning. Moore's Talking Typewriter— or, to be exact, the cognitive idea behind it—was an explicit inspiration. He started with the problem of overhyped educational technology. To overcome that problem, Kay argued, one must begin by addressing the learning process itself. The explicit subject of his memo, then, was this, not the machine itself. Its defining concept was of the child. And here he insisted on Piaget's view: a child was an explorer and model-builder, or, in Kay's terms, "a 'verb' rather than a 'noun.'" His approach would be opposed to pettifogging standardized tests, because, as he appealingly put it, he was more interested in Sistine-Chapel-Ceilings-per-lifetime than in any minor, incremental improvements that might be measurable. And this kind of autonomous creativity was not necessarily a matter of technology at all. After all, the traditional printed book served the need quite well, and any new technology should at least accomplish the (surprisingly difficult) task of replicating a book's properties. But there was scope to do more: it could also be "active (like the child) rather than passive." To make such a design and give it purpose, one needed the learning theories of Piaget and Bruner, the approaches to practical creativity of Moore and

Papert, the vision of media of McLuhan, and the algorithmic sensibilities of Minsky and Herbert Simon.

Kay followed this with a brief piece of science-fiction, giving an uncannily perceptive—if a little too hopeful—portrait of everyday life with networked laptop computers. It was too hopeful, however, for the same reason that Kay devised it at all. He believed that the Dynabook should, and therefore would, serve primarily as a tool for a creative, educative process of "reflexive communication." He recognized its likely use as a mere consumption device, but he assumed that people would want to be at least as actively creative as they were already with pen and paper. If the point of Kay's prospective machine was to teach "thinking" and "models," he assumed that people at least did want to learn those things. His plan would assist learners in developing "interface" skills like seeing. And this would be done, among other things, by facilitating reading. In that light the account Kay gave of his ideal machine was revealing in its emphases. He envisaged a $500 device with a screen, keyboard (perhaps a virtual one), stylus, removable storage, and rechargeable battery. It should use an operating language close enough to the concepts of a lay user that nonexperts could program it, and it should be networked. Even that basic specification reflected convictions about the nature of reading. For example, the storage capacity of a Dynabook was defined in terms of printed pages. Beyond that, he insisted that the screen display characters with a resolution and contrast approximating that of a printed page (something we have arguably still not achieved). A user must be able to change fonts on the fly, and even design new ones. His lab had in fact researched "the art of character generation" quite extensively, he added, and he included a discussion of this work. Characters on the screen looked better than expected, he commented, and he guessed that a "noise reducing filter" between the eye and the brain might be acting to smooth them out— something that sounds like it was adopted straight from recent accounts of perception. As an example, he showed renderings of the first page of Irving Stone's biographical novel about Michelangelo, *The Agony and the Ecstasy*, in different fonts. The choice was surely significant. Stone's 1961 work was not only a bestseller but also a milestone in portrayals of the heroic, creative individual in a turbulent world.

In the future, Kay suggested, readers would not buy books like Stone's on paper at all. Instead books would be "instantiated" in the Dynabook by downloading them from a library or vending machine via a "global information utility" and rendered in whatever font the reader desired. He confidently predicted that this would not harm the publishing industry, just as photocopying had not, because "people are not interested in acting as a source or bootlegger; rather, they like to permute and play with what they own."[98] He then added that one reason for this focus was that the Dynabook was designed with a specific purpose in mind: to replace school textbooks. And because textbook purchases generally cost around ninety dollars per year, over the lifetime of the machine the savings would almost cover its purchase price. That educational economy, it turned out, was a principal reason why he had set himself that target of five hundred dollars. If he hit it, maybe the machine could even be given away free.[99]

By this point, Kay's group was already hard at work on a nonportable prototype based on this sketch, with software informed by Bruner's ideas about cognitive learning. But Kay obviously recognized that more than technical expertise was required. A real Dynabook, he declared, would demand "a corollary inquiry" into "why good readers do so much better than average readers"—which, remarkably, had been perhaps *the* defining question of the science of reading for generations. At much the same time as his "Personal Computer" memo, moreover, he expatiated on this theme in a presentation to the National Council of Teachers of English at the University of Minnesota.[100] The filmed presentation would be far less renowned than the memo, but it provided a fascinating insight into the reasoning that lay behind that document.

At the Minnesota meeting Kay gave a freestyle, slide-based talk before a small but clearly riveted audience. Noting that he was inspired by the ideas of Piaget, Moore, Papert, and even Dewey, he promised to introduce an "environment" in which children would learn skills such as reading, French, or mathematics by exploration. For that to happen, he continued, the central technology had to be portable. He likened it to the pocket calculator, which became revolutionary simply by virtue of being sized to fit in a pocket. So his engineers were going to "beat the technologies into submission" in order to make a device the size of a paper notebook. As in his memo, Kay insisted that it must have all the attributes of paper but be "active" too. He then demonstrated a couple of examples of what this meant. One was the use of different typefaces, which Kay illustrated again with the example of Stone's *The Agony and the Ecstasy* (which, he remarked offhandedly, his lab had "typed in"). He noted that one could also design one's own fonts, making this, in McLuhan's terms, a "hot" medium. A reader could also substitute different kinds of characters altogether—he suggested Pitman's Initial Teaching Alphabet, a simplified system that McLuhan had also highlighted in *The Gutenberg Galaxy*. Or one could even substitute an image for a word, such as replacing the phrase "Winnie the Pooh" with a picture of the bear. Every act of reading would then become an opportunity for creative experiment, he suggested: what if one changed every *he* in A. A. Milne's book to a *she*? Copyright law might be an obstacle, he conceded, but the Dynabook would be a "new medium," and like all new media it would radically disrupt existing ones.

At this point Kay's audience interjected, and on significant grounds. He had mentioned in passing that the acquisition of language proceeded by words rather than phonemes, and he had implied that learning to read occurred similarly. That was one of the most controversial of all the claims of the science of reading, and his audience knew enough to question it. In addition, was he not making the same mistake as countless earlier researchers by assuming that a beginner's reading should be modeled on adult reading? Far from being just a demo of a pioneering vision of personal computing, then, what emerged in this setting was a dialogue about "research about reading." In disagreeing, Kay remarked that he himself had learned at the age of two—so much for "reading readiness"—and that he had done so by watching his mother run her fingers down the page as she read bedtime stories. That sounded very like the speed-reading tactic championed by Evelyn Wood's notorious program, which the Carter administration would shortly introduce into the White House. Thanks to it, he said, he now read at two thousand words per minute. Kay then further insisted, in good cognitive-science fashion, that learning linguistic or reading skills was not a matter of *instruction*, per se, but a matter of getting students to *become* speakers or readers. Experimenting and exploration were essential to that process. Unlike contemporary devices for "computer aided instruction,"[101] then, which were self-contained and prescriptive, the Dynabook would be networked, open, and dynamically responsive. Each machine would act like a gateway to a library through which users could roam freely. Moore had proved that computing of this kind attracted children, after all. Such a machine would attract adults in the same way, making everyone into a tool-using creator. So when Kay predicted programs that would allow people to compose and edit their own texts in real time—in essence, word-processors—that vision was an immediate and natural consequence of this notion of learning. What would someone do with his imagined machine, he was asked? "As much as you possibly can." It would engender "a renaissance man effect."

As it happened, Kay's Dynabook project never did generate a portable computer. But his "interim Dynabook" did have consequences. It evolved into the Xerox Alto, the first personal computer to rely on a wholly graphical user interface, which could be networked together to make a collective of twenty to thirty machines.[102] From the Alto such interfaces were appropriated by Apple to make its Lisa system, and from there they were to be reiterated across the personal-computing world. When we experience computers today, by and large we do so through mediating conventions that are descended from the project that Kay initialized. And that means that, at a further remove, they come from the ideas of cognitive science, the Talking Typewriter, and, ultimately, the debates that the science of reading sparked in the 1950s. We shall see more of this history in the next two chapters. So the story will have to be left at this point for now. But it is worth stressing how much Kay's work at Xerox PARC on Dynabook has always been regarded as a canonical step, and how little we have appreciated the centrality of reading to his vision.

Many of the elements of Kay's project—integration, miniaturization, portability, networking, storage, and the use of symbolic communications to allow everyday users to exercise autonomous, open-ended creativity without the need for programming skills—would become definitive of "personal" computing as it developed. To be sure, he was not the only person thinking along such lines. We have become familiar with how countercultural ideals of the late 1960s came to be repurposed into foundational principles of personal computing in much the same period.[103] And Kay did acknowledge the existence of communitarian culture in the computing world explicitly in his famous memorandum. But he did so in order to insist that what he was doing was something different. What he had in mind was explicitly much more Moore's vision than, say, Stewart Brand's. It is salient that in 1981, coinciding with the launch of the Osborne 1 as the first commercial portable PC, the progressive computer pioneer Edmund Berkeley reprinted Kobler's report on the Brooklyn success of the Talking Typewriter in his *Computers and People* (the oldest computer magazine, previously *Computers and Automation*). He did not have to choose that work. Berkeley's rationale was that cheap microcomputers could finally make possible the realization of the Talking Typewriter's aim of teaching reading to everyone.[104] By contrast to the path via the counterculture, this other trajectory is now all but completely unknown. True, Joy Rankin tells a compatible story in her recent *People's History of Computing in the United States*, centering in her case on the PLATO system developed at the University of Illinois. PLATO was a counterpart and sometime rival to the Talking Typewriter, although its pedagogical ethos was rather more top-down and it demanded more technical skill of its users.[105] But in general the history reconstructed here has remained in the background. There are reasons for that. It lacks "culture heroes"—there are few Pirates of Silicon Valley here to tempt the higher journalists of the *New Yorker* or the *Atlantic*. Worse still, it involves the history of education, which is notoriously a backwater in modern intellectual culture. And it is, undeniably, a complex, multithreaded, reticulated tale, resistant to rendering as a conventional narrative. It runs through a range of locations, from the high cultural atmosphere of Toronto, via the disciplinary anxieties of MIT and Harvard, to the dangerous precincts of Cabrini Green. Beginning at the peak of commercialized reading programs based on eye-movement traces, reading accelerators, and Samuel Renshaw's tachistoscopic boosterism, it passes through the cognitive revolution and the invention of information science, before arriving at a wholesale effort to transform American education—and each one of those steps had its agonistic politics as well as its scientific complexity. Something with so many venues, aspects, and complexities was never going to be easy to track. But perhaps this is just what human culture is like. It is as multifarious and irreducible as the cognitive revolutionaries liked to believe.

8

Reading Wars and
Science Wars

An informed American reader of the mid-1950s might well have believed that the future of work and leisure in an information age looked rosy. The essential skill for the postwar world, that of reading quickly and accurately, was in scientific hands. America's government, forces, and corporations were all forging ahead with programs to refine the abilities of the professional classes to manage information at ever-greater speed—and this, as Marshall McLuhan would endlessly repeat, was the new heart of the nation's industry. Experimental scientists were using sophisticated technology to understand ever more precisely how reading worked, and they were teaming up with educators to help the next generation do it better. The newest approaches in cognitive science and information theory were both drawing on this cutting-edge research and, in turn, contributing suggestions for how it might advance in the future. Military uses of this science had helped win the Second World War and would surely do the same in Korea. And the most American of scientific prophets—the Robert Heinlein who would soon forecast a Hayekian version of 1776 in *The Moon is a Harsh Mistress*—was predicting that Renshawed superheroes would spread American values through the galaxy. All this was eminently within reach, and your children could look forward to it confidently. What was more, you too could share in the experience. You could become a turbocharged reader yourself, for the investment of just thirty to a hundred dollars at the Foundation for Better Reading. There was probably a branch in your neighborhood.

But beneath all the boosterish rhetoric of the science of reading there was always an undertow of skepticism. Champions of the science had pointed to awful test scores among America's children and servicemen since well before the war in order to support their work, but those scores did not seem to be rising. What if your children would not, in fact, enjoy such blessings? Middle-managers sent to reading-improvement courses (sometimes at their own expense) and military officers trained in the army and air force programs did not always emerge enthused by the experience. What if the promises made by tachistoscope-wielding tutors were all hot air? Doubts about the more extravagant claims on behalf of the science of reading had always existed, and they could certainly be unearthed now if one looked hard enough, even though on the surface they seemed to be outgunned by Renshaw's military record and the stentorian consensus of America's governmental and corporate leaders. Lurking behind the marketing hype of speed-reading accelerators, tachistoscopes, and eye-movement devices lay nagging questions. Did they actually do any good? Even if they did, was faster reading necessarily *better* reading, with greater comprehension and retention? And did their benefits, if any, last? At almost the same moment that McLuhan arrived at Renshaw's doorstep in Ohio, the future literary great Wayne Booth published a short but acute satire on the whole issue entitled "*Lady Chatterley's Lover* and the Tachistoscope." Booth imagined how people coached in the Special Tachistoscopic Army Reading Training (START) would hyperquantify their emotional reaction times to texts, much to the despair of the rival Society for the Lowering of Wasteful and Anaesthetic Reading Speeds (SLOWERS). Having seen in the *New Republic* that Lawrence's novel was intended to "evoke erotic responses . . . in your very glands," one academic reader steeped in START's ideology tries to measure his Very Gland Reaction Time (VGRT, pronounced *Vagrant*—soon renamed PRT on the basis of experience) while reading *Lady Chatterley* at 6,731 words per minute, and bolts when the results come in.[1] It was hardly subtle. But there was never a shortage of rollicking, Johnsonian skepticism of this kind in the pages of America's middlebrow press. And in the end, it hit home.

The Science of Reading in Crisis

Just as McLuhan vaunted the importance of Renshaw-style insights to the multisensory future, the science of reading lurched headlong into a crisis of credibility. Three currents seem to have converged to trigger a public backlash: a new appreciation among researchers of the importance of reflexive effects in psychological experimentation, which led to a fear that the scientific character of the enterprise was shaky; a reaction in the general public against the crasser claims of commercial speed-reading programs; and, above all, a furious backlash at the perceived failure of public schools to produce improvements for America's children. For the commercial institutes, the effects were serious enough. A real downturn in the enterprise seems to have set in around 1955, and although it did not disappear—it still exists today—it permanently lost the unquestioned association with a bright scientific future that had previously been so alluring. But for schools the results were much worse. Practices and conceptions that had held sway for decades were suddenly cast into radical doubt, in a context of unremitting cultural conflict. The effects were immediate and lasting. And they continue to affect the lives of countless Americans to this day.

The first factor undermining the science of reading was a new awareness of the complexities involved in psychological experiments in general. It had become clear in many branches of experimental psychology that the actions of human subjects could be affected by aspects of the laboratory environment that scientists might not be aware of. Harvard psychologist Martin Orne, for example, coined the concept of "demand characteristics" in the late 1950s to indicate how a person might respond to subtle perceived cues in the experimental setting, estimating that the experimenter might want certain results and seeking to be a "good" scientific subject. Recognition of the reality and implications of such reflexive interactions characterized the human and social sciences of the period. It made the positivism inherent in techniques like Renshaw's—not to mention all those miles of film containing eye-movement traces—suddenly seem dated and naive. To be sure, some of the leading scientists of reading who had performed those experiments had explicitly acknowledged similar problems; Guy Buswell's project on the viewing of pictures had placed them center stage. But the broader enterprise was ill-equipped to handle experimental reflexivity. The problem was relatively minor for the commercial science of reading, because at first it was largely practicing scientists who found it a compelling question. But educationalists could not ignore it, especially as leading figures in the new cognitive sciences were seeking to reshape education itself in ways that would embrace reflexivity as an element in the learning process. Before long, then, it was being invoked as a general issue of concern that ought to give anyone pause before committing to actions based solely on reading experimentation.

This epistemic conundrum took on greater significance in the later 1950s, in the context of a gigantic societal shudder at the sheer hucksterism of the science's commercial practitioners. Competition between profit-seeking programs had grown intense, leading to claims that applied science could increase reading speeds by orders of magnitude. These claims were increasingly questioned in the press as self-evidently ludicrous. Psychologists and other expert authorities pointed out that legitimate scientists had almost never made claims for quite such extravagant improvements (Renshaw, apparently, had been quietly forgotten). They defused the confrontational character of their charge, at least a little, by raising the possibility that something real might indeed be happening, but that it was not *reading* that people were mastering but a different skill that they called "skimming," or occasionally "scanning." This was the practice of moving one's eyes over a page very quickly and at a controlled rate, taking in almost nothing as one did so. The validity of their charge may seem obvious in retrospect, especially to those who remember the saga of Evelyn Wood's specious "Reading Dynamics" enterprise in the later 1960s, which involved exactly this kind of practice.[2] But we should pause before assuming that the earlier commercial ventures were prototype versions of Reading Dynamics. In fact, they were indeed generally based (if a bit selectively) on reputable scientific practices and ideas. Moreover, unlike Wood, they employed versions of the instrumental repertoire that academic researchers used. But that very scientificity turned out to be a problem. Members of the public were supposed to realize such dramatic improvements by using what turned out to be a confusing array of often shoddily produced and overpriced "instruments." Some operations were even out-and-out frauds, like the company called Study Systems, of Hackensack, New Jersey. Not long after the Talking Typewriter's success in the borough's public-school system, Study Systems tricked priests and nuns working in Catholic schools in Brooklyn into issuing endorsements for its proposed "reading laboratories," and then sent salesmen to tell parents that they were required to purchase reading accelerators at the absurd price of $400 each. New York's Consumer Affairs Department forced the company to refund the money, in the largest rebate the department had ever won.[3] Most ventures were probably not quite as unscrupulous as that, but they were not exactly scrupulous either, and a lasting reaction set in against their combination of implausible claims and tacky devices. By 1956 even the Better Reading Foundation had been ordered not to use the term "foundation" any more.[4]

By the late 1950s, the "speed mania" whipped up by marketing was being blamed for a general reaction against reading machines of all kinds. During the preceding decade, one commentator reported, "a bumper crop of 'reading programs'" had been "widely and sensationally thrust before the public," most of them centered on some kind of device. The result was a feeling that "all use of aids is suspect."[5] Colleges and corporate training programs started to question the value of "mechanical approaches" rather more intently too: a survey of the time found that although virtually all industrial programs used tachistoscopes and most used reading accelerators, many were discovering that straightforward reading materials were more effective than either.[6] It was precisely to avoid the stigma that machines had acquired that Wood's Reading Dynamics venture discarded instrumentation altogether. It relied instead on the reader moving a hand down the page as he or she read, acting as a human counterpart to the reading accelerator. The strategy seemed to work, as the program temporarily avoided the notoriety by then attached to reading devices; and the practice was exactly what Alan Kay later remembered his parents doing. But the abstention from instruments ultimately left it vulnerable, and it was when academic scientists of reading publicly challenged Wood to let her students be tested in their eye-movement cameras that the enterprise started to collapse.[7]

The third problem that afflicted the science of reading in the mid-1950s was much broader and more damaging than these. It was a massive public backlash, not against the commercial enterprises, nor directly against the science of reading itself as a laboratory practice, but against its involvement in the everyday educational experiences of America's children. That involvement had been a reality for perhaps three decades by this point, having been secured in the 1920s in the generation of Judd, Buswell, Gates, and Gray. When children in American schools learned to read, they did so by using the books written by scientists of reading (or at least under their direction), in programs devised according to the theories of the scientists of reading, and on calendars established by the experimental traditions of the scientists of reading. The alliance between science and pedagogy in this domain was one of the fixed points of almost all Americans' lives. So it came as a tremendous shock—and, to some, a long overdue one—when in 1955 a book was published that roundly declared the whole thing to be a scam, and when that book immediately became a national bestseller. The book revealed a scandal. The very science of reading that promised to lead the free world to victory over conformist Communism was in fact guaranteeing its rapid and absolute defeat. Not only were America's children not being taught the fast, efficient, and creative reading techniques that would underpin a brilliant techno-democratic future: the awful truth was that they were not being taught to read at all.

Johnny and the Red Scare

The herald of this message was a forty-four-year-old erstwhile legal scholar who had fled the rise of Nazism in his home country of Austria in the mid-1930s and made a new life in the United States. Like many of his generation, Rudolf Flesch had latched onto the question of "readability" posed by the social science of reading at that time. He chose to investigate what he called the "marks of readable style" and enrolled in the PhD program in library science at one of the centers of the science of reading, Teachers College at Columbia University. He participated in the readability laboratory's work, and Flesch concluded from his experiences there that he could design tests to provide an objective measure of how readable any given work was. Armed with his measurement system, in the immediate postwar years he became prominent among the advocates urging "plain English" on the midcentury bureaucratic and administrative establishment. Soon he had several books to his name, with titles like *How to Test Readability*, *The Art of Clear Thinking*, and *How to Write Better*—all three of which appeared in the one year of 1951. He took on a market identity prominent enough to warrant a discussion that same year in Marshall McLuhan's voyage into the maelstrom of commercial print, *The Mechanical Bride*. Popular as they were, though, his books were also respected: of all the competing scales of readability, his was the most highly regarded. When J. C. R. Licklider and others in the cognitive revolution appropriated the science of reading to investigate perception and creativity, they routinely used Flesch's scale as a viable measure of read-ability.[8] Having proven himself in this field, Flesch then put his expertise in readability to powerful use by turning his attention to the other side of the issue—that of the skills of readers. Before long, he came to the con-clusion that something profoundly wrong—in both the epistemic and the moral sense of that word—had taken root at the heart of American intellectual culture. Somehow, the most advanced country in the world had stopped producing good readers. It was, in his view, nothing less than a national emergency of existential importance.

Flesch's warning cry about the science and pedagogy of reading, *Why Johnny Can't Read*, appeared in March 1955 and was an immediate sen-sation. It was widely serialized in the press, drew critical commentary across the media, and had to be reprinted four times in a month. The book remained on the bestseller lists for almost the rest of the year. It sparked debates that not only lasted for years—in some ways they still continue today—but did so at a consistently ferocious level of intensity, sustained partly by an incendiary advertising campaign. *Why Johnny Can't Read* became one of the best-known titles of the century; it is tell-ing, I think, that when I told colleagues in 2020 what I was working on, almost all of them referred to it spontaneously, although none had ever read the book. What drove that success was in large part the urgency and brilliance of Flesch's prose. He had learned his own lessons about clarity and readability well: the book was lean, clear, fluent, and punchy. It mounted a rhetorically devastating case against a culture of teaching that, Flesch claimed, was failing American children by the millions.

Flesch's charge that schools were not teaching children to read at all rested on his view of what reading itself was. He claimed that pupils were no longer being educated in the recognition of letters one by one but were being exhorted to "guess" words from their overall shapes. To that extent, his claim was an exaggerated version of one that had been made before; others had long disparaged the "whole-word" or "look and say" style of pedagogy. What set Flesch's polemic apart was its explanation of what had gone wrong. The disaster, he reported, was the outcome of a histori-cal process. That process had played out since around 1920, in what was virtually a conspiracy by experts against citizens. He largely blamed Ed-mund Burke Huey for it. Huey, Flesch maintained, had taken the relatively moderate claims of earlier researchers like Dodge and Cattell and, writing "as if in a fever," ramped them up into a crusade against the teaching of "what each letter stands for." But he had not been alone. The textbook industry had embraced his crusade in the mid-1920s, with Gray's Dick and Jane series in the vanguard. Since about 1925, the whole-word ideology of Huey had taken over, thanks to this industry. At much the same time, the "fad" for teaching silent reading—again, on the grounds that because adults did it, children must be taught to do it too—reached its peak. And Arthur Gates of Teachers College—who, incidentally, had still been teaching there when Flesch was a student—had delivered the coup de grâce by devising the cunning expedient of relegating phonics to "incidental" status and designing a test to prove the relegation justified.[9] Eventually, by the 1930s it was simply impossible to buy elementary books that were *not* written on the whole-word principle.

Children were not only subjected to the non-teaching of reading, then: they were also restricted to the use of "horrible, stupid, emasculated, pointless, tasteless little readers," in which "dozens and dozens of totally unexciting middle-class, middle-income, middle-I.Q. children's activities" occurred simply in order to give the books a chance to recite phrases like "See the funny, funny animal" over and over again. These books—and the industrial system that gave rise to them—were to Flesch "the nub of the whole problem." An oligarchy of companies made millions every year by selling such books, and the whole-word system meant that schools had no choice but to buy them, because in that system students were restricted to specific lists of words drawn up by experts in the science of reading. Pupils could not be given just any books to read. And an unholy alliance endured between capitalist publishing and spurious science, because the industry's books were almost always compiled under the well-remunerated authority of some credentialed academic expert (especially William Gray, in the case of Dick and Jane) who oversaw teams of authorial drudges.[10] It was all very neat. The whole-word method created a closed market that the companies and the experts could exploit forever.

So Flesch did not simply allege that the prevailing practice of teaching was inane and corrupt: he provided an account of how it had become that way. The credibility of the charge rested on the plausibility of the account. In that light, it is important to note that he was careful not to place initial responsibility on the science of reading at all, perhaps partly because he held it in such contempt that he wanted to avoid granting it efficacy of any kind. Flesch argued that the whole-word method had not in fact originated in laboratory psychology—it simply suited the experts to claim that. It had actually come into being thanks to a homely episode recounted in John Russell Webb's *New Word Method* of 1846. Webb told how he had hit upon the method by chance, when he and a young girl learning to read had encountered a cow just as she was looking at the printed word *cow*. Flesch preferred this tale to the stories about kymographs and eye cameras told in earnest research monographs. As Webb's approach subsequently gained traction, he explained, the traditional letter-by-letter method had become encumbered with defensive "scientific" measures to hold off the threat, such as elaborate diacritical markings. The paradoxical result was that whatever it gained in scientific rigor it lost in intuitive appeal, and by the 1920s the battle was almost lost. The psychologists—Dearborn, Dodge, and above all Huey—had merely delivered the coup de grâce.

But it did not need to be this way. Flesch claimed to have been inspired to start his investigation of the whole issue by coming across the preface of an unpublished alternative primer drafted by a renowned prewar structural linguist, Yale's Leonard Bloomfield. Bloomfield's work proceeded on quite different principles—which meant, on Flesch's account, that it was no wonder that he had been unable to find a publisher for it.[11] Flesch particularly seized upon what he saw as his unalloyed embrace of teaching by the sounds of individual letters. He reckoned that this approach—"phonics"—represented genuine science, as opposed to the self-interested and superficial impostor found in prevailing pedagogy. In point of fact, his enthusiasm was a little excessive. It was true that Gray had used part of Bloomfield's argument to exemplify the dangers of a crude phonics approach, but in fact Bloomfield was much more equivocal in his recommendation than either he or Flesch registered (and Gray was much more equivocal in his critique of it).[12] But Flesch required a Manichean state of "warfare" between "entrenched 'experts'" and "the advocates of common sense" for his polemic to work. He was determined to distinguish categorically between real expertise, which he reserved to linguists and psychologists, and specious quackery, which he attributed to the scientists of reading and their followers. He claimed that the latter both dominated the pseudo-discipline of educational psychology and filled the faculty ranks at its pseudo-institutions, the teacher-training colleges. There they indoctrinated future teachers by means of textbooks written by the same soi-disant experts who profited from the primers.

Perhaps the biggest sign of how spurious this pseudoscience was, in Flesch's eyes, was its assumption that the results of experiments based on adults should be applied to children. Again, this was exaggerated—the early researchers had often included schoolchildren among their subjects—but when it came to pedagogy the charge was accurate enough to be damning. Its point of application was the central principle of the science of reading ever since Cattell: the claim that reading involved grasping whole words. The science of reading asserted consistently that fluent adult readers proceeded this way. Even if they did, Flesch maintained, there was no reason to expect that children could or should leap straight to the same practice. But he also refused to believe that even adults, if properly taught, really did seize on whole words. He argued that a child trained to read phonically would then acquire a habit (he agreed with the psychologists that far), and would go on reading words letter by letter throughout adulthood. "He does this fantastically fast," he conceded, "and quite unconsciously," but this was what happened. He simply denied flatly the claim that mature readers operated by larger units. A capable adult reader had "never in his life" read a single word as a unitary whole, Flesch argued. So the method of "word guessing," as he pejoratively called it, might well teach a certain habit, but it was not a habit of *reading*—for children, and for adults too.[13]

But what about eye movements? Flesch recognized the richness of that tradition, but in answer to it he took up the skeptical arguments that had been mounted by the psychologist of typefaces, Miles Tinker. For any child, he wrote, the eye movements simply corresponded to whatever reading method that child had been taught. Regressive movements were symptoms, not causes, of poor reading (Buswell had made a point of saying that too, of course); and he implied that they arose from a training in word-scale guessing. Remedial work, he then continued, typically involved "retraining the eyes" to reduce regressive movements, on the assumption that "the mechanical gadgets, the films, the tachistoscopes, the reading accelerators, and so forth" would "teach the eyes to do a better job." But, far from helping, on Flesch's account tachistoscopes and the like did the wrong thing. They sought to increase fixation spans, when increasing spans merely meant more and faster word guessing—that is, it strengthened a bad habit rather than remedying it. Even if it worked in the sense of speeding up the process a little, it would produce less accurate reading than before. Regressions were much more important, but in truth *any* improvement in reading would reduce the number of those, because regressions were symptoms of confusion. An improvement brought about by phonics would not affect the fixation span at all, then, but that did not matter. What it would do was reduce the regressions, because it removed the confusion, while at the same time also lessening the durations of pauses. The student would "learn to *see* the letters on the page more quickly."[14]

Even if one accepted that there was a crisis, that it had historical roots, and that phonics provided a better remedial strategy than the tachistoscope, there still remained the problem that the whole scientific community—well beyond the ranks of those profiting directly—was united in consensus on the whole-word approach. Why? Flesch claimed to have spent two days at Teachers College trying to track down the answer. He was looking for the scientific basis for the whole-word method. In essence, he found that no such basis existed. Every empirical study cited in the literature, he announced, either confirmed the superiority of phonics or shriveled under his skeptical analysis. On the other hand, there *was* evidence that his preferred strategy worked: he was able to tout a number of Catholic schools in the Chicago area that were apparently succeeding by means of phonics methods. Yet somehow an entire scientific community—not to mention the truly massive industry that was public-school teaching—had managed to explain away negative experiences and conjure positive results out of thin air. Not that this was necessarily a product of dishonesty as such: the disconnect had created anxiety, to be sure, but scientists and teachers had managed that problem by inventing concepts to explain experiences away. The most significant of them was the much-touted notion of "reading readiness."

There was a large literature by 1950 on the concept of reading readiness. An early and influential foundation had been established by Arthur Gates and his friends, who by the mid-1930s had already deployed "every type of test . . . which the authors could think of" to try to measure it. Flesch accounted the concept "the holy of holies, the inner sanctum of the whole 'science' of reading." There was a reason all those efforts to measure it had failed to come to a definitive result. It was in his view simply a kludge, adopted on an ad hoc basis to get around the awkward fact that so many children did not do well in the whole-word system. In no other country in the world, he claimed, did the concept of reading readiness even exist. He insisted that there was no such thing in nature—except in the sense that it referred to the readiness of the teacher to let a child read. In the end, he conceded that while absolute certainty was unachievable—there would always be demands for more experiments and more tests—in practice the evidence was unambiguous. "The record is perfectly clear," Flesch concluded. "Our 'scientific' educators simply don't *want* to know the truth."[15]

More to the point, the scientists did not want *you* to know the truth. A major reason for that was that they were not real scientists anyway. They were nervous of being found out. Only two kinds of experts should be paid attention to when it came to reading, Flesch told an imaginary schoolteacher: linguists (he claimed that all of them were on his side) and psychologists. And by psychologists, he emphatically did not mean "teachers' college professors who happen to be members of the American Psychological Association," but "scholars whose main work is the study of the human mind." The pseudo-experts who generally informed pedagogy were in his view charlatans. As evidence, he remarked that they tended to assert that the whole-word method was based on Gestalt psychology. Real psychologists, he said, would know that this was completely wrong. A genuine expert would know that the point of Gestalt psychology was not to focus on wholes as such, but to understand the moment when the mind appreciated how things fitted together into structures. A genuine Gestalt approach to reading would emphasize precisely the joining-together of letters into words. The whole-word method was starkly opposed to this. What it was really based on was a very different, and in Flesch's portrayal much more sinister, kind of science. And here he got to the central point of his book—the real reason why he was so indignant. After a passing reference to Rousseau's Émile, he revealed the truth. Whole-word teaching was, he announced, Pavlovian conditioned-reflex psychology in action. Just as Pavlov had trained dogs to salivate at the sound of a bell, so psychologists in awe of him had realized that children could be trained to associate words with things. And the horror of it was that they had had some warrant, because in fact it *was* possible to get a child to associate the word *chicken* with a chicken. But the crucial point was that in Pavlovian methods the child was never told *why* that link was made, any more than Pavlov explained associationism to his dogs. "Conditioning," then, was inherently and ineluctably "an authoritarian process." In essence, it was not *teaching* at all, but "animal training." It was, Flesch declaimed, "inhuman, mean, stupid." In short, "the word method consists essentially of treating children as if they were dogs."

In effect, then, Flesch's book should be seen as a bravura demotic performance of a certain kind of sensibility. That sensibility was of a piece with cognitive scientists' declarative rejection of behaviorism. It was in this context that he insisted that his stance not be identified as reactionary. He did not accuse the scientists of reading of being Communists, he pointed out. On the contrary, he reminded them that he himself had a PhD from Teachers College, virtually the epicenter of progressive educational policies, and he avowed that he admired Dewey as much as anyone. But the Dewey he invoked had held that education should be "democratic, free of senseless formalism and drill, based on interest and meaningful experience, and inseparably joined to the real life that goes on around the child." And so Flesch insisted that he was a "liberal" and that his position should be counted as progressive. "There is a connection between phonics and democracy," he continued, and this was "fundamental." The right of equal opportunity for all was supposed to be inalienable in America, and yet "the word method interferes with that right." Nor was this merely a question of moral philosophy. In practice, he pointed out, what happened in the 1950s was that wealthy or cultured parents surrounded their children with tutors and books, and those children learned to read at home. The children not fortunate enough to have that home life were the ones who had no other option but the whole-word method taught in schools, and they fell by the wayside. The "American Dream" of equal opportunity via free education for all—the dream with which we began this book—was consequently vanishing, and the whole-word method was entrenching inequality. In the end, the reason to rebel against the whole-word method—and against the pseudoscience of reading that upheld it—was that it was "destroying democracy in this country."[16]

That concluding flourish suggests the reason why I have spent several paragraphs on Flesch's argument in *Why Johnny Can't Read*, a book that is generally assumed to be distinctly one-dimensional. There was a lot more to it than that. Oddly enough, the most appropriate counterpart to the abrasive Flesch may actually have been the congenial, urbane, and erudite Jerome Bruner. Although Flesch was certainly no cognitive scientist himself, his diatribe had a good deal in common with the rejection of prior authorities common to the postwar generation of social scientists like Bruner. In a sense, he should be added to the list of intellectual reformers given in the previous chapter, although he was clearly a very different kind of cultural presence from Bruner, Moore, and even McLuhan, and even though he was angry about a very different level of institution. Like them, he was rebelling against a doctrinaire scientism; they called it behaviorist, he called it Pavlovian. They all associated their targets with authoritarianism, conformity, mediocrity—to co-opt Alexander Pope's goddess of Grub Street, with Dulness. Moreover, a consideration of the politics of Flesch's campaign complicates any account of the politics of the social sciences and cognitive sciences too. To place his diatribe in their company forces us to recognize that the consequences of information theory, cognitive science, and linguistics extended to the everyday formation of American citizens in schools across the country.

Those stakes were clear from the outset. Flesch's polemic immediately caused a stir in the publishing industry, which had long enjoyed the profits from selling elementary reading primers. In particular, in spring 1955 the education division at the publisher Houghton Mifflin was headed by William Spaulding. Spaulding was the son of the inaugural dean of Yale's Department of Education, Frank Ellsworth Spaulding, who had long been a participant in the debates over children's reading; in the early years of the century Spaulding senior had devised his own series of Aldine Readers, a key principle of which had been the use of rhyme to teach beginning readers, and he had published a guide to assessing the quality of rival primers. In the immediate wake of Flesch's sensational success, William Spaulding seems to have remembered that emphasis on rhyme. He summoned an old army friend to Boston to talk over an idea he had had for a response. Spaulding wanted Ted Geisel—whose pen name was Dr. Seuss—to write an elementary book that would answer the challenge mounted by Flesch. It must have a limited vocabulary (he suggested a list of 225 words) and be so captivating that a child would refuse to put it down. Geisel went home and considered the idea. He spent months poring over Spaulding's word list. At a loss for ideas, in the end he fixated on the two first words on the list that came to his attention as rhyming:

cat and *hat*. The result was *The Cat in the Hat*, which appeared to immediate and universal acclaim in 1957. As the book's jacket declared, children would "discover for the first time that they don't need to be read to any more."[17] In terms of its sheer influence in fomenting reading habits across the population, *The Cat in the Hat* would have a fair claim to be considered the most important poem published in the twentieth century.

Rudolf Flesch was certainly not the first critic to question the scientists of reading. Nor was it unusual for the methods of schoolteachers in this field to be called into strenuous doubt. After all, many of the contributors to Buswell's and Waples's versions of the science of reading in the Depression had justified their efforts precisely by pointing to statistics showing that adult Americans' reading rates were dismal, and they had blamed poor teaching, in part, for a problem that they warned endangered the national well-being. And the rapturous reception that *The Cat in the Hat* enjoyed proved a pent-up appetite for something new. Yet somehow Flesch's polemic touched off a paroxysm of anxiety far greater than anything sparked by any earlier revelations. Part of the reason for this was surely Flesch's own rhetorical skill. Another part was the fact that his account treated the whole-word method as not merely false but authoritarian. That claim spoke loudly to Cold War audiences primed to recognize un-American infiltrations. But perhaps the greatest single reason for the furor was that the Cold War suddenly became a much more immediate presence in American lives.

On October 4, 1957, the Soviet Union successfully launched the first artificial satellite to orbit the Earth. Sputnik was probably the USSR's greatest Cold War triumph. The experience of having a potentially hostile object pass overhead seven times every day, without any possibility of defending against it, was profoundly disquieting to Americans across the political spectrum. And its implications for geopolitics were scarcely less momentous. The USSR seemed suddenly to have "taken over world leadership in science," as the *New York Times* put it. And at the same time the United States could not even produce a literate populace, let alone a scientific one! Such fears took popular shape thanks to repeated expression in the press, as well as in bestselling books such as Arther Trace's *What Ivan Knows that Johnny Doesn't* (1961), which argued that in the US system "a textbook is in fact a tyrant" worse than anything in the USSR.[18] As it accelerated efforts to launch America's own spacecraft, then, the federal government suddenly found itself faced by intense and multiply-ing demands to transform a sleepy and complacent education system, and to do so from top to bottom. The very day that Sputnik reached orbit, moves began in Washington to recast education as a matter of defense. In 1958 the major legislative result of the crisis, the National Defense Educa-tion Act, was passed.[19] The NDEA drastically increased the involvement of the federal government in higher education, in the name of improving science in the face of Soviet competition. But if America had too few sci-entists and mathematicians, it seemed obvious that the problem must lie not only—or even mainly—with its universities, but with the people, places, and practices that educated children earlier in their lives. So pub-lic anxiety also focused on schools. Suddenly everyone agreed that major reforms were essential here too. If Johnny could not even read, how could he ever become a space scientist?

Flesch's fusillade soon found supporters in this new environment. One of the most vociferous was Charles Walcutt, whose *Reading: Chaos and Cure* (coauthored in 1958 with Sibyl Terman, daughter-in-law of Lewis Terman of IQ testing fame) and *Tomorrow's Illiterates* (1961) laid out par-allel critiques. Walcutt then went on to trump Flesch by creating a com-plete reading program for the publisher Lippincott. By 1959 the National Conference on Research in English, prodded by William D. Sheldon, was convening a committee on research on reading, which met at Syracuse for three days in 1959 and then again in Chicago in 1960. The participants agreed that the state of knowledge on how children began to read (as opposed to that on mature reading) was lamentable—so much so that nobody could even tell whether Flesch had been right or wrong. At that point, conversations between a linguist, James Soffietti, and Jeanne Chall and Ralph Staiger of the International Reading Association led Chall to propose a critical analysis of the entire body of existing research on read-ing pedagogy. Designed in part to address the contentions that Flesch and his followers had made, the resulting survey would be the most im-portant account of educational impact of the science of reading ever undertaken.

Like Flesch, Chall was of central-European origin. She had been born in Poland in 1921 and her family had brought her to New York seven years later. She had worked in the Bureau of Educational Research at Ohio State in the immediate postwar years, where she became a convinced advocate of "readability." (We have come across her already in that context.) She and her supervisor, Edgar Dale—the same Dale who had worked on the study of African American readers in Ohio that had been so influential in establishing the need for a science of readability in the first place— formulated a rival test to Flesch's to measure the readability of any given text.[20] By 1961 she was teaching at City College in the CUNY system and had become a director of the International Reading Association. Before long Harvard would hire her to create a Reading Laboratory for the university. The Carnegie Corporation agreed to fund her proposed investigation, and with its support Chall embarked on more than two years of interviewing teachers, reading assigned books, and analyzing existing research in the science of reading. She also traveled to Britain to explore practices there. The outcome was a two-volume mimeographed report that she first submitted to the corporation in 1965 and then published in revised form two years later as *Learning to Read: The Great Debate*.

Chall's *Learning to Read* immediately became the standard authority on the ways in which the science of reading affected teaching practices, and the de facto starting point for all sides in later struggles. It included a survey of some thirty experimental studies published in the twentieth century on rival teaching approaches, which showed that Flesch had a real point. Overall their consensus was overwhelmingly in favor of including at least some phonics instruction. Her major conclusion was that as a method for *beginning* readers, perhaps until third grade, teachers should embrace a greater emphasis than they currently did on letter-by-letter work. She called the latter the "code-emphasis" approach, and in fact it was endorsed by a decorated NSA cryptologist, Mitford Mathews, who published a detailed history of teaching approaches at around the same time.[21] But there were several different kinds of code work, and which was best remained unsettled. Nor did Chall advocate eliminating meaning-based approaches altogether—fearing, as she put it sardonically, that that would inevitably lead to the publication in 2010 of a book called *Why Robert Can't Read* that eviscerated the artificial and ideological code approach in favor of a "natural" focus on meaning. That fear reflected the other essential point of her report: that the Manichaean binary that Flesch had portrayed was a myth. In practice, schools almost never adopted a method that was either all phonics or all "look-say." The question, therefore, was one of balance.

Yet Chall discovered that virtually no work had been done to assess where to strike that balance, or even by what criteria to make the determination. In place of that, the whole debate had taken on an "ideological" character, reminiscent to her of debates in religion or politics rather than "science and scholarship." She blamed the moral panic over Sputnik for most of the passions, in addition to fears that automation would destroy blue-collar jobs. The shenanigans of the publishing industry were another factor. Publishers' sales representatives could be extraordinarily forceful, adopting tactics such as the spreading of rumors about neighboring schools stealing a march by adopting their latest curriculum. But in general charges and countercharges flew so incessantly that it was impossible to know where a responsible solution lay. "Critics of the prevailing methods use such accusations as attention-getting devices," she warned, "but few are willing to commit themselves to a face-to-face evaluation." She suggested that publishers at least should be required to test their textbooks' outcomes, in the way that pharmaceuticals companies were obliged to test new medicaments.[22] But the problem ran through the entire culture. Teachers, scientists, and parents alike were reluctant not only to concede that different approaches had virtues but also to notice that in practice they were not all that different in the first place.[23]

The most important thing to know about the science of early reading, Chall warned, was that it was so "shockingly inconclusive" that statements could be plucked from it to "prove" anything that a combatant wanted. It was not that there was any lack of research as such: by the mid-1960s more than a thousand studies were being published every year. Indeed, one major problem was the sheer mass of it. As experts had warned about a number of fields of research in the postwar era, some kind of computerized storage and retrieval system would be essential to deal with it all. Chall suggested that it be combined with a published digest to be issued every month or so. (Indiana University was about to launch such a system under the federal government's aegis, she announced.) Lacking such a system, it was impossible for anyone to read the science of reading.[24] Another factor was institutional. Schools of education were not research establishments, and they lacked the networks and resources to sustain sophisticated investigatory traditions, which meant that earlier generations of experiments had often been better—more theoretically grounded, more sophisticated—than recent work. She respected

Buswell's work, for example, despite its statistical naivety, because it was more attuned than later studies to the specifics of how individual readers and their teachers acted. But by the same token, that earlier work had been "parochial," and its practitioners had made claims beyond anything their experiments could possibly have warranted. She chided Huey in particular for recommending a later start to reading and an emphasis on meaning, decades before psychologists might be able to settle those questions. However, recent research too tended to be directed at solving some very particular problem but then invoked as if it proved much larger points. Since Huey, there had never been a synthetic account of early reading that might have served as the fount of a progressing research tradition. The result was that there were no shared criteria, and it was common to see options airily recommended as "more interesting" than rivals. Authorities like William Gray had felt able to disregard what others counted as experimental evidence, in the way that Thomas Kuhn said that scientists wedded to old paradigms could never be compelled logically to abandon them. But on the other side, although supporters of phonics tended to claim that any improvements in classes that adopted their ideas were consequences of those ideas, such improvements also correlated with reductions in class size, increased provision of books, team-teaching, and more. In effect, *no* method of teaching had *ever* been tested, at least in any properly scientific sense of the term "test."[25]

By this time, contests over the teaching of reading had become highly politicized. The movement to reform pedagogy was strongly aligned with conservative groups advocating lower taxes and a leaner, narrower, more traditional idea of education. Interest groups like the Reading Reform Foundation (started in 1962) were associated with private schools and opposition to progressive education. And it was primarily middle-class and college-educated parents who were active on that side of the controversies, largely because they were fearful that their children might not get into good colleges: activists generally resided in suburbs rather than in the poorer cities. Even the faddish Montessori system, Chall noted, had changed from an innovation in Italy meant for poor families and handicapped children into a favorite of the rich and favored in the United States. It was no accident, she added, that Omar Khayyam Moore's "computerized typewriters" had been installed first in a suburban private school already reputed for high standards. (And we may recall from the previous chapter that Richard Kobler had reported on the system's success to the Reading Reform Foundation.) These social and political alignments, she suggested, had been counterproductive. Public-school staff and parents had certainly noticed them, and they had therefore grown inclined to resist changes that they understandably saw as of a piece with aggressively conservative policies.[26]

Chall devoted considerable attention to Moore and his autotelic environments, as perhaps the most notable example of a novel approach that might, in principle, supervene such divisions. She described the Talking Typewriter laboratory and the child's experience in it from witnessing it at first hand, and declared that she found the abilities it inculcated to be among the most impressive of any group she had encountered. Unlike most experts Chall interviewed, Moore built his venture on the principle that a child was naturally an explorer, and that rewards must come straightforwardly from discoveries, not from teachers recognizing successes. She also noted that he was unique in rejecting completely any notion of "reading readiness," believing instead that one should start to read as early as two or three. She associated the approach with that of linguists like Bloomfield who favored her own "code-emphasis" approach, albeit as part of Moore's broader "theory of problem solving and social interaction."[27]

As Chall described it, then, Moore's idea of combining a code emphasis with an autotelic environment offered a way forward, even as the "great debate" was polarizing into these stubborn political camps. The debate was not only about learning to read, now, but about what Chall called "philosophies" of education. If one squinted, then one might see those philosophies as progressive and traditional; casting the distinction that way served their alignment with established political factions. But a different possible mapping would render the differences in terms of cognitive psychologies on the one hand and training regimes on the other. And the cognitive approach had the potential to appeal to teachers and parents alike, precisely because of its deep-seated, foundational association with freedom, individual autonomy, and creativity. So Chall was somewhat optimistic in spite of what seemed a debilitating and intractable conflict. The basis for her sanguinity was that the terms of the debate were changing to those of cognitive science. The commitments of information science, cognitive psychology, and psycholinguistics—partly developed, as we saw in the previous chapter, out of a reappropriation of experimental practices from the old science of reading—were intrinsic now to both technological ventures like Moore's Talking Typewriter and educational initiatives like Bruner's and the MACOS program. Diehard phonics warriors (including Flesch himself, who would restate his convictions in a 1981 book) might see any innovations on these lines as a retreat, conjured by a "conspiracy for sight methods" masterminded by textbook publishers and the International Reading Association. But she felt confident that intransigence would not endure in the face of such exciting innovations. The combination of liberty, creativity, and technology sounded like one that would appeal across otherwise impervious political boundaries.[28]

Needless to say, she spoke too soon.

Psycholinguistics and Guessing-Games

In 1969, shortly after moving from Harvard's Center for Cognitive Studies to Rockefeller University, cognitive psychologist George Miller addressed the American Psychological Association as its president. The times were critical and revolutionary, he told them, and psychology should be playing its part. In Miller's view, what that meant was that academic psychologists should give up the relationship with the public that they had clung to since the formation of the discipline. No longer should they see themselves as authoritative founts of knowledge, from whom lay communities should take guidance. That had been the old way—the way associated with aspirations to control human behavior, and with political authoritarianism. Instead, from now on psychology must become something practiced by nonpsychologists. "The people at large" would generate and use knowledge for themselves, in the forms they needed. Only when this happened could a "peaceful revolution based on a new conception of human nature" take place. So the responsibility of the professional psychologist in this moment was simply to take disciplinary expertise and "give it away." And the best way he could think of to do that was to democratize the science of reading. It was time for the old model, in which august psychologists decided on methods and authored the books that teachers must use, to be jettisoned. If the role of science in this field could be got right, then its role in society at large might well follow.[29]

As a veteran of the cognitive-science movement, Miller had long been a combatant against behaviorist science's autocratic associations. By 1969, however, with America stuck in the Vietnam War and the high-command of social science represented in the RAND Corporation and Robert McNamara's meticulously soulless planning exercises, those associations seemed more consequential than ever. In the face of this, his example of the science of reading as a good field in which to forge a re-calibration of the place of social science in American culture was well chosen. A new consensus about reading seemed to be forming, in the wake of the conflicts launched by Flesch and surveyed by Chall. As Miller no doubt knew, the new approach sought to supplant the old way of Gray and Gates, which it decried as high-handed and personally damaging. In their place, it emphasized the kind of "autotelic" experiences that Miller referred to. And it used Miller's own discipline, psycholinguistics, to articulate its role in realizing the kinds of hopes he voiced.

In the mid-1960s a new cognitive account of the acquisition of reading skills came to the fore. It was often cited against Chall's preference for a "code emphasis," particularly by those who sensed in Chall's advocacy a surrender to the Flesch camp. The new account centered on what were (inauspiciously) called "psycholinguistic guessing-games." That phrase originated in a talk presented at the American Educational Research Association in New York City in February 1967 by Kenneth S. Goodman, a politically engaged educationalist then based in Detroit.[30] Goodman had not met Miller, but by this point he had become a convinced acolyte of Miller's one-time colleague and ally, Noam Chomsky. His notion of reading would become first a vehicle for a thorough reform and then the target of a vitriolic and sweeping reaction.

Goodman had earned a doctorate in education at UCLA, where he had started to wonder about the linguistics of reading. From 1962 he held a faculty position at Wayne State University in Detroit; he would later move to the University of Arizona and become perhaps the most prominent advocate of "whole-language" methods through the end of the century. While he was in Detroit, Goodman undertook a study involving one hundred children of various ethnicities. He became fascinated by what he called the "miscues" that these children made while reading. These small and swiftly corrected "deviations" from the text were constantly being made and almost immediately qualified. Such endemic phenomena should not be dismissed as errors, he decided, let alone treated as mistakes to be corrected by a teacher. They must be seen as part of the learning process itself. So he concluded that both the decades of science involving eye fixations and what he called "our mania for devices which eliminate regressions in reading" were based on a shared and "lamentable" scientific failure. Far from being errors to eradicate, regressions and pauses were the major means by which a child learned. Goodman reasoned that teachers should abandon the concentration on letters and words—phonics and look-say—and instead develop a pedagogy based on this active engagement with language.[31]

Helen Robinson of the University of Chicago—heir to Judd and Buswell's program—heard about this idea and forthwith got Goodman appointed to a new Committee on Linguistics and Reading founded under the aegis of the National Council of Teachers of English (NCTE) and the International Reading Association (IRA). As a result, in 1966 Goodman spent the summer at Cornell with Harry Levin's federally funded Project Literacy, an endeavor to use cognitive tools to understand the psychology of learning. There he encountered Chomsky, who was visiting from MIT. Chomsky had an immediate and radical impact, as he often did when he first met other scholars. He suggested in conversation with Goodman that he reconceive of reading as what Chomsky called "tentative information processing," and it was this notion that Goodman then glossed as a "psycholinguistic guessing game."[32] The point came from a combination of the cognitive revolution with Shannon-style information theory. Chomsky had earlier presented an account of listening on these bases, and Goodman felt that his story could be told just as well about reading. A reader, like a listener, was always trying to make use of a "signal" to reconstruct an original "message." What happened when one read, he concluded, was nothing like the sequential, letter-by-letter or word-by-word perceptive acquisitions that scientists of reading and their enemies had portrayed for decades. Instead it was a process of rapid informational feedback. It involved constant anticipating followed by confirmation or qualification. The anticipating itself might feel instantaneous, automatic, unconscious, and intuitive, but it was in fact shaped by experience, language, and "thought development" over time. And it was about "phonological or oral output," not meaning. What a reader "guessed" was a linguistic entity and not, initially at least, a semantic one.

Meaning was something arrived at by a distinct "decoding" process that involved Chomsky-style "sampling."

Here, Goodman decided, was a model that could explain reading far better than all those tachistoscopes. Tachistoscope experimenters adhered to what he called an "end of the nose" view of reading, according to which a reader took in information from a small space around the point of the eye's fixation and moved sequentially, picking up more meaning at each pause. Dismissing this as inadequate because it could not explain such everyday events as an adult author repeatedly missing the same misprint while proofing a portion of text—a notably Bruneresque point—Goodman retorted that real reading involved not one, but three kinds of information. Readers certainly relied on "graphic input"—the characters on the page. But they also used syntactic and semantic information, drawn from short- and long-term memories, to project dynamically an account of each sentence. A reader was constantly "sampling from the page just enough to confirm his guess of what's coming." What constituted *skill* in reading was therefore not greater perceptual precision, as such, but the ability to make good initial guesses, so that one required fewer visual cues to arrive at stability. The process of learning to read well, then, was one of learning which cues to select, and how best to produce guesses based on those cues and one's prior experiences. Again, this resembled very closely how cognitive psychologists like Miller understood the process of learning spoken languages.

Goodman came to represent a new generation, which dismissed both candidates in the previous decade's conflict on the basis that they rested on a false, if commonsensical, idea of perception. The practice of reading was now to be seen as a "selective" exercise, in which the reader acted as a very fast-moving experimenter along the lines of the students Moore envisaged in developing his autotelic environments. Taking in "minimal language cues" from the page, the mind processed them as "information" and made tentative decisions about meaning; those decisions were then bolstered, rejected, or refined as the reader proceeded. And despite Goodman's disdain, tachistoscopes and eye cameras could be called into action in support of this account. Paul Kolers of MIT—champion of the reissuing of Huey's work—seized the opportunity to use both, showing rotated letters in a tachistoscope to confirm that correct perceptions were not sufficient, and using eye-movement cameras of the Buswell type to demonstrate that fast readers did not fixate in any regular pattern on the page. The combination proved to him that a good reader "tells himself a story" based on "clues" on the page. The key to skilled and fluent reading, consequently, was not precise perception of every character or word at all. It was almost the opposite. Learning to read was a matter of developing "the ability to anticipate that which has not been seen." The best readers had the ability to select the *fewest*, most productive "cues" from the page, and to make the best "guesses" with them. So the practice of teaching should focus much less on perceptual accuracy and much more on "the clue-search and information-extracting characteristics of reading"—which, Kolers argued, was more or less what Huey had recommended back in 1908. Thanks to the cognitive scientists, he concluded, the teaching of mathematics had recently been transformed; it was high time, after five hundred years, that the teaching of reading have its own "revolution."[33]

Psycholinguistics thus suddenly became the most important trend in the science and pedagogy of reading alike. Unfortunately, it was also, as the president of the International Reading Association lamented, the most confusing. Psycholinguistics was a highly technical and abstruse field, which could generate extraordinarily complicated models. It is unclear, for example, what a hard-pressed elementary-school teacher was supposed to make of Berkeley scientist Robert Buddell's diagram summarizing the learning process in reading (fig. 8.1).[34] There was consequently enormous scope for simplified and commercialized versions. Just as the reading accelerator fad retreated and Evelyn Wood's speed-reading enterprise lost credibility, so schools and homes found themselves "flooded" with "materials labeled 'linguistic.'"[35] Goodman and others contributed to this inundation. So did the British-born and Australian-educated Frank Smith, who, even more emphatically than Goodman, took approaches from Harvard's Center for Cognitive Studies and applied them in this domain. Smith earned his PhD from Harvard in Psycholinguistics in 1967, and four years later in *Understanding Reading*—a title that echoed McLuhan's *Understanding Media*—he provided the most influential single treatment of the subject. He frankly confessed that the book was a product of his having been "indoctrinated" by the "heady psycholinguistic atmosphere" around Chomsky, Miller, and Bruner, and he dedicated the book to Miller. And Smith enthusiastically endorsed Miller's point about the need for everyman to become a psychologist. The way to learn to read,

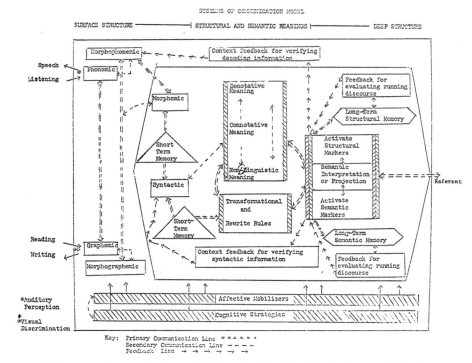

8.1 A "systems of communication model" for the teaching of reading, based on psycholinguistics assumptions. Robert B. Ruddell, "Psycholinguistics Implications for a Systems of Communication Model," in *Psycholinguistics and the Teaching of Reading*, ed. K. S. Goodman and J. T. Fleming (Newark, DE: International Reading Association, 1969), 73.

he argued, was by actually reading—in the same way that the best way to learn how to ride a bicycle was to ride a bicycle. The professional scientist could no more produce a set of rules for the one than the other. So he was quite reticent when it came to recommending specific methods of teaching. Smith rehearsed the long canon of literature on eye movements—all the way back to Cattell—to inform teachers of what, he believed, was their most important task: that of providing rapid, considerate feedback to pupils as they learned. But beyond that he said little beyond noting the complexity of the problem and the legitimacy of teachers' experiences. Convinced as he was that psycholinguistics did represent a clear advance on prior approaches, he expressly disclaimed any intent to start "a classroom revolution."[36] In fact, the advantage of psycholinguistics was precisely that it did *not* imply some specific technology or prescription in the manner of the Gray readers.

The psycholinguistic science of reading championed by Smith, Goodman, Kolers, Miller, Chomsky, and others seemed set to dominate at the start of the 1970s. It benefited from a third great expansion in educational funding. The "Right to Read" became a ubiquitous slogan for the new decade, as Smith remarked, "replacing moon landings as the focus of a monstrous national effort."[37]

The phrase *Right to Read* had long been used to resist censorship efforts, and it deliberately echoed civil rights campaigns for African Americans to gain access to libraries. But in 1969 it was taken up as the label for a grandiose national campaign. The New York Commissioner of Education, James E. Allen, moved to Washington that May and became the federal Commissioner. He immediately began agitating for a national reading program. Convinced that it would be popular, President Richard Nixon gave his approval, and Allen announced the initiative in September in a speech in Los Angeles. He likened the "Right to Read" campaign to the Apollo program—a link that would become a cliché over the next few years—and with good reason: the administration's promise that by the end of the 1970s "all children would be reading properly," depending on what one meant by *properly*, might well be as hard to realize as a lunar landing. Patricia Nixon, the First Lady (and a one-time high-school teacher), became honorary chair of the campaign.[38] But the initiative was slow to get going, and in June 1970 Allen was forced to resign after criticizing US actions in Cambodia. Two months later, a forty-member National Council was finally appointed under the chairmanship of Walter Straley, a vice president at AT&T. Soon, Ruth Love Holloway, an energetic African American educator and administrator, was appointed as director of the overall effort, which she too likened to the Apollo program, while rebutting perceptions that reading itself was obsolete in "the McLuhan world." More than one million dollars was assigned for the first year of work. Right to Read "Task Forces" were then established in states like New York, and it was announced that a "national reading center" would be founded, under the leadership of Louis Gonzaga Mendez Jr., a decorated World War II paratrooper and educational administrator in Virginia. The new national center would conduct experimental research and coordinate scientific programs. Its first call, Mendez and Straley agreed, would be to devise "a yardstick to measure reading deficiencies"; after that it would move to create TV programs and publications "to help people read better."[39]

The new program was an entirely characteristic move for its time. In the wake of physics, cancer medicine, and the space program, the science of reading, newly equipped with a modern, prestigious, and socially consequential theoretical basis in cognitive psychology, was finally set to achieve a political and institutional milestone that had never seemed possible before. To be sure, the amounts of investment involved were minuscule by the standards of those mammoth endeavors. But that was not the point. For the first time in its history, the science of reading was becoming a national endeavor with the full support of the federal state to bolster and unify it. And it had the ear of the president. In that sense, at least, it was about to become Big Science.

The Reading Wars

In 1969 the science of reading became one of the branches of the scientific enterprise that the federal government looked to for guidance in forming social policy. For the next fifty years, successive initiatives (roughly one per decade) would be launched from Washington, DC, with grandiose aims of solving the reading "crisis" that, almost everyone agreed, afflicted America's children and blighted their prospects in an information age. After the Right to Read program would come the Nation of Readers venture, and then No Child Left Behind. The success of each program, it is fair to say, was at best mixed. But that was probably inevitable, given the rhetoric that accompanied each at launch and the continuing disagreements surrounding educational approaches. What is more notable is the extent to which reading, like other objects of state attention such as diet and pollution, became an object of explicit, sustained partisan politics. Like those other entities, it had of course never been apolitical—Flesch's concern about being typecast as a reactionary was justified. But from this point the science of reading had to be a political science as well as a psychological, experimental, and pedagogical one. That is, its practitioners, if they were to flourish, were well advised to think anthropologically and recognize that the old "modern" binary of nature and society had broken down. Or perhaps it would be better to say that they had to be Machiavellian, in the true (amoral) sense of that word: realists about the realms in which they operated, and prepared to build alliances between formerly disparate communities, interests, and methods in order to make progress. It is not that reading pedagogy became *more* controversial after 1969. It had always been agonistic, and often passionately so. But the controversies did change in character as they became inextricable from public debates about other areas of state policy such as privatization, healthcare, and social programs. This was to enter into a domain in which, on the one hand, real, large-scale changes could be engineered, but, on the other, public conflict was incessant and irresolvable. By the early 1990s, the media were starting to refer to the resulting arguments routinely as "the reading wars," taking their cue from the "star wars" phrase that the Reagan administration had embraced in the mideighties for the Strategic Defense Initiative. The title became commonplace around 1995. It was the same period in which the media identified "breast cancer wars," "memory wars," "pollution wars," "science wars," and more, all of which ballooned in the next decade. There were a lot of wars going on, rhetorically at least, and reading was a key battleground.

The Right to Read program did chalk up one major early success. The prospective creation of a well-funded national campaign in 1970 brought a moment of relative peace in what the *New York Times* was already calling the "battle" between whole-word and phonics proponents, and the Children's Television Workshop, having just created *Sesame Street*, took advantage. It appealed for support to create a new series aimed at older children who lagged in reading. The campaign agreed, and approved a new series slated to cost seven million dollars—a huge amount for the time. "The reading show," as it was at first known internally, was devised as a 130-episode-long series of televised lessons in learning to read, based on Chall's code-emphasis method. It eventually became *The Electric Company*. Fronted by Rita Moreno and Bill Cosby, it included a fast-changing series of scenes in which actors dramatized the fun, creative, and rewarding nature of decoding printed letters. Among the regulars, for example, was a young Morgan Freeman, who played the character Easy Reader—"a hip enthusiast for print who bursts into excitement every time he reads a word aloud."[40] By the time the second season was under way it was being described as "the world's most complicated and most expensive television series," with all those resources being used for "the single narrow purpose of teaching 7-year-olds to read." And it was a huge success, at least in terms of viewership. More than 50 percent of American elementary schools that possessed TV sets showed *The Electric Company* to pupils.[41] Chall herself came to feel that the greatest achievement of her work was not any advance in the science of reading per se, but the creation of *Sesame Street* and *The Electric Company*.[42]

That success notwithstanding, Right to Read struggled to make any impact on the problem it existed to address. The program was slow to get off the ground, and by September 1971 it was already said to be moribund. The council was supposed to have one hundred members, for example, but the Nixon administration struggled to recruit forty, and many whom Allen had recommended were reportedly rejected for partisan reasons. The National Reading Center lacked a president until January, after Mendez had moved on.[43] Even in the middle of the decade the *New York Times* could report that "Johnny Still Can't Read Very Well." Whatever doubts might exist about the tests used to measure such things (and they were real and widespread), it was generally agreed that scores were still "sagging." The evidence was especially dismaying when it came after such a long period of reform—dating back, the *Times* suggested, all the way to Flesch. Reading teachers themselves might point to increased recognition of problems like poverty, and "growing medical evidence that poor nutrition in the early months of life may cause irreversible brain damage." The Right to Read campaign was focusing on training teachers, in the belief that teachers with high levels of expectation would succeed "regardless of method." But the aim was still to ensure that by 1980 every child would leave school with "the skills to read to the full limits of his desires." It was a "herculean task."[44] And it was not achieved. In 1979, the Right to Read campaign was quietly wrapped up into the general Bureau of School Improvement, and no more was heard of it.[45]

In the wake of the failure of Right to Read, a curiously disconnected status quo emerged. The political and evidential balance was shifting strongly against cognitive-science approaches, as part of the general tilt to the right that occurred after the election of Ronald Reagan to the presidency. And yet, at the same time cognitive-science and whole-language methods were the practical orthodoxy in classrooms across the country, bolstered by a sense among practicing teachers that they corresponded to their experiences in day-to-day encounters with children. As a result, when another iteration of reform agitation started, it gave rise to clashes that were even more partisan and even more intractable. The process originated in 1983 when the National Academy of Education and the

National Institute of Education established a Commission on Reading under the leadership of Richard C. Anderson, a professor at the Center for the Study of Reading at the University of Illinois. The Anderson Commission issued a report in 1985 entitled *Becoming a Nation of Readers*. The report rejected once again the old Manichean dualism of phonics versus whole-word methods, and insisted that a balance between the two (which might well vary with the age of a child) was appropriate. At the same time, it acknowledged that teachers played a critically important role, and that erosion in the income and public status of the teaching community had had a real impact: the relentless barrage of reform agitation in the 1970s and 1980s had "oppressed the spirit" of these professionals. In its wake, however, the US Department of Education issued a guide, *What Works*, that recommended approaches for public schools.[46] *What Works* largely elided the qualifications that the Anderson Commission had registered, instead beginning with a declaration, in tune with the administration's controversial call for a "curriculum of the home," that parental conduct, not family poverty, was what affected children's success.[47] It went on to recommend phonics wholeheartedly. But the breezy pamphlet took *all* learning, not just reading, as its province, so these observations stood alongside, for example, a stress on learning science by doing experiments. Children might learn "science in action," it seemed, but they should learn reading by rote.

Yet any debates provoked by *What Works* were insignificant compared to the reading wars proper that broke out in the 1990s. Their main battleground was California. Generally regarded as a dynamic, technologically advanced society, in 1995 the state was dismayed to find that it ranked fourth from bottom (tied with Louisiana) in reading skills. Sixty percent of Californian fourth graders fell below the minimum federal reading level. How could this have happened? In the charged political atmosphere fomented by Governor Pete Wilson's Proposition 187 campaign to deny public services, including schooling, to undocumented people, immigration was for some all too tempting an explanation. If overall test scores in reading were low, the charge went, it must be because of greater numbers of nonnative speakers in the classroom. Lack of resources was another favored explanation. In the wake of the Reagan-era tax revolt, California's public spending had been radically curtailed, and by this time the state ranked forty-third nationally in expenditure on education. On an annual basis it now dispersed some $30,000 per classroom less than the average American state. It was also possible—if perhaps impolitic—to maintain that there was no state-specific problem at all, because scores for reading had declined everywhere. But the focus of public anxiety, fostered by conservative interests in Sacramento, fell on teaching. In the early 1990s, California had abandoned grades, standardized tests, and—"most destructive of all," according to the African American columnist and psychological expert Brent Staples—phonics.[48] It had moved to a whole-language approach, and the shift was all too easy to correlate with the decline in scores. It was certainly possible for defenders of the whole-language method to point to the stringent budget cuts that the anti-property-tax Proposition 13 had caused, to call test scores into doubt, or to point out, as Goodman did, that it was hardly surprising if teaching methods designed to target test scores resulted in higher test scores.[49] But such ripostes were readily cast as special pleading. Antagonists to "whole-language" teaching treated the issue from the start as a matter of popular politics.

As has happened almost every decade since the early 1900s, in the 1980s and 1990s the nation faced a "reading crisis." In 1985, a "staggering" twenty-seven million adults—one in five of the population—were reportedly "functional illiterates," and the total was estimated to be increasing by 2.3 million per year. The situation was decried as "a national disgrace." As in earlier iterations, too, private companies saw the crisis as an opportunity. Some five hundred American Learning Corporation "Reading Game" centers sprang up nationwide. But if the purported crisis was a national one, it was largely a matter for the states to address, and California became the first and largest to do so decisively. The result was what the book editor of the *San Francisco Chronicle* called "one of the most fascinating controversies to hit the reading scene since 1955, when Rudolph Flesch kicked American education in the right side of its brain with his bestseller *Why Johnny Can't Read*."[50] For about fifteen years, from the early 1980s to the late 1990s, elementary education was to be found on the front pages of California's newspapers on an everyday basis.

The Anderson Commission's report had been for the most part a measured document, revisiting as it did themes that had been articulated repeatedly for decades. It largely rejected the old notion of "reading readiness" and embraced phonics as an essential part of early education, at least to the extent of "getting children off to a fast start." In later years it urged a greater focus on comprehension and on improving reading in the course of classes on other subjects such as history.[51] None of that was particularly controversial. But the report was received in an environment where administrators, parents, and teachers—like Bill Honig, for example, then working in a San Francisco elementary school—were already looking for answers of a certain kind. They feared that existing methods were turning children into drones—"human computers who can't." The report was received in California, particularly, as vouching for a clear change of course. The shift would be away from a strategy of teaching that was associated with a scientistic trend to identify "sub-skills" and inculcate them discretely, in much the way that the "new math" curriculum was supposed to do. As in the new math venture, contemporary reading pedagogy disassociated the skill into a large number of subskills (some 290 in the case of the program taught in Chicago's public schools), each of which could be tested. The approach involved a lot of mechanical work, or so Anderson's report was read as concluding, leaving children bored and alienated. It also laid great emphasis on certifying banal achievements (like being able to circle the consonant that appeared most in a sentence) rather than the ability to parse and understand prose as a whole. So although it had a patina of scientific rigor, critics complained that children were not actually *reading*; and it turned out that pupils who initially tested well on elementary skills often did poorly in later years, presumably because the emphasis on piecemeal mechanical abilities led to neglect of "comprehension." As the *San Francisco Chronicle* put it, thirty years after Flesch, his attack had succeeded all too well. "Johnny can't read precisely because he's phonics'd to death." In that context, the new report was widely seen as urging a "return to the more traditional approach." Beyond absolute beginners, children should be reading "stories and books." In tune with the notion that this would be a restorative move, the conservative US education secretary William J. Bennett expressed his approval for the shift alongside increases in the teaching of robust history, science, and literature.[52]

The initiative to transform the teaching of reading in California's public schools came with the election of Honig as state superintendent of public instruction. Having experienced the existing approach himself, Honig regarded a transformation as his priority. He forthwith launched a reform that came to be called the California Reading Initiative. At first, the move was widely popular. A "reading renaissance" was at hand, which would see the state's children educated as "connoisseurs of fine literature." Moreover, the venture was part of a wholesale reshaping of arts and humanities in the Californian education system. And it was self-consciously intended to be a national model.[53] It amounted to a dramatic "swing of the pendulum" away from the "hyper-rationalism" of the 1970s and away from the micromanagement of skills according to diktats of "behavioral scientists." Tests and drills would be replaced by real books, and classrooms would become places for "active," engaged learning rather than sites for dreary exercises.[54] It was as though teaching practices in this area were finally going to embrace cognitive science. "We are in the midst of a revolution," Honig's broader initiative flatly proclaimed—one finally based on real understandings of "how the brain works, how we acquire language, and how we construct meaning in our lives."

California's education system was enormous, so any shift of this degree would be a major endeavor that would take years to implement. In 1988 the State Board of Education started the process by rejecting textbooks based exclusively on phonics (which led to a three-year lawsuit by the publisher).[55] The detailed new program arrived in Californian schools a year later. Training teachers to put it into practice took still longer, and some were still reporting years later—by which point the tide had turned—that they had never been trained in the new techniques. But in the early 1990s it seemed that California was in the lead and forging ahead successfully. When the Clinton administration took office in the wake of the 1992 election, the way seemed set for a national cognitive renaissance along the lines pioneered by Honig.

But by that time the initially auspicious environment in California itself was changing. In 1990 Pete Wilson defeated Dianne Feinstein in the gubernatorial election, beginning a period of Republican leadership in Sacramento that would become increasingly conservative in tenor. In addition, in 1994–95 national and state test scores appeared that showed the state's children to be, if anything, worse at reading than they had been a decade earlier. The state's own CLAS test system was one thing: defenders of the reform could point to many problems with the tests' "misleading" methodology and flawed implementation. (In fall 1994 Wilson himself canceled the system, in a move seen as a sop to the religious right.)

But national rankings were harder to contest, and in 1995 California appeared thirty-ninth out of thirty-nine participating states, again alongside Louisiana. Whatever the situation might be in the rest of the country, the *Los Angeles Times* complained, "Johnny can't read or add very well in California."[56]

California had aspired to be a "national model" for its 1980s reforms, but now it risked becoming a trailblazer of a different kind. Wilson blamed the adoption of the whole-language approach for the humiliation, and he was not alone. Already whole-language methods were under scrutiny in other states where results seemed to be poor and where political climates (not all of them conservative in a partisan sense) furnished sympathetic audiences for calls to revert to older norms, such as Texas, Minnesota, Missouri, Colorado, Indiana, and Maryland. Proponents of action—described in the media as a combination of "respected researchers, minority parents, Christian fundamentalists and conservative politicians"—demanded that the newer approaches be dismantled in favor of a return to the phonics that they recognized and preferred.[57]

Inevitably, these divisions became both commercialized and partisan. The new demand for phonics was both grassroots and driven by advertising. On AM talk-radio stations, commercial phonics programs were marketed extensively; but at the same time ex-teachers disillusioned by the whole-language approach also founded their own small companies to distribute teaching materials.[58] Parents, both fearful of what they saw in their own districts and persuaded by media campaigns, embraced homeschooling—which in turn led to suspicions in the schools about indoctrination. In June 1995 a letter circulating among teaching staff in the Fullerton area cautioned that parents pulling children out of school to learn phonics (among other things) were seeking "to establish a far-right presence in our district."[59] Suspicion about motives was bolstered when it became clear that lobbying firms had realized the potential in artificially cultivated popular anger for mounting pseudo-grassroots campaigns.[60] And in any case there was good evidence to support such fears in the case of reading. The hoary Eagle Forum, Phyllis Schlafly's populist conservative network, certainly weighed in. At seventy-one, Schlafly announced that her last campaign would be to return reading education to sanity. Although no longer the public presence that she had been in the 1970s, with an audience of eighty thousand members, aligned with some fifty college chapters and many spin-off organizations, she was still a force to be reckoned with.[61]

The combination of populism and commercial interest was manifest in the rise of "Hooked on Phonics," probably the most intensely marketed of the commercial programs in these years. The enterprise was founded in 1985, during the national reading crisis that stimulated California's own policy, by a one-time composer of advertising jingles. It charged $230 for a set of readers, workbooks, flash cards, and cassette tapes. By 1993 it had sold more than a million of these kits and had revenues of about $130 million per annum. It expanded into retail stores, and launched efforts to design programs for mathematics, history, and other subjects—to create, in the end, "a whole classroom in a box." The success was driven by relentless AM-radio advertising, as well as by testimonials of successful use. But did it work? Experts and laypeople disagreed, partly because—as ever—they disagreed on how to define reading itself: professionals who stressed "understanding" were less likely to be impressed than parents looking for the skill of parsing individual words. But the company claimed that it could produce extraordinary successes even for learning-disabled children, and this led to more sustained questioning, not least by the TV program *Dateline NBC* in late 1994. The Federal Trade Commission then investigated, and eventually determined that the advertising line had no sound basis in science.[62] The ruling threw the company into a crisis, but, tellingly, it also generated a problem for the FTC itself. It prompted thousands of letters, calls, and emails of protest, much to the bafflement of an agency that had never experienced such resistance before. The onslaught apparently came from parents convinced that the commission's action was in reality a maneuver to suppress not merely the company itself but homeschooling in general, and to do so in the interest of an inept and self-serving public institution. In the end, Hooked on Phonics' parent company went bankrupt, owing more than fourteen million dollars to various media companies for all that advertising. But it had certainly demonstrated the potential for a populist campaign favoring an alternative science of reading.[63]

In the meantime, after winning three elections as superintendent of public instruction, Honig was ousted from office in early 1993 after a scandal involving conflict of interest. Until the next round of elections the position was held by an acting superintendent, but in 1994 Delaine Eastin, a Democrat, was elected. Faced with public dismay, Eastin wanted to move quickly. But what to do? Reducing class sizes was clearly one priority, but otherwise the choices were murky. Phonics proponents were loud, but professionals were suspicious of them and advocated some kind of "balance." That could easily sound like special pleading. Worse still, as one angry teacher put it, for some protagonists reading was only the start. "There are also 'progressive' frameworks in science, physical education, history, etc.," this writer complained, all of them generated and backed by "so-called experts." The conservative columnist Debra Saunders, who spent much of this period lambasting what she called "the edutocracy," likewise claimed that "'the experts' too frequently are the problem." If all these experts—a category that included Department of Education staff, superintendents, school-board members, principals, scientists, "and even some teachers"—could not be trusted, and all the more if they would not take responsibility for failures, then on whose knowledge should action be based? An indication of how confused the situation was came when one pro-phonics school-board member in Ventura cooperated with a technology company to propose using an eye-movement camera to resolve the debate: one hundred children would be asked to read texts on a computer screen "to determine whether there is a connection between the way children move their eyes as they read and how well they comprehend." It was as though a century of scientific work on reading had never happened. The idea won support from Caltech's Jet Propulsion Laboratory, but what became of it is unknown.[64]

In the event, Eastin appointed two panels to examine the issues in reading and mathematics. Ironically, one of the most influential figures in shaping the discussions of the reading panel was to be none other than Bill Honig.[65] After leaving office, Honig had spent time doing community service as a teacher of elementary reading, and the experience had helped convince him that his own policies of the mid-1980s needed serious qualification. But perhaps the greater persuasive factor was his acquaintance with one of the other members of Eastin's reading panel, Marion Joseph. Joseph was a veteran Democrat and erstwhile chief adviser to Honig's own predecessor in office, Wilson Riles. She had retired when Honig defeated Riles for reelection but remained an active figure in local and state issues. She had then become attuned to the issue of reading education when her grandson began having difficulties. Starting at a 1994 meeting in San Francisco, she became an energetic campaigner for change, making common cause with the now-repentant Honig. Joseph developed a powerful and extensive networked political operation, active both inside Sacramento and in the broader Californian population. She cultivated what she called "moles" in the state government to give her advance notice of likely proposals, and she encouraged journalists to report on stories that would alert readers to important issues. She also used a fax machine to link a wide network of allies among parents, teachers, and researchers, based at first on connections she had made years before while working with Riles. And Joseph cajoled Honig into writing a book, *Teaching Our Children to Read* (1995), which circulated widely as a samizdat typescript before formal publication. She also persuaded him to set up a consultancy to aid school administrations keen to move away from whole-language approaches.[66]

The Eastin panel moved impressively quickly, but even before it had done its work there were signs of legislative action. Drafts of an "ABC bill" were soon being drawn up in Sacramento (its sponsors were two Democrats and one Republican, their surnames beginning with the letters A, B, and C). It would uproot the 1987 reform and in its place make the teaching of phonics and spelling a legal requirement. So obvious did the principle seem by this point that the *Sacramento Bee* likened it to drafting legislation that would mandate the law of gravity. Sure enough, the resulting "back to basics" measures received broad political backing and passed. Wilson signed them into law in early October 1995.[67] Meanwhile, the panel's report appeared in September, while the ABC bill was in process. It had the title *Every Child a Reader*. Eastin immediately accepted its recommendations, calling the California Reading Initiative "an honest mistake." She announced that she would empower a combination of parental organizations, university educators, and others to plan changes in detail. The "crisis" demanded action, she declared, to ensure that "all children are reading independently no later than the end of third grade." Once again, the state was going "back to basics," as the media reported—and with Texas having already acted to reintroduce code-emphasis approaches, California's shift meant that the nation's two largest textbook markets would become a new orthodoxy, inevitably reshaping the teaching of reading nationwide.

Once again, teachers could be forgiven for pointing out that the kinds of practices being mooted were in fact already happening in their classrooms—classrooms that were otherwise overcrowded, multilingual, and deprived of books by scant funding levels.[68] But in any case, matters lagged, partly because of such disagreements about the relationship between policy and practice, and partly because a simple need for resources. With 7,700 schools, and 59,000 teachers just in grades K–3—the Los Angeles school district alone had 632,000 students—the Californian system was gigantic. Steering it in a new direction would take time. One of the problems with Honig's previous reform, indeed, had been that it had never been completely implemented, and now another wholesale change was suddenly being demanded. It meant creating new materials, manufacturing new books, and retraining teachers. Just retraining K–3 teachers alone was projected to cost $100 million. Ultimately, however, after at one point proposing to hand over $50,000 to every school in the state, Wilson put just under $200 million into the reading effort. Teachers would be retrained at the same time as class sizes were cut. His state budget also exercised force majeure by channeling funds solely to schools that embraced the new approach.[69]

By this point, in mid-1996, a relieved Eastin could announce that California had an armistice in hand that would finally end the "famed Phonics Wars." The new plan would combine phonics for beginners with attention to comprehension in succeeding years. Its watchword was to be that much-maligned term, "balance." And in fact it does seem that the "wars" faded somewhat once the new programs had started to be implemented in earnest and emulated in other states. That is not to say that Eastin's hope was realized and that they ended altogether, however. Indeed, terms such as "phonics wars" and "reading wars" continued to increase in familiarity over the next decade. That was partly because they were used retrospectively, to make sense of the earlier clashes. But it was also because, in defining a past era, they also carried a warning about the present and future.

The decade of California's reading wars might have been a period of endless ferment, radical hopes, and desperate fears, but on one view it amounted to a lot of sound and fury that signified little.[70] A one-sentence summary of that view might run like this: what started in 1985 with a "back to basics" initiative to replace ill-judged scientism culminated in 1995 in a "back to basics" initiative to supplant ill-judged scientism. The two scientisms were not the same, to be sure. But that perhaps helps us grasp what was truly significant about the whole sorry sequence. It was a key process in the emergence of a new kind of practical politics of culture, in which campaigns came to be mounted as popular resistance, grounded in empiricism, against scientifically authorized public institutions. Neither of the two sides in the reading wars was anti-science per se. Both in fact claimed to be *defending* science. But they had different conceptions of what science was and how it should be upheld. A magnanimous outsider might have observed—and some involved in the debates themselves did observe—that the resulting multiplication of perspectives was, potentially, a good thing. The fact that agreement was unachievable might have conveyed the irreducible complexity of reading itself. It could also have been pointed out that it was potentially helpful for public understandings of science too. All sides began convinced that science and truth were to be equated, and that empirical observation (at some level) could clarify where that truth lay. It was hard to emerge at the end of the process without accepting (again, at some level) that science was something more complex than that. Perhaps it was a set of enterprises that were irredeemably human, situated, provisional, and fallible, while at the same time being responsible, self-critical, skilled, and knowledgeable.

In practice, this does not seem to have been the outcome. What contemporaries—and especially protagonists—seemed to learn from the experience of the reading wars was that scientific authority could always be challenged by adopting the mantle of science itself. They came to appreciate not that the diversity of science made it productive and useful, but that that diversity—when set beside a simplistic account of science as a structure built from uncontested facts—made it vulnerable.

The Reading of Science

By the early 2000s, the reading wars had already become conflicts over public scientific authority—what it was, how to recognize it, and who could claim it. This seemed to many a good thing. Science, after all, meant a body of knowledge that was objective and robust; so popular and governmental understandings alike led people to expect that the authority of science would be the best source for a consensual conclusion to this otherwise endless dispute. In place of repeated calls to start again from scratch, an agreement to cleave to experimental research ought to provide a firm foundation for progress—"normal science," in Thomas Kuhn's formulation. But in practice it was never so. Rival camps disagreed radically on where the true science could be found, on what grounds it could trump local expertise, and which scientific conclusions to endorse and apply. In effect, they disagreed on what authoritative science *was*. And each side could appeal to reputable accounts—accounts by people and institutions characterizable as "scientific"—to back up its argument. Science might be the output of rigorous, designed, and technically proficient experimentation; or it might be the quantitative data produced out of local, nuanced, professional expertise. There were no clear quacks to denounce.

In the new century, the federal government made an emphatic point of the centrality of science to the issue. A flagship of the George W. Bush administration's "No Child Left Behind" legislation was a five-billion-dollar reading initiative (that is, roughly four hundred times as large as Right to Read a generation earlier) that gave a pivotal place to science in determining questions of educational policy and practice. "Reading First," as it was called, explicitly required states to show that they were helping schools by reference to "scientifically based reading research."[71] For the first time in its history, it seemed, the science of reading had been granted formal, federally warranted authority to be the decider in cases of teaching controversy. Yet it was not quite the mandate that it might seem. In fact, the new law was based on the report of yet another preceding National Reading Panel, and that panel report had defined only an approved subset of research as "scientific." Such research had to be *experimental*—or it could not be counted at all. This had the effect, as the one teacher on the National Reading Panel had complained, of discounting a lot of what would otherwise count as empirical experience. In particular, it discounted experiences in places, and with people, that were underrepresented in formal scientific institutions—classrooms and ethnic minorities being the most obvious examples.

How, then, were parents, teachers, and supervisors to decide what "science" endorsed? The Reading First program had an answer to that question. It circulated a short pamphlet by two University of Oregon professors entitled *A Consumer's Guide to Evaluating a Core Reading Program*.[72] The pamphlet opened with a declaration that deciding upon a strategy was too complex and important a matter to be left to individuals; it had to be decided by objective evidence. So it led the "consumer" through a brief flowchart of steps to exclude unworthy candidates. "Is there Trustworthy Evidence of Program Efficacy?" was the first and easily most important question. The criterion of trustworthiness, it stipulated, was that there be "prior scientific" research to demonstrate effectiveness, through "carefully designed experimental studies." Whether that term meant only *laboratory* experiments was left obscure. And in any case, a vast amount of what were now being questioned as unscientific dogmas had in fact originated in experiments done with every appearance of scientific orthodoxy: precision instruments, exacting protocols, laboratory settings, and formal publication by top-tier presses. After that, a candidate program must also provide systematic and explicit instruction in "phonemic awareness," "phonics/decoding," vocabulary, and comprehension—these being elements that were presumably above and beyond scientific question. So, again, any initiative that did not prioritize phonics was ruled out. Only after a program had passed this test could it be further considered, which the pamphlet declared should be done by filling in a multiple-options questionnaire. Short as the process was, the impression the tract sought to create was that this was a judgment to be made via an algorithmic decision tree, each branch of which reflected a putatively consensual scientific verdict. In fact, as critics were not slow to remark, the boughs of the tree were not all grounded in the science of reading as it then stood. The superintendent of Madison, Wisconsin's school district, for example, claimed to have quantitative proof that the district's "balanced" program hit all the targets required, while the Reading First officials could furnish "not a single ounce of data" in favor of their alternative. The criteria, as Richard Allington put it bluntly, looked "made up."

The real problem facing policymakers and school officials in the fraught and politicized context of the first decade of the twenty-first century was that science, like reading, was not one thing but many. All sides could claim to be standing up for unbiased experience—for true science—against partiality, and all sides could claim to be representing common sense and disinterested citizens against elite professional interests. Campaigners for what might be described as neo-phonics purported to uphold nonmetropolitan and middle-class objectivity, against blinkered and self-interested teachers' colleges. Whole-language advocates for their part claimed to represent honest, hard-working, and professional teachers, in plucky opposition to an organized and well-connected pressure group funded by the wealthy. Any appeal to objectivity was therefore bound to be compromised and complex—as was any charge that one camp or another was a "special interest." In consequence, the reading wars turned into an irresolvable sequence of efforts to insist upon boundaries, of the kind that often happen in public debates involving the sciences.[73] But at the same time, the various camps were not, in their eyes at least, engaged in work to build such boundaries. All sides insisted on regarding the boundaries that they observed as *obvious*.

As with any broad-based political clash involving sciences, there was an organizational aspect to this. In general, protagonists in such episodes tend to downplay the importance of management, because in their view the explanation for winning is the truth of their position rather than anything more mundane. But in fact the organization of letter-writing drives, the lobbying of congressional representatives, and the raising of funds are all vitally important, as the Californian experience in the reading wars implies. More than that, they are not readily separable from the constitution of credibility for the positions being advanced. The exemplary case is the "memory wars" of the 1990s, in which one side effectively constructed a new natural phenomenon, "false memory syndrome," in order to seize the initiative from psychiatrists and litigants who based sensational claims on recovered-memory testimony. The new "syndrome" came into existence with its own foundation to study it, a network of expert scientists and laity to undergird the research, and an organized campaign to make the phenomenon visible and viable to the media.[74] The versions of "simulacrum science" described by Naomi Oreskes and Eric Conway in their exposé of the tobacco and energy industries' efforts to undermine public policy were similar endeavors, better funded but in a sense less sophisticated. In the lingo of modern social movements, they were Astroturf rather than grassroots campaigns. The campaigns about reading showed similar traits, but they occupied a middle ground between the cases of recovered memory and simulacrum science. By and large, no side was arguing in obvious bad faith. All sides could plausibly maintain that they were acting in accord with science, that research was in their favor, that the theories and models they worked with were more sophisticated and mirrored reality better than those of their opponents, and, for good measure, that the historical trajectory of the science of reading bent toward them.

The closest thing to a massive corporate interest acting surreptitiously behind the scenes—if one excludes the teaching profession itself—was the textbook industry, which was certainly large but not particularly surreptitious. And, as we shall see, the reading campaign did promote its own science of reading to place in competition with that emerging from Columbia, Chicago, and other universities. The combination makes the reading wars a subtler case of a science conflict than that of climate and tobacco wars, if a less consequential one—though that too is debatable.

The alignment of skepticism about the science of reading (in its older forms, at least) with ideological conservatism was well known by Flesch's time, which was why he took such pains to disavow it. But there is a difference between an alignment and an organized, managed affiliation. In the 1990s and early 2000s, as opposed to the 1950s, what the teaching profession and the scientists of reading faced was a coherently engineered movement. Because literacy had long held such a prized place in America's self-image, the science of reading was always liable to become a major political football; but it was in this time and place that that finally happened in earnest. That explains why, even after the phonics side essentially won, the wars continued. Fresh skirmishing has been reported every few years since the 1990s, and continues to break out in the 2020s.[75] But one aspect that the reading wars in these years did share with other conflicts involving science was the tendency for private and state agents to merge. For example, Robert Sweet Jr., a former congressional aide, wrote most of the Reading First legislation. A convinced phonics advocate, Sweet also became president of the National Right to Read Foundation, "a pro-phonics group." And it was not long before teachers were encountering pressure from federal agencies, which had been co-opted by "contractors" (educational publishers) and pressure groups (phonics campaigns) to impose new policies on the schools. This was the context in which the mandate to take only scientific knowledge into account was supposed to be put into practice. No wonder the mandate ended up guaranteeing prolonged discord over what it meant by *scientific* in the first place.[76]

The reading wars have never really ended.[77] But one can still ask whether they achieved anything in terms of the anxieties that originally motivated them. And the answer seems to be that they changed very little. Statistics tracing the reading abilities of American children did not show much improvement, but remained fairly stable from 1992 to 2017. A global survey in 2016 found that the US still lagged internationally in reading. Perhaps the new world of information technology had rendered that point obsolete—"Who cares if Johnny can't read well," asked the *New York Times* at the end of 2016, "so long as he can multiply?"—but the country actually did even worse in mathematics. "Something has to change," the paper concluded, adding, with an awfully familiar rhetorical flourish: "A national reading push would be the moonshot that makes all others possible."[78]

The Forever Wars

If the reading wars ascended into struggles over the character and role of science and scientific authority, that was no more than had been predicted at the outset. In 1961, the great historian Jacques Barzun contributed an introduction to Charles Walcutt's *Tomorrow's Illiterates*, one of a number of books to rehash Flesch's case. Barzun made the claim explicitly and presciently that the real issue, in the end, would be the public reputation of scientific expertise. His point, however, was not that the public was too skeptical about science, as one tends to fear nowadays. It was that the public was too credulous. "Every parent," he noted, was familiar with the "cluster of ideas and catchwords" that characterized contemporary pedagogy in reading—the notion of "reading readiness," for example, and the ideal of speed. Such notions had the cachet of scientific terminology and as such tended to go unquestioned. But he agreed with Walcutt and Flesch that they were spurious. What was needed, then, was not just a reformed reading program, but a sustained public reflection on expertise and a reengagement of science with society. To correct the course of increasing separation between scientific specialists and the general populace was, he maintained, "one of the fundamental tasks of society." Barzun was not anti-science as such: his point was that citizens had the right and responsibility to choose *which* science—which experts—to attend to. Qualifications provided only an initial presumption of competence, and that any particular expert was actually competent in a particular point of contention must be demonstrated. Scientific advisers, he wrote, should be selected "with the same care we bestow on choosing a car or a mate." In other words, the public must embrace "the same suspicion of 'scientific' conclusions that the scientists themselves do." Nobody should accept the value of IQ testing on faith, and the same skepticism must be applied to claims about "the television curriculum and the teaching machine." In short, the campaign for reading reform would only be worthwhile if it were carried further, to create a permanent shift in the relation between the public and science itself. "If this great awakening is in the end to give us better schools, we must take time first to cure our gullibility."[79]

Barzun's declaration was extraordinary in both its diagnosis and its prognosis. In the event, it did indeed transpire that the reading-reform movement triggered by Flesch contributed to a new kind of public skepticism about scientific expertise. That is one reason why the rebellion against the science of reading may seem disconcertingly resonant today. The specific allegations that Flesch, Walcutt, and their followers leveled against a scientific community—of complicity in a commercial racket, of willful obduracy, of condescending elitism, and of surrender to an autocratic ideology—were remarkable for their character as much as for their content. Coming as they did at the apogee of Mertonian representations of science as an enterprise defined by a morally tinged ethos, his allegations effectively amounted to a charge that the science of reading was not science at all. Or, even worse, Barzun implied that it had once been science, but that every single one of its practitioners had betrayed the scientific ethos and that it therefore deserved expulsion from Olympus. And it was the right and duty of citizens to discriminate between true scientists and false prophets; that distinction could not responsibly be delegated irredeemably to the sciences themselves. Such skepticism seems to have struck a chord, because from Flesch and Walcutt it spread broadly through the reading-reform movement that occurred between the 1970s and the first decade of the twenty-first century. It was initially restricted to the particular science of reading. But it was soon extended: to testing, to other realms of pedagogy like mathematics, and before long to matters of the common good in general. Arguably, we live in the aftermath. While Barzun identified excessive deference to science as a major problem of his age, today, two generations later, we inhabit a world in which the public questioning he encouraged has metastasized. Of course, Barzun himself was not responsible for this, but the campaign inspired by Flesch did constitute a long-term, well-funded, consistently populist social movement, with broad appeal across classes and regions, and as such it almost certainly helped secure the credibility of such questioning in general. Public skepticism is now widely felt to be a debilitating problem for policymaking in many areas of concern to which scientific research is germane. The reading wars may not have had transformative results for reading itself, then, but in the broader area articulated by Barzun their consequences may have been all too great. The critique of the science of reading had purchase, but at a high price.

To see the broader implications of this, we need to go beyond the bounds of reading itself. Since the 1990s, America has seen the growth of a public culture of doubt and disbelief about science that has repeatedly crippled efforts to establish policies for public health, climate change, and tobacco regulation. Since its publication in 2010, Naomi Oreskes and Eric Conway's book *Merchants of Doubt* has held sway as the most plausible explanation for this otherwise puzzling development. Oreskes and Conway demonstrate that powerful corporate interests have fostered various kinds of "simulacrum science" for decades, with the express aim of cultivating a public perception that reputable scientific findings are dogged by uncertainty, that the real science is "unsettled," and that the status quo must be left in place pending the resolution of (manufactured) controversies. Their account is authoritative and damning, both of the CEOs involved and, perhaps more, of the credentialed scientists who have lent their authority to these efforts. At the same time, though, their approach suggests that we now need to ask a question that they themselves have somewhat sidelined. Theirs is essentially a "supply-side" story: it focuses on the interest-driven efforts of particular parties to manipulate public opinion and stymie policy initiatives. What is not explained is why those efforts fell on fertile ground.

To the extent that Oreskes and Conway engage with this question, their approach is that of classical media sociology. They point to the end of the "fairness doctrine" in US broadcasting during the Reagan years, and they ascribe eroding public confidence in genuine science to the propagandistic programming that flourished in the wake of that shift.[80] They are right, of course, as any degree of exposure to such broadcasting will soon confirm. Still, perhaps the good history and righteous politics of *Merchants of Doubt* have distracted attention from its view of communication. Oreskes and Conway are not naive about this, and they show that the doubt-merchants have always been engaged in a complex feedback relationship with public responses. Yet there is a lot more to be said about the consolidation of beliefs at the receiving end of such media messages. We could even learn from the pioneers whose work represented a conduit running from the social science of reading to the new field of media and communication studies in the mid-twentieth century. As we have seen, veterans of propaganda like Waples, Lasswell, and Bruner argued from experience against simple models of media efficacy.[81] When Lasswell closed his classic study of propaganda by quoting Anatole France's dictum that all society was "run by an unseen engineer," he did so precisely to assert a contrast—that such engineering was no longer possible in a complex, dynamic, interconnected culture. Readers were agents now. That is something of a commonplace today, of course, but it means that a different question may need to be asked about the roots of our "crisis" in public attitudes to science. Take the time when the tobacco industry began its efforts at simulacrum science, which seems to have been in the mid-1950s. How and why were such large populations, starting in around 1955–60, disposed to credit such unfamiliar notions as science's multiplicity and dubiety? That was the period when conventionally reassuring representations of the scientific ethos were at their height, after all. Somehow Americans seem to have found the fragmenting of what one may call capital-*S* Science unremarkable. What made the erosion of science's credibility itself credible?

We are probably not yet equipped to answer that question definitively. But a key element may well be that at the same time millions of American families were confronting a drastic problem of science in action— one that was not about relatively distant questions of epidemiology and biological causation but immediate matters of their children's welfare and futures. It was a problem that no responsible parent could ignore. Decades of public anxiety told them that, even if the experiences of their own children did not. It demanded prompt attention. And it directly connected the most intimate anxieties, via perceptions of science, to high political questions of institutions, taxes, and public goods. When it reached schoolrooms across the country, the debate over reading that Flesch launched meant that millions of citizens *had* to think critically about science and deal with multiple conflicting claims to scientific authority. To reiterate, Flesch's original argument was not in itself antiscientific. On the contrary, he claimed that the problem with the science of reading was precisely that it had devolved into non-science, and he called real science into the lists against it. But that was the point: the argument over reading split the authority of Science. And Flesch and the reading-reform campaigners then insisted that lay citizens had not merely the right but the responsibility to come to judgments about what was a real scientific claim and what was not, and about who was a legitimate scientist and who was not.

At a barely implicit level, Flesch and the reformers appealed for the revival of the prized ideal of the Jeffersonian public citizen. Their whole case rested on this point: that an educated, reasonable American, equipped with common sense and a decent library, was entitled to appraise some half a century of experimental science. All the more so if the American involved happened to be a parent, which Flesch was. And that was very much the contention enhanced later by Barzun. He and Flesch were perhaps pushing at an open door. The long-standing tradition of robust skepticism about overweening experts (a tradition only partly explicable in terms of frontiers and farmer-citizens, because similar currents have also existed in other cultures) may have been due for a revival anyway. But in the scientific and technological context of the postwar era, that tradition took on a new form. Suspicions about the phenomenon of "big" science were certainly part of that, and they existed within the scientific community itself. But another, more positive, element was a panoply of novel forms of experimental "science" conducted as exciting enterprises in that realm itself. Past-life regression and Velikovskian mythography were probably the two most prominent examples, but one could also cite Wilhelm Reich's orgone accumulators and various movements in alternative medicine. Such "sciences" took place outside the walls of conventional research establishments, and they depended on lay citizens' involvement. For the first time in a generation, Americans found that the most interesting science was plural and local—and it was something to which they themselves could contribute.[82]

The achievement of reading-reformers like Flesch, Walcutt, and Barzun was to seize upon an issue that was at once more down-to-earth than any of those notorious cases, equally close to citizens' own identities, and of massively greater immediate consequence. And they did so at exactly the right moment. Every parent in the land was likely to have experienced doubts about how reading was taught to their children, many would have

seen test scores (and media jeremiads about them), and the vast majority would be horrified to learn that teaching was being done in systematically the wrong way. Flesch's book was pitch-perfect in that regard. It
hardly mattered that experts dismissed it as ill-informed and one-sided—
which they did, repeatedly, persuasively, and indignantly. His persona of
the baffled and indignant citizen-at-large, conveyed not only in the book
itself but in his publisher's cannily hectoring advertisements, was resistant to such countermoves. It succeeded with the readers it had to. One
is reminded of Merton's monitory point that to many in the general public
the claims of advanced sciences are often less plausible than the populist
arguments of firebrands. But there was more to it than that. As if he were
a Lilliputian counterpart to the giants of big tobacco, Flesch advanced his
own pedagogical science of reading—not quite a simulacrum science (its
foundations were firmer), but a rival science nonetheless. About a third of
Why Johnny Can't Read was given over to a practical guidebook for teaching Johnny to read on the basis, as Flesch said, of this real science—and
the whole section was soon reprinted as a separate volume called *Teaching Johnny to Read*. Parents were exhorted to try it out over the summer
holidays. They could and should place this science in competition with
the official version practiced in schools, and they could and should see
for themselves which one "worked."[83]

In concert with its rival science, the reading-reform campaign also pioneered the strategy of treating a conventional research community as
just another interest group, an institution no more deserving of public
deference than any other. Authoritative scientists of reading might well
dismiss Flesch's recommendations as specious, then, but in the context
of this strategy the natural response of his readers was that that was exactly what they *would* say. At least their own science was not the creature
of some corporate mastodon—one to which parents were obliged to pay
hundreds of dollars every year. And that is the light, too, in which to appreciate the parallels between critiques of the science of reading and suspicions of government-funded science in general. Although Flesch's campaign targeted one particular field rather than science itself, it did bear a
resemblance to qualms being voiced at much the same time both within
and about that larger domain. More prestigious sciences were starting to
see concerns expressed about their worldly and conservative tendencies,
not least by scientists themselves. Even physics, the acknowledged key to

victory in World War II and now to security in the Cold War, was vulnerable. When the leader of one of the national laboratories expressed public doubts about the managerial and conformist character of "big science," he could hardly be accused of undermining science itself for ulterior motives. So the campaign against the science of reading drew on growing unease about the worldly nature of even the most prestigious of all scientific fields.[84] That was another reason why Flesch's charge that a particular science had become complacent, clannish, defensive, corrupt, and authoritarian found purchase.

As Chall put it in 1967, when it comes to a practice that is second nature to a vast proportion of the population, "the existence of a controversy is itself a phenomenon deserving serious study."[85] The campaign to reform how reading was taught came to stand as an extraordinarily successful model. It became a prototype, not only for future iterations of the campaign against the science of reading, but also for polemical crusades against other branches of science. Essentially similar strategies would be rehearsed in later years in many arenas of debate around scientific issues, including climate change, vaccines, and census statistics. But perhaps that understates the case, because it was more than a model or prototype. Recall that question about why Americans found the diversification of science congenial. Perhaps all those corporate strategies in the domains of energy, tobacco, pharmaceuticals, pollution, and more worked, to the extent that they did, because they could exploit attitudes that were *already* immanent on a large scale in the American middle- and lower-classes. If so, then it seems likely that one forger of those attitudes was the campaign against the science of reading. The campaign arguably reached more people, earlier, for longer, and with greater personal impact than any other about a scientific issue. It encouraged a broadly dispersed population, extending across classes, regions, genders, and races, to see itself for the first time as a popular collective, defined by its position vis-à-vis scientific authority and possessing a responsibility to judge such authority against alternatives. The real achievement of reading's forever wars was thus not any new view of schools, teaching, or reading. It was a public endemically liable to hearken to the sirens of scientific skepticism.

9

Readers, Machines, and an Information Revolution

For all that the reading wars became conflicts over the relationship between science and the public, the actual techniques that had been central to the science of reading for about a century were conspicuously absent from the fray. With the exception of one odd proposal to measure the eye motions of a small group of schoolchildren in Ventura County, California, at no point did traditional instruments such as eye-movement cameras and tachistoscopes play any role. The status and meaning of scientific claims were certainly at issue, but they either involved discrete sciences like statistics and testing or they were somewhat downstream from the practice of experimental research. There was a certain irony to this, in that the G. W. Bush administration enacted policies intended to push school districts nationwide to shun methods and approaches that were not based on attested results from expressly *experimental* science. What "experimental" meant in this context was never made very clear, but it did not seem that the federal government had experimental psychology in mind. To a large extent, the contentions of both sides in the continuing tussles over reading in America's schools did adduce evidence that was founded *historically* in some version of the experimental science that had used tachistoscopes and eye cameras. If, as John Maynard Keynes once said, people who prided themselves on seeing the world as it really was were usually "the slaves of some defunct economist," then apparently pragmatic thinkers about reading in these struggles were generally reflecting some long-forgotten trial by Buswell, Huey, or Cattell. But the instrumental culture of the experiments themselves seems never to have been attended to in any of the wars' various skirmishes.

There was something rather appropriate about this. The 1960s through the 1990s were also the generation that believed itself to be undergoing an information revolution, culminating in mass access to computing and the Internet. As we have seen, both casual commentators and more dedicated analysts speculated then that reading itself might be obsolescent, doomed, like print, to irrelevancy in an age of dynamically interactive visual and auditory media. The future lay in moving images and multimedia experiences, not in texts—let alone in books. Perhaps it was no wonder that all those elaborately intricate experiments on readers were disappearing from view. Maybe the science of reading was ceasing to be a going concern because the information revolution was gearing up.

As a claim about the science of reading itself, this conclusion would have been resoundingly false. Not only has the experimental science of reading—eye cameras, tachistoscopes, and all—survived into the twenty-first century; it has also flourished, diversified, and even undergone its own version of democratization. One can get an impression of its continuing importance simply by hefting in one's hand either of two so-called handbooks to survey the field recently: Blackwell's *The Science of Reading* (2005), which weighs in at 661 pages, or *The Oxford Handbook of Eye Movements* (2011), which at 1,027 pages is almost twice as long. The overwhelming preponderance of the chapters in both volumes address scientific work done after 1990. The real question, then, is not why the science of reading fell into abeyance, as it seemed to do in the years of the reading wars, but how it managed to endure and prosper in such different technical, scientific, and political environments. Yet at the same time there was something worth noticing about how this science was perceived—what its role and purpose were taken to be. In this sense, major changes did occur with the advent of computing and new media, but the changes were not only to the science of reading. What "information technology" was to be emerged only through a process of engagement with the question of reading. That meant that the information revolution was substantially shaped by virtue of its arising in dialogue with this science.

The large question of the changing fortunes of the science of reading in the information revolution may usefully be resolved into three more manageable ones. First, what was the science of reading *for*? Second, what did it *do*? And, third, what was its *subject*? In fact, these three questions have dogged the pages of this book since the beginning, albeit generally in a rather incognito fashion. So it is fitting that as we approach the end they should be addressed directly. It transpires that there are three principal reasons why the science of reading endures today. While at first blush each of them is new and peculiar to the contemporary environment, all in fact have deep historical roots, extending back to the early twentieth century at least. That is another reason why it will be helpful to explore them now. By leading us back over the decades to the 1890s as well as forward to the 2020s, they allow us to arrive at an appropriately historical understanding of a set of scientific claims that today's scientists of reading are generating. Those claims are circulating widely in contemporary media, where they have been received with largely unalloyed acceptance. Together they constitute a promise to furnish radical new insights into culture and cognition, which will allow for both a resolution to conflicts like the reading wars and the generation of a novel constitution of reading publics. Knowing the history of the science of reading will help provide us with means to calibrate such claims. It may also, therefore, help us move beyond the simple binaries of credulity and dismissal that have characterized the public fortunes of this field for so many decades.

Consuming Readers

The first reason for its survival was that the science of reading found new constituencies. At the same time as eye cameras seemed to be falling into disregard and desuetude for the study of reading in general during the 1960s—largely because in an era of psycholinguistic explorations the questions they were designed to answer were no longer deemed to be cutting-edge—they reappeared in different venues. In the late sixties, the device became a key instrument for a newly thriving "science" of market research. It was a simple matter to take instruments originally designed to show where in a line of text a reader paused and took in information and repurpose them to show where in an advertisement (or a filmed commercial) a viewer's eyes lingered, presumably imbibing influence. It is even possible today that the device has had greater public influence through this channel than by virtue of its role as an instrument in experimental science itself. So one important component to an understanding of the science of reading in the information age has to be an account of how market research and reading science converged. That account needs to go back to the origins of both.

Although the sustained use of eye-movement devices in market research seems to have begun in the late 1960s, the record of affiliations between the science of reading and attempts to understand consumers extends back much further. That should not be surprising, because the modern advertising industry originated with the same boom in print media that led Cattell and others to believe that the future of society depended on a scientific understanding of the practice of reading. The distant roots of market research lay in that era, with late nineteenth-century efforts to define the effects of industrialization and urban immiseration through the use of statistics, mapping, and surveys. Just as these social-scientific techniques could be used by companies to demonstrate the extent of their impact, so in the years around 1900 there were also scattered efforts to use psychological techniques to measure how well people recognized and remembered advertising. And there were some new departures, too, such as the idea of Ralph Tilton of the *Century Dictionary and Encyclopedia* to include an identifying code as part of a mail-in coupon incorporated into advertisements in particular publications. The company could assess how effective different advertisements and venues were by recording how many coupons were returned from each. By around 1910 independent firms were emerging to pursue such strategies. They emerged from the world of experts in addressing reading communities—that of the publishers.

The first market-research company seems to have been Business Bourse, founded in New York by J. George Frederick, formerly of the trade journal *Printers' Ink*. Business Bourse would later publish the path-breaking book on "consumer study," *Selling Mrs. Consumer*, by Frederick's spouse, Christine Frederick. At much the same time, Charles Coolidge Parlin established a Commercial Research Division within Curtis Publishing, in Philadelphia, which undertook large-scale studies for a number of industries and mapped the "climate of consumption" across American cities. And R. O. Eastman began systematic reader research for Kellogg's, again before launching his own independent firm. Such initiatives became prototypes for many more research operations, both within corporations and as autonomous companies in their own right. One early follower was the Psychological Corporation, the firm that Cattell, Thorndike, and Woodworth founded in 1921, which had its own Market Surveys Division from 1930. The division produced a regular survey of consumers' responses to advertising that became known as the "Psychological Brand Barometer." Like Cattell, other academic social scientists and psychologists found themselves in demand, and several of them took advantage. Most famously, the founder of modern behaviorism, John B. Watson, left Johns Hopkins in 1920 to join an advertising agency. At the same time, units within academia—Harvard Business School to the fore—moved to embrace the opportunity. The trend continued beyond World War II, when sociologists at the University of Chicago created Social Research Incorporated. And other hybrid ventures—half academic, half commercial—emerged at Stanford, Michigan, and elsewhere. The designation of such enterprises as "market research" became conventional in the late 1920s. The field's first professional association, the Market Research Council, was formed in 1926, and a trade journal, *Market Research*, appeared in 1932. The new profession sought, in various ways, to generate reliable, salable knowledge about the extent and impact of reading. It made information about readers into a commodity, to be sold to the corporations sponsoring the research.[1]

All these ventures, to varying extents and in different ways, proposed to use social and psychological sciences to understand the environment within which sponsoring companies worked. The act of reading itself—how an individual made meaning with the help of print on a page—was not at first their prime concern, although it did appear in their projects. They focused instead mainly on reading populations considered in the aggregate, in the manner of Douglas Waples. As a result, in the 1930s their techniques correlated with those developed in the social-scientific research carried out under the aegis of federal agencies to discover large-scale trends in reading alongside those in unemployment, poverty, and health. At the same time, researchers in this hybrid field embraced a diversified and active concept of the reader, partly on the basis of such research, but more directly as a result of the diversification of media. The electromagnetic ether seemed set to replace ink as the principal vehicle of human communication and culture. As what had initially seemed a brightly utopian prospect—countless articles in the 1920s forecast the end of war by virtue of radio, and the emergence of a unified (perhaps even telepathic) global community—took on a much darker hue, so methods inherited from an older science of reading had to be repurposed and reshaped. They were used now not only to tackle problems of marketing but also to address the political questions posed by broadcasting itself—and the two issues were seen as intimately linked. Hence such early market research became a critically important element in the emergence of a field of media and communications studies.

We can see something of the importance of these questions in a duel between methods and institutions that arose in the mid-1930s between two radio networks, NBC and CBS. Both of them were reliant on advertising, and both therefore needed good evidence for the appeal and efficacy of their programming. NBC was the larger, more prestigious, and more prosperous business. It had enjoyed great success with an audience research methodology developed by perhaps the best-known of the first generation of American market researchers, Daniel Starch. Starch himself had started out as a collaborator with Walter Dearborn in work on reading and "efficiency." His classic *Principles of Advertising*, originally published in 1910, contained detailed treatments of how typefaces and line-lengths affected influence, tackled by means of records of a reader's eye movements.[2] He held a PhD in psychology from the University of Iowa, and he taught at Wellesley, Wisconsin, and Harvard before leaving academia in 1926 to run his own market-research organization. While at Harvard he invented a method for measuring the impact of print advertisements by interviewing readers of magazines. Through the 1930s, his firm refined that technique, generating a commercial "Starch Advertisement Rating Service" based on the prodigious feat of interviewing 100,000 people every year, in communities scattered across the country. (How the firm selected and trained interviewers and then calibrated their techniques are interesting questions for a future sociological history.)[3] And Starch himself continued to work on the subject for decades, becoming probably the clearest case of an individual whose work linked the science of reading to market research. In 1928, however, the then-nascent NBC hired him to survey its radio audience—a task with both obvious similarities and treacherous differences from his magazine work. The result was the first audience survey of radio listeners ever carried out. Starch essentially modeled his method to gauge listening on that to gauge reading: he sent out thousands of interviewers, who asked the members of some

17,000 households about their habits.[4] The feat was impressive enough that in succeeding years the network continued to rely on the Starch model. Through the 1930s, as a result, NBC conceived of its audience expressly in terms of "circulation," and it told its advertisers that it designed commercials, "even to the smallest details," in order "to produce the same effect on the listener as he obtains from the printed page."

NBC's rival, CBS, therefore had to come up with a rival science. It found the answer when a graduate student in psychology at Ohio State University wrote to a senior executive to tell him about his research. The young Frank Stanton had become interested in the question of radio audiences some years earlier, as an undergraduate at Ohio Wesleyan University. He had spent the interim period teaching typography in Dayton, before, as a first research project at OSU, returning to one of the oldest questions in the science of reading. Stanton asked whether paper texture and color affected reading speed, apparently hoping to find an answer that would attract the interest of potential employers in the print advertising industry. Like researchers in the tradition of Huey, that is, he was interested in how to make reading move at the speed of thought. As it turned out, however, he found that there was no effect.[5] At that point he returned to radio, a medium in which perception did at least move at the speed of speech. He wondered whether people who *heard* advertisements retained them better than people who *read* them. And here he did get a result. Stanton found a significant difference in favor of radio—one that he suspected might even be greater if the readers concerned were not skilled university students.[6] Radio beat print. And this, he realized, offered an entirely new kind of answer to the question of efficacy being asked of broadcasting companies by advertisers.

Stanton wrote excitedly to executives at both NBC and CBS to tell them of his discovery. NBC sent back a form letter—why do more, when the company already had what it needed?—but a CBS manager responded in earnest. Encouraged, Stanton went on to do his PhD at Ohio State on radio-listening research. (We do not know if he ever met Renshaw.) He invented a small device that would record details about when a household radio receiver was switched on and how it was tuned, and he persuaded fifty-eight householders in Columbus to let him install them in their homes, telling them, quite misleadingly, that he was doing an experiment involving electric current usage. He found that the information garnered from his machines differed markedly from householders' memories as gathered in interviews. That had major implications for broadcasting companies' claims about their reach and impact—so much for Starch's surveys—and their attractiveness to advertisers stood to be transformed. It was, in fact, something of a Holy Grail for radio broadcasters, who otherwise lacked any hard data corresponding to newspaper circulation figures to prove their prowess to advertisers.[7] Stanton immediately wrote again to CBS in New York. The network promptly hired him as a researcher (only its third), and immediately began to promote his work in corporate pamphlets distributed to advertisers.[8] "Do people remember as well what they *hear* as what they *read*?" the pamphlets demanded: psychological research had now proved that hearing was not only different from reading, but distinctly better. In effect, a gauntlet had been thrown down. Starch's methods of audience research (derived from the social science of reading) would duel with Stanton's (defined by their departure from that science). The contrast drew wide attention both within academic psychology and in the public realm. The mainstream press highlighted it, too—and no wonder, because the entire commercial basis of the newspaper industry was at stake too. An "audimeter" descended from Stanton's crude surveillance instrument would even feature in Marshall McLuhan's *Mechanical Bride* as the commercial counterpart to a spy's bugging device.[9]

Perhaps even more than corporate supremacy was at stake. Radio and the press were huge issues of public and political concern in the 1930s, when the potential for determinative media effects frightened Americans all too aware of developments in Europe and the USSR. Since the early 1920s, media and middlebrow representations of radio had tended to highlight utopian forecasts of the advent of a harmonious, peaceful world, in which old animosities would be dissolved by the unifying force of the media technology. But in the thirties, this first started to look like complacency and then collapsed altogether, to be replaced by fears of propaganda and mass ideology.[10] In that context it is significant that the first influential scientific study of radio listening was by two psychologists, Hadley Cantril and Gordon Allport. Their *Psychology of Radio* (1935) argued that broadcasting was indeed creating a "new mental world," but that it needed to be understood scientifically. For Cantril and Allport, the important questions surrounding radio listening were all in the specific terrain of psychology—something that came naturally when all the important questions surrounding book-reading had been seen in that light for a generation, and when so many assumed that the reason radio mattered was that it was destined to take the place of print. The central chapter of their book, in which the two modes of reception were compared, was in fact authored by a recent Harvard graduate named Merton Carver, who had done his doctoral research on the experimental psychology of learning by listening and reading. The distinction that Carver drew between the cultural effects of radio and those of print was consequently a direct extension of the science of reading into this new domain. He pointed out, for example, that one experienced printed words not in isolation but as "related items in larger groupings," which a reader made "as his eyes play about the words in the normal process of reading." Any given reader determined "the range and tempo of his own perceptual experiences" in order to extract as much meaning as possible, and fluency and speed came with the ability to group words included those *following* the one directly before the eye. Listening was different. In listening, words were separated by time; to the extent that they were elided, this was done by the speaker, not the listener. A listener could not listen ahead in the way that a reader could read ahead. Carver thus identified a sharp contrast between "stepwise" auditory experiences and "interlocking" visual ones. The cultural differences between radio and print were to be explained by this distinction. So, for example, Cantril and Allport inferred (somewhat counterintuitively) that citizens at "lower cultural levels" got more from reading than listening, because they had the power to pore over difficult passages. "Untrained minds are frequently quite unable to listen intelligently to material which they could, perhaps with difficulty, understand and reproduce if they read it," they argued. At least, that was true for the moment, given current reading practices. For they added the enigmatic remark that radio was "gradually training the masses to listen."[11]

The concerns of advertisers that Stanton and Starch tackled therefore converged with broader and deeper fears about the potency of media in general—print as well as wireless—and psychology was seen as the way to understand and master them. As Stanton put it in a speech in late 1936, he himself wanted to use psychology to arrive at robust, scientific knowledge not only *that* people listened to certain programs, but *why* and

how they did so. He wanted, he said, "to quantify influence." This determination soon took institutional form. Encouraged by Hadley Cantril, he
submitted a proposal to the Rockefeller Foundation suggesting an initiative to create the research methods to tackle such questions. The methods would need to be interdisciplinary by the standards of contemporary
social science, and they would have to attend to different locations and
subjects too. The foundation agreed to fund the idea. But the original
plan that Stanton and Cantril had concocted soon fell apart for personal
reasons. They had intended to collaborate in directing the program, but
Stanton decided to stay at CBS rather than divert into academia, and
Cantril had too many other duties. They turned for advice to Robert Lynd,
whose classic study *Middletown* (1929) was an important point of origin
for Depression-era social science. As it happened, Lynd had recently encouraged a Viennese sociologist named Paul Lazarsfeld to remain in the
United States at the end of a period of research there, rather than return
to face the perilous political climate of his own country. Lazarsfeld had
been instrumental in creating a new kind of research institution in Vienna
designed for nondisciplinary subjects; he was currently at the University
of Newark, trying in vain to interest Cattell and Thorndike in his ideas for
their Psychological Corporation. Lynd suggested that he would make a
fine director of Stanton and Cantril's new program. They concurred. So
Lazarsfeld began as director of the new Office of Radio Research, and for
the next few years he and Stanton would meet weekly to chart out the
course of what was fast emerging as a discipline in the making.[12]

The Office of Radio Research was in fact based in New York City, and
it was eventually taken over by Columbia University and morphed into
what became known as the Bureau of Applied Social Research. On the
face of it, such a drastic change of name would seem odd. But in fact it
was fitting, because the research that both the Office and the Bureau undertook, although focused on media and communications, and although
it included market research for major US corporations, did aim to address
major social questions from the outset. To that end Lazarsfeld cultivated
contacts with the second Chicago school of scientists of reading—in
1944 he coauthored *The People's Choice* with Waples's protégé Bernard
Berelson—at the same time as recruiting colleagues who included Robert K. Merton, Theodor Adorno, and Max Horkheimer. They developed
new techniques for gauging the reception of programming, including the

use of a "continuous panel" of radio listeners. These listeners were not only interviewed repeatedly over time but also subjected to home visits by CBS staff to see which products were passing through their households. The technique was soon co-opted to create the Nielsen Radio Index. In addition, to address the problem of *why* people listened to certain programs, Lazarsfeld, his wife Herta Herzog, Marjorie Fiske, and Merton put to use what became known as the "Lazarsfeld-Stanton Program Analyzer." Given its resemblance to what Stanton significantly called "Ohio State behaviorism" the instrument was probably more his invention than Lazarsfeld's. (It came, he was saying, from the culture of Pressey, Renshaw, and Sherman.) In theory, it allowed researchers for the first time to record reactions to media messages in real time. A listener would turn a knob to signal liking or disliking parts of a broadcast as it was happening, while the instrument recorded the gestures as a trace. From this mix of sociology, psychology, and interviewing, Lazarsfeld and his partners generated a corpus of work that defined the early field of media and communications studies.

The first volume of research from the radio project to be published, Lazarsfeld's *Radio and the Printed Page*, appeared in mid-1940. It was, as Michael Stamm has argued, a landmark in the history of media and communication studies. Respectfully diverging from the stance of Cantril and Allport, Lazarsfeld maintained that radio listening had to be interpreted as part of social life, experienced differently by different communities. The interdisciplinary approach that he applied followed from and reinforced that contention. Although it has tended to be remembered as a work exclusively about radio, in fact the book's title was accurate, and it had much to say about readership too. That much was also indicated by Lazarsfeld's list of helpers, which included not only Stanton and Cantril, but also Mary Carter and Miriam Thompkins (library scientists at USC and Columbia), Alberta Curtis of the New York Public Library, Harold Lasswell, Samuel Stouffer (the University of Chicago sociologist of symbolic interactionism), and especially Douglas Waples. Their collective effort showed that both radio and print had to be viewed as social, historical achievements. Active measures to build listenership were essential to the newer medium, and without them radio would fail to reach the audiences that print did not address.

The questions that Lazarsfeld posed in *Radio and the Printed Page* derived from those early hopes and fears about the new medium. He began with the conventional attribution of "seriousness" to reading. Ever since the Reformation, print had been identified with epochal historical transformations. But for a democratic society facing historic problems it had a major drawback. "Students of reading," he noted, had shown that "a considerable part of the public has not developed sufficient reading skill, nor enough of a habit of reading, ever to read a text on a serious subject." The reason why radio had initially been hailed as such a marvel derived from this anxiety about reading as well as from its own properties: here was a tool for communication that recipients did *not* require a special skill to use. The outstanding question, then, was whether it was realizing this hope. Did people who refrained from reading "for information" now listen to "serious" programs? That in turn implied that significant research on radio must depend on first understanding the reading habits of people—which was a notoriously difficult matter, as Buswell and Gray had shown. One component of the book, therefore, was an attempt to tackle that problem. The Office of Radio Research collaborated with the Book of the Month Club to furnish a new classification of reading habits in terms of a hierarchy of five "levels," each of which was further subdivided into "types." As Lazarsfeld noted, this was an important contribution to the social science of reading because it represented "an unusually refined classification of reading level," which was "beyond anything tried so far in the nature of reading classification."[13] But for him the point was that it allowed for a direct comparison between listening and these "levels" of reading. Lazarsfeld reported a clear correlation. Listening was certainly more common at lower "levels" than higher, but those who listened more were not tuning in to "serious" programming. In fact, lower reading levels corresponded to fewer serious listeners. And Cattell and Thorndike's Psychological Corporation furnished more data that buttressed this. The corporation conducted extensive interviews in Buffalo, New York, which showed that lower-"level" listeners tuned in to serious programming about half as much as higher.

What this attention to reading showed was that customary anxieties about radio were misplaced. The real issue was not the sheer availability of broadcasting, Lazarsfeld said, but the interest of citizens in actively seeking out programs. That was a problem, but it was also an opportunity. "Print did not raise the intellectual standard of living just because it was invented," he pointed out, "but because it was used by educational institutions such as schools and promoted by cultural agencies such as libraries and publishers." This was very much the kind of point made by Waples and Louis Wilson. Were the equivalent done for radio, it might generate new constituencies with serious interests that print had never catered to. But that would require cultural institutions to step in, to provide programming and build audiences. Radio alone, in and of itself, could not "contribute substantially to the enlightenment of the American community," but radio managed properly could. To understand how to achieve that, one needed to ask *why* people listened as they did. And so the central chapter of *Radio and the Printed Page* analyzed why people decided whether to read or listen. Here the existing science-of-reading research on readers' abilities came into play in a novel context—to tackle radio's role "in the fight for democracy."[14]

The lynchpin of Lazarsfeld's book was this central chapter, entitled "To Read or Not to Read: The Place of Reading Skill in Radio Listening." It argued that there was a principal reason why people at higher cultural levels preferred reading to listening: they were more skillful readers. In other words, print was for them "a more efficient form of communication" than radio. Not only was it faster, though it was that. Skilled readers could fix their own "rhythm and rate of speed," and they could select which parts of a page to attend to closely. Unskilled readers, on the other hand, tended to complain in interviews about the "energy" it took to read, to the extent that for such readers the very process of reading itself seemed to demand such concentration that attending to meaning became impossible. Whether a corresponding "listening skill" (the kind of thing Allport and Cantril had envisaged) even existed was not clear—Stanton suggested developing tests for it, along the lines of those used for reading skill—but it seemed much less consequential. But in any case Lazarsfeld could now explain an otherwise remarkable fact: that people who preferred print and those who preferred radio both gave the same reasons for their preference. Both groups were really referring to the same thing, he concluded: the "experience of efficient communication." Any subjective qualities that good readers might attribute to their experience of print—ease of comprehension, absorption, concentration, and more—were in truth qualities of their skill in reading.[15]

Skills, however, were not adamantine. They were learnable, and they were always exercised in contexts. At one point Lazarsfeld even recommended that broadcasting be used to help such learning, by informing people about "reading as a process": he wanted radio to convey the conclusions of the science of reading over the air, to dispel popular misconceptions such as the belief that slow reading was better for retention. In general, his emphasis on the dynamic and contextual was one of the major distinctions between his work and that of Cantril and Allport, who had relied on laboratory experiments to compare experiences of radio and print. Lazarsfeld wanted to change the "frame of reference," he said, and to pursue "the choices and habits of people under real life conditions." While he hoped, then, that his brand of research might help clarify what he called "the much confused controversy regarding 'eye versus ear'"—a reference to the relevant chapter in Cantril and Allport's book—he also suggested that the "the physiological means of perception" did not matter much in the communication of ideas. Much more important was the conjunction of subject-matter, social situation, and "the reading and listening habits of the respondent." So, once again, his approach discounted simplistic notions of media shift. Radio alone was not making the newspaper obsolete.[16]

Lazarsfeld ended his book with a cautionary flight of fancy. He invited his reader to imagine taking a time machine back to the year 1500, when information spread through Europe by means of the newsletters produced by the powerful Fugger merchant family of Augsburg.[17] What predictions might have emerged from research sponsored by an imagined Fugger Foundation into the impact of print at that time? Wrong ones, no doubt. His point was not that such a project would have made foolish predictions, however, but that the question itself would then have been ill-posed. The importance of print was undefined at that early stage. That importance emerged only as a *consequence* of the Reformation and subsequent revolutions; print itself was not their cause. The same would be true of radio. "We can be fairly sure only that radio will not of itself mold the future," Lazarsfeld concluded. "What we, the people of today and tomorrow, make of our social system: this will define the place of radio in history."[18]

Lazarsfeld's conviction that radio's place in history was indeterminate might sound deflating, but in practice it enabled the emergence of an ambitious and open-ended field of inquiry. In that sense it was of a piece with the distinctive principles that made the Office of Radio Research and the Bureau of Applied Social Research into the most important early centers for communications and media studies. Here the fact that both radio and newsprint depended on advertising meant that the need for a science of *reception* was critical in a consequential sense as well as a conceptual one. Not least, it meant that research would get funded. A science of reception would henceforth be central to attempts to understand both contemporary politics and a purported revolution in the making. Researchers treated it as an active, learned, and changeable skill—another notion modeled on views of reading. To understand reception, then, the science of reading could be repurposed. But there was no one way to do that, and in practice it was pursued variously by psychologists, surveyors like Starch, and others. Lazarsfeld's multidisciplinary researchers built a practical tradition by doing so in their own fashion; it is often forgotten that his own work on listening was carried out in concert with what he called "reading research," primarily on magazine usage. Lazarsfeld even endorsed the possibility of developing a reading-measuring instrument, the equivalent to Stanton's audimeter, in the form of a device that used special chemicals to record how long a page had been exposed to light.[19] The machine seems never to have caught on. But, even so, the science of reading itself was changed by the coalescence, in ways both evident and subtle. It now came to be seen as one branch of a more general science of reception. That paved the way for accounts in later decades that would emphasize the contingent, open-ended, active, and even elusive nature of reading as a skill.

As they worked to repurpose the science of reading in this context, Lazarsfeld and his colleagues developed and refined a generalized concept of *media* as well as that of reception. It was both general and internally variegated, such that the researcher interested in media must pay attention not only to technical affordances and psychophysics but also to context, change over time, training, and diversity. But in the meantime, the existence of the broad-based "frame of reference" pioneered by the new discipline (to use a phrase both Lazarsfeld and Allport liked) had one other major consequence. It meant that when Claude Shannon devised his theory of "information" in 1948, the conceptual and practical grounds were already in place to recognize it, and to welcome it as something significant. Conceivably, this was one reason why cybernetics initially took off as a social phenomenon—as something that experts across the academic landscape immediately found conceptually compelling—rather than remaining just another of Norbert Wiener's brilliant ideas.[20]

Eyeball Marketing

Another consequential convergence of market research, media studies, and the science of reading commenced as cybernetics started to disintegrate in the 1960s. This was the era when market research itself became truly an international big business. One reason why Madison Avenue was able to vaunt itself into such prosperity was that it could claim social-scientific warrant for its claims about the reach, power, and impact of advertising. And one part of that was that it rediscovered eye-tracking.

The notion that eye-movement cameras might be used to refine advertising was an old one, as we have seen. In the 1920s Guy Buswell had taken Judd's two-dimensional version of the instrument used by scientists of reading and used it to research the viewing of images. Buswell pioneered the use of density-mapping and other representational techniques to display where a viewer's eyes paused in examining a picture. He and his followers identified these pauses with the fixations of the eye in reading, during which all meaningful perception happened. As Buswell remarked in the conclusion to his pioneering *How People Look at Pictures* of 1935, the obvious inference was that researchers could use eye-tracking devices to record the parts of an image that led to reception and response. And he concluded the book by applying the technique to three pictorial advertisements, remarking that the advertising industry ought to look further into it. In the midcentury years his advice was largely neglected. Marketers were preoccupied with different sciences: statistics, survey research, focus groups, and mass psychology. But in the 1960s they returned to the eye-movement camera in a bid to produce persuasive data about the relative powers of different graphic designs to attract a particular consumer's gaze. The mechanism became a key tool in their repertoire. Advertising firms adopted the technique with alacrity, especially when combined with other methods (focus groups and the like) to realize the potential in Buswell's approach. With its aid they claimed to be able to record attention, and thereby to produce scientific recommendations for advertising designs. It was perhaps the most important single step on the way to what would eventually become known as an "attention economy."

One inspiration for the marketers' use of eye-movement cameras was the rediscovery within psychology itself of the earlier science of reading. As we have seen, the midsixties cognitive scientists saw themselves as reviving questions asked by Edmund Burke Huey and never really answered.[21] But another, rather ironically, was the importation of a curious piece of Soviet science. Born in 1914 in Moscow and trained as a physicist before World War II, Alfred Yarbus had completed his PhD on visual illusions in psychology after the war. He then took a position at the USSR Academy of Sciences' Biophysics Institute, where he devoted himself to eye-movement research, inventing an eye-movement instrument similar to Dodge's. In 1963, at the peak of Soviet cybernetics, the Biophysics Institute became part of a new Institute for Problems in Information Transmission. Yarbus made the transition and participated there as part of a diverse group that included pioneers in AI, pattern recognition, communications, sensory physiology, and neuroscience (he even coauthored with neuroscientist Alexander Luria). In this very exotic context he returned to Buswell's questions about the viewing of images and used his new instrumentation of eye-tracking to extend them. The resulting book, *Eye Movements and Vision*, was initially published in Russian in 1965, and then reissued in English in 1967 by Basil Haigh, a British translator living in Cambridge who had made a specialty of translating Soviet neuroscience. It made its way to the United States under the patronage of a specialist in visual psychology at Brown University. Yarbus's work caused a stir when it arrived, not least because it revived a point that Buswell had made, characteristically, in passing: that the places where the eye paused in looking at an image changed if the preceding instructions were different. In other words, this rather mechanical technique of eye-movement tracking proved that image-reception was not itself mechanical. It could not be explained as a simple matter of stimuli, and had to involve prior history.[22]

Facial-recognition work would later draw on Yarbus, but in the shorter term his work became a point of reference for claims to be able to steer visual perceptions in the advertising world. It was from the mid-1960s, then, that marketing companies began regularly to take groups of a hun-

dred or so people, seat them in a theater, and have them view advertisements on a screen while their eye movements were tracked. The resulting
traces were then buttressed by personal interviews. The technique over-
promised at first and provoked skepticism, but it was refined over time,
and by the 1980s the use of computer recording and analysis had made it
more effective. By that point eye movements would be recorded digitally
sixty times per second, and the trials would include "through the book"
experiments in which users would be asked to "turn" pages on an onscreen simulation of a magazine. (When they did this, apparently, they
missed brand-names in ads some 40 percent of the time.) In the hands of
a newly confident "marketing science," the science of reading had given
rise to a versatile and impressive resource for linking corporations to consuming publics.[23]

The next step in this history was also one that Buswell had foreseen.
He had designed the first "portable" eye-movement device, which became
the ophthalmograph used in educational research, remedial reading institutions, and speed-reading clinics in the 1930s, 1940s, and 1950s. But
that instrument was, to use a phrase from the computer world, more lug-
gable than portable. It certainly could not be used to record the eye motions of a reader as he or she moved around during the day. Efforts to create a truly portable device seem to have started with Norman Mackworth,
the British colleague on whom Jerome Bruner depended for help in his
cognitive psychology experiments. Mackworth made a head-mounted
instrument in the late 1950s using television cameras. In 1962 he tried
to use his distinctly Rube Goldbergesque device to record the eye movements of car-drivers, but the machine was still too ungainly. The potential
was clear, however, and a generation later miniaturization had gone far
enough to make the idea viable. While sophisticated, high-resolution eye-
tracking devices remained static installations (in hospitals, for example),
lower-resolution versions became small enough to be mounted on the
head—perhaps even as spectacles. By the 1990s, companies were using
head-mounted units to record where a consumer's eyes lingered as he or
she moved through a supermarket.[24] Wearable eye-movement recorders
at length became small enough to become inconspicuous to the user.
They remain to this day a mainstay of market research, often now coupled
with some purported "neuro" element. They are also employed routinely
in virtual-reality systems, as well as in augmented-reality gadgets such

as Google's intermittently available "Google Glass" spectacles. Popular reactions to Google Glass focused on the antisocial implications of the device's ability to record others willy-nilly, but if it incorporated technology to track the wearer's eye movements then that would create a quite distinct issue, and possibly a more serious one.[25]

Meanwhile, the question of reading had become even more complex with the shift from paper to screens and the advent of the Internet. Was screen-reading different? What about online reading, with all its distractions and diversions? The first researchers to tackle these questions using eye-tracking systems were probably the designers and critics of what had previously been called "human-computer interaction" and later became known as "user experience" (UX for short). The following section will deal with them. But market researchers showed almost the same alacrity. As soon as the first Web advertising appeared, analytics corporations sprang up offering to analyze its effectiveness. They turned to the technology and research methods that they were familiar with in order to get a grip on this otherwise disconcertingly novel medium. They employed human subjects and used eye-tracking devices to observe which areas of a web page attracted the most attention, and which remained, as it were, visual deserts. For marketers in this movement, what Buswell had called density maps became "heat maps," put to use to assess—and, recursively, to create—the attention economy.[26] Success was measured in terms of how many "eyeballs" a site was reported to attract, and after that in those of where on the page the eyeballs lingered. In the monumental *Oxford Handbook of Eye Movements*, there seems to be no mention of the use of eye-movement tracers in designing computer information interfaces, or in online marketing and other capitalist enterprises, yet these have certainly been among the most consequential uses of the techniques in the twenty-first century.

But they are far from alone. The technology can be found not only in healthcare but also in aerospace and entertainment (think of its applications in gaming, and in augmented and virtual reality). The automotive industry is a major prospective venue, too—today, Cadillac's "Super Cruise" system uses an eye-movement camera to make sure that a driver is paying attention while its semiautonomous AI handles the car's controls, and virtually every major carmaker has a similar system either operational or in the pipeline.[27] There are also, naturally, military uses.

McLuhan was impressed by the marriage-like intimacy that a fighter pilot developed with his aircraft, but now what sustains that intimacy is not just training but a helmet that knows moment by moment where the pilot is looking. And then there are more outré applications, such as lie-detection (where its utility is, of course, dubious).[28] The ubiquity of these devices is greater now than ever before, and it is going to grow further. Altogether the global market size for eye-tracking technology is currently predicted to exceed two billion dollars by 2026.[29]

One reason why this element of the science of reading in particular is suddenly such big business is that it is perfectly suited to the world of "surveillance capitalism" with which we have become all too acquainted.[30] Thanks to the development of sophisticated tracking technologies that could be deployed *ad distans*, the old research domain of tracking that had previously been confined to the laboratory or schoolroom was now dispersed through the Internet. And it became part and parcel of the system by which, for example, a vast range of websites record every mouse movement and every keystroke we make while facing them, and Amazon puts every act of reading we ever do via Kindle into a surveilled database. But to see where the trend is really heading, we need to go back once more to the beginning—to the ambitions of Wundt and Cattell and Huey, and to their puzzlement at what reading, in the end, actually *was*. The reason for this is that today's readers are often not human beings at all.

Perhaps the most successful branch of artificial intelligence to date is that known as "machine reading." This has its roots in a shift away from explicitly rule-based algorithmic techniques in natural-language processing that took place in the late 1990s and early 2000s, and then in the embrace of artificial neural-net methods in the succeeding decade. Alongside data mining, text mining, and machine learning—terms that are sometimes used interchangeably, though technically they refer to discrete practices—the purpose of machine reading is to ask questions, which ideally can be posed in normal language, of massive and cross-field data sets.[31] Such techniques have been used successfully in pharmaceuticals research, for example, to explore potential new uses for compounds that are described in discrete corners of the enormous published scientific literature that a human scientist would be extraordinarily unlikely to encounter. At the time of writing they have been deployed to search for medicaments that may be worth testing empirically for efficacy against COVID-19. Machine-reading *comprehension* resembles the human activity of reading a little more closely, at least in its visible aspects. As a demonstration by Microsoft Research puts it, MCR "scans documents and extracts meaning from the text, just like a human reader."[32] In what amounts to an interesting extension of the famous Turing test, "the reader," aspirationally at least, may now be either a human or an algorithm, and it should be difficult for a scientist of reading to tell which it is from the meaning it extracts from a text.

It should not be too surprising, therefore, that eye-tracking researchers have extended their remit to explore how digital "eyes" scan digital "texts." In the 2020s a fast-growing corporate research industry has arisen that offers *simulated* eye-tracking. This industry bypasses human readers altogether. It uses machine-reading algorithms instead, to produce "heat maps" indicating where human viewers *would* pause their eyes if they were to be tracked by traditional technologies. The advantage is that the approach is fast, cheap, and scalable. You can even try it out for yourself for free, at the time of writing, by accessing the websites of companies like Feng-GUI.[33] Feng-GUI's algorithm practices what the company calls "predictive eye-tracking," showing the fixations a representative group of human eyes in a focus group would manifest in the first five seconds of encountering an image or text. It can be used for all forms of presentation, not just Web pages; for example, one could show the algorithm a poster design or a restaurant menu. Given such a source, the system identifies what the company tellingly calls "attention leaks," apparently on "the neurobiological basis of visual processing." Compared to traditional eye-tracking experiments, its benefits are supposedly clear. The AI version works in seconds as opposed to days, costs roughly one-thousandth the price of conventional experiments, and claims greater than 90 percent correlation with eye-tracking sessions using fifty real people. In addition, it requires of the corporate user no expertise, research staff, or equipment. And those figures will improve if the system attracts clients, because the technology not only tracks simulated reading but *is* simulated reading. As a standard machine-reading algorithm, the more it is used, the more it trains itself. It should, then, become ever more accurate in its representations of human reading—or at least, of human readings carried out in circumstances defined by the system. In any case, champions point out, even existing versions, while not perfect by any means, are generally good enough. Fine precision is not needed in this domain. Targeting advertisements does not require the precision of a latter-day version of Norbert Wiener's anti-aircraft system of the 1940s, perhaps targeting missile-defense batteries at incoming ICBMs. It's not, as they say, rocket science.

So far, we may suppose, so dystopian. As actual, human readers, we may even end up feeling neglected in this technological future; at least someone once thought we were worth tracking. But there is another aspect that deserves at least passing consideration. In the face of so many contemporary fears about surveillance, the proponents of machine-reader analysis speak in winning terms about a democratization of the research process itself. Up to now, they point out, the means to understand influence have been the preserve of scientific and commercial elites. In the future, that will change. We shall be able to conduct studies of cultural influence ourselves, on the fly. And our subjects will not be restricted by ethnicity, class, gender, or location, in the way that they were for Cattell, Dodge, and their successors. Such technologists aspire, at least in their publicity materials, to make the science of reading into a citizen science, emancipated from the constraints of the modern world at the levels of researcher and subject alike. And perhaps they have a point, for all the inevitability of the hype. It is certainly true that in the twentieth century not many people had access to laboratory-grade eye-movement cameras and tachistoscopes, while the science created with such devices shaped their educations and hence their destinies. To be sure, the version of the past being conjured here is too monochromatic. Those earlier machines were hardly the equivalent of particle accelerators, after all, and schools and commercial reading institutes did use both. Moreover, some at least of the earlier researchers were acutely conscious of the need to look at all kinds of readers and be sensitive to their differences. Buswell used the science of reading, monolithic as it could be, precisely to argue that existing assumptions about reading propensities on grounds of race were prejudicial. Yet the relative ease, cheapness, and rough-and-ready sufficiency of the virtual eye-tracking algorithms are good reasons to believe that they will catch on, and a forecast that they may permit new kinds of distributed, socially contextual research is not altogether implausible in that context. Puffery notwithstanding, the moral claims for citizenship made about this new, democratized knowledge of attention and influence in process are at least worth considering seriously. What will the dynamics of culture be, when we are all scientists of reading?

Designing Information

Part and parcel with the adoption of concepts of media and information in the era of World War II was a recognition of the importance to reading of information design—that is, the manner in which information in general was presented on a page, oscilloscope screen, cockpit instrument panel, or other surface. By the end of the war, researches by Fitts and many others had made clear that design was certainly not an inconsequential "factor," as the term had it, in the efficiency, speed, and accuracy with which humans grasped information in contexts ranging from submarines to schoolrooms. In the 1960s, 1970s, and 1980s, their efforts would be remembered and become reference points for the creation of a computational world. To follow the ways in which this happened is to see how the science of reading helped shape our mundane experiences of information to this day.

One of the most striking common features of the biographies of those figures who played key roles in the cognitive and information sciences in the later 1940s and 1950s is that so many of them had worked in fields involving informational displays. As we have seen, Paul Fitts was engaged in using the science of reading to understand why bombers crash-landed so frequently. J. C. R. Licklider was putting his training in experimental psychology to use in Harvard's Psycho-Acoustics Laboratory, before moving to MIT where his continued interest in what was becoming known as human-computer interaction saw him doing eye-pursuit experiments with oscilloscopes—updated, fun versions of the trials that the student Samuel Renshaw had done in the thirties. He then looked at the uses of screens in the SAGE air-defense system.[34] Mackworth, later to pioneer wearable eye-trackers, spent the war years investigating fatigue and boredom among radar operators. The examples could easily be multiplied. In several institutions, including Bell Labs, lasting research traditions of human engineering emerged out of these experiences.

Licklider spent much of 1961–64 working on what may be thought of as the third in a pivotal trio of reports on reading, media, and information to be composed at a key moment in the shaping of a new information age.

The other two we have already met: Marshall McLuhan's report on "Understanding New Media" for the NAEB, and Jeanne Chall's on "Learning to Read" for the Carnegie Corporation. Licklider's work, which eventually appeared in print as *The Library of the Future*, is not usually considered alongside the others, but doing so helps us see what was remarkable about both it and them.

The report was commissioned by the Council on Library Resources, a unit established by the Ford Foundation in 1956. Its purpose was to address the problems of information storage and retrieval that had been identified by Vannevar Bush in his much-discussed *Atlantic* article of 1945, "As We May Think"—the piece that, as the foundation put it, had "opened the current campaign on the 'information problem.'" Bush had argued that the biggest problem to face the sciences after the war would not be the terrible threat posed by atomic weapons but the sheer amassing of information decoupled from any retrieval technology capable of keeping pace. We have seen that others had similar anxieties. His "Memex" idea—later often identified as the *fons et origo* of networked and personalized information access devices—was a thought experiment that highlighted the need for some creative response. In effect, it dramatized an insoluble problem. The council sought Licklider out to produce an analysis of how this "information problem" might be characterized and tackled in the coming decades. At that point he was working at Bolt, Beranek, and Newman (BBN), a technology company launched by two MIT professors that would win renown for its developments in computing, networks, and information technologies. He put his contacts to work and quickly corralled together a group of engineers and psychologists. They then spent two years brainstorming about the question of how information storage and retrieval might best operate in the year 2000.

Licklider's initial insight was to redefine the question. The problem was not really one of libraries at all, he decided, if by that term one meant collections of physical books, but one of designing "man's interaction with the body of recorded knowledge." He then started to address that issue with the notion of reading as an active, learned skill—the conception that had come to fruition in the 1940s and was foundational to cognitive science. But whereas all his contemporaries (without exception, as far as I know) valorized the skilled reader, Licklider did not. He regarded the need for such skill as a fundamental *problem*. It was something that must be overcome if a future information age could come into being. What was needed was a form of information-institution that did not need such a refined, restrictive skill to use. His group thus commenced by accepting the possibility of redesigning everything from scratch. But then they proceeded by thinking in terms of another cognitive-science notion: that of "schemata," or the experiential concepts that people adapted or combined to form new concepts. Simply abandoning everything already familiar about libraries ran the risk of leaving no schemata to think with, so Licklider decided to retain what he called "lower-echelon, component-level" elements. The notion of pages stayed, for example, even though those of traditional books and libraries could be jettisoned. "As a medium for the display of information," he conceded, "the printed page is superb." The page was thus to be the unit of future informational experience. It worked because it afforded "enough resolution to meet the eye's demand" (an intriguing way to put the matter) and it contained enough information to occupy a reader for "a convenient quantum of time." The page also gave the reader autonomous control over the rate of reading—the point that midcentury media researchers like Cantril and Allport had made. But many of these advantages were lost, Licklider continued, when pages were bound together into books. Books were heavy, bulky, expensive, and slow to circulate. They were "not very good display devices"—except when used for the kind of leisurely "consecutive reading" that was not his concern, and for which he seemed to feel some distaste. Current libraries, therefore, were a bad starting point too: they inherited the book's "passiveness" and demanded too much of the user in searching out and gathering information. The absence of "active processors" within books, Licklider argued, stood at the heart of the problem of human interactions with recorded knowledge.

Behind that claim lay a further question. What was *thinking*, as an activity? To answer that, Licklider decided to monitor himself actually doing some of it. To his disappointment, when he did he found that most of his time was spent on tasks that were essentially "clerical or mechanical"—seeking, managing, and storing information. Preparing to cogitate was far more time-consuming than actually cogitating. Moreover, he found that he made choices about what to cogitate *about*, in the main, according to banal considerations of "clerical feasibility." Most of the tasks conventionally grouped under the term *thinking* were therefore ones that could be better performed by automated systems such as his imagined "active" library.[35] In a world where it was already impossible for any one person to keep abreast of a technical discipline by the practice of reading, what Licklider devised to address the problem that Vannevar Bush had articulated was a new kind of informational interaction. Licklider's vision was that human and machine would work together "in intimate association"—at least, until the machines took over completely. He had in mind, interestingly, the kind of close cooperation that he himself had experienced in "good multidisciplinary scientific or engineering teams." That meant that his proposal was to be distinguished categorically from any concept of technological "extensions" of human senses, of the kind so often invoked by McLuhan (although in this context Licklider actually cited not McLuhan himself, but the chair of British aircraft manufacturer Boulton Paul). His own notion was one of "symbiosis."[36] In this vision, rather than being expected to do the detailed work themselves, as readers, people should be reconceived as managers of the processing of information. He called it the "control-and-monitor approach." Reading would not be eliminated altogether—you would still need to read *something* in the end, after all the management—but that task would be radically downgraded.[37]

The trouble with this idea was that it meant externalizing into technologies the processes by which readers operated, and that required understanding what those processes were. Licklider suggested rendering readers' practices into algorithmic instructions for selecting, reformatting, and the like, which could then be recoded so that some hypothetical future computer could replicate them. That sounded simple enough, but the science of reading had hardly generated an algorithm-ready model of the practice at hand. Even in his speculative version, rendered in a kind of pseudocode, the details that Licklider sketched out soon became elaborate and confusing (fig. 9.1). The proposal was clearly not only beyond the technology of the time, but beyond its conceptual apparatus too—not least because it would require kinds of natural-language processing that were then almost unimaginable. Still, he passed handily over that problem by a radical extension of the principle of recursion that had so captivated the cognitive theorists. Licklider represented the whole system—user, machine, storage, technical staff, institution—as an "adaptive self-organizing process" that would recursively improve itself

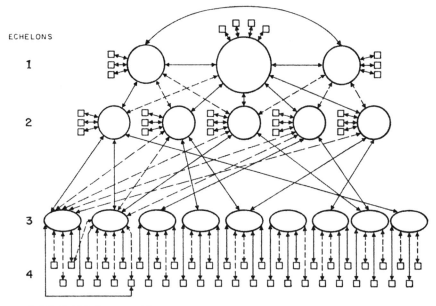

9.1 J. C. R. Licklider's proposed "structure of the procognitive system" for a future library. Echelon 1 represents centers devoted to sustaining the "total fund of knowledge," echelon 2 those organizing corpora, and echelon 3 those devoted to processing required by networked users. Echelon 4 represents consoles at which users interact with displays of information. Circles and ellipses represent "advanced and specialized computer systems," with squares standing for "man-computer interfaces." Solid and dotted lines represent connections established during operation. J. C. R. Licklider, *Libraries of the Future* (Cambridge, MA: MIT Press, 1965), 41. © 1965 Massachusetts Institute of Technology, by permission of The MIT Press.

as it was used. How that would be programmed remained unclear, but he predicted that about twenty years should suffice for the future library system to be achieved, or perhaps for it to achieve itself. At that point it would be possible to ask a computer to "mull over" a technical subfield and organize it for a person's use — "if we can define precisely what 'mulling' should mean," Licklider cautioned.[38]

So Licklider maintained that the book and the library should be discarded in favor of what he called "precognitive systems." These systems could retain certain organizational concepts from the traditional library, ranging in scale from words to chapters, indexes, authorial attributions, and even (as classes of information) "books" themselves. But they would not be physical buildings full of bound volumes. At the same time, too, the institution of the "computer center" must likewise be abandoned. What would replace both would be one concoction that Licklider dubbed the "intermedium." By this he meant everything in the human-computer symbiosis. In terms of experience, it meant individual desks or consoles at which users would sit in order to interact with computing systems by means of screen and light pen—a conception clearly indebted to radar and air-defense system consoles. The desk itself would be an "active" device, which the reader of the future would come to regard as a personal "intellectual Ford or Cadillac." Through a time-sharing system its console would have access to a mass of recorded information. (Licklider rehearsed Shannon's old account of the entropy of printed English here to argue for the possibility of the mass storage of literary works.) The computer would be able to parse and sieve this store for utility to particular tasks, in a manner akin to what we would now recognize as data mining. And the user would then manipulate its responses, in a way suggested by an experimental system Licklider had designed. He called it "Symbiont." Symbiont displayed technical documents to a "student" user and allowed for reading, note taking, and other activities.[39]

What leaps out in the twenty-first century about Licklider's proposal, especially when compared to those of other technology and media commentators of the time, is how little attention it paid to the human activity of reading. In fact, Licklider seemed to marginalize the practice altogether, replacing it with various kinds of gestural, acoustic, and graphic mediation. Where the need for it persisted, he sought to rebalance the task between the human and the computer. And it was the computer, not the human, that he expected to "introspect" in order to reveal what was salient about the process and hence help improve it. Licklider's relative silence about human reading stood in strong contrast, not only to contemporaries like McLuhan and Chall, but also to fellow pioneers of human-computer interaction like Ivan Sutherland, Douglas Engelbart, and Alan Kay.

This was reflected in Licklider's curious gestures at demonstrations of these ideas, such as Symbiont. Of all the technical components needed for his envisioned symbiotic system—memory, processing power, languages—the least-developed, he felt, were its "displays and controls." These were the means by which a human and a computer would interact freely and intuitively enough to achieve his desired marginalization of human reading. But in 1960 displays meant oscilloscopes, and controls meant electric typewriters, or at best a light pen (itself descended from

the light-guns used in radar consoles). He envisaged extending the latter to employ a shared "display surface" on which the machine would
show prompts and the human user could handwrite—his ideal computer
would read handwriting, though he recognized that this was a massive
hurdle. It is significant, incidentally, that Licklider's imagined future computer would be able to speak and listen as well as interpret handwriting,
because, as he put it, to expect a corporate executive or military officer to
learn typing was clearly unrealistic.

The kind of graphical interaction Licklider had in mind was indeed feasible at the time, handwriting recognition largely excepted. It was demonstrated in Ivan Sutherland's famous MIT PhD project, a design-oriented
program called *Sketchpad*.[40] Sketchpad replaced textual input with diagrams, symbols, and images; for example, it would translate loosely drawn
boxes into a flowchart design. The idea was not only to get around the
problem of computers reading handwriting, but also to circumvent the
need for humans to produce such writing. As Sutherland himself put it,
prior to this people had been "writing letters to rather than conferring
with our computers," and his system inaugurated "a new era of man-
machine communication." He also envisaged it recursively creating an
ever-expanding library of design elements, while also incorporating details about materials and costs so that they would be included automatically in a design process.[41] Much could already be done, therefore. So it
is striking to turn to Licklider's own efforts, manifested in an educational
device that he designed at this time along the lines of Omar Khayyam
Moore's Talking Typewriter. Licklider's "Tutor 1" used an electric typewriter to interact, as Moore's did, but in stark contrast to Moore's vision
(and to Sutherland's) it was very much *not* an exploratory "environment."
Tutor 1 was a linear question-and-answer machine; it even required the
user to learn arcane control codes to make it work at all.[42]

What constrained Licklider's much-touted vision, it seems, was not so
much technology itself as an insufficiently intensive reflection on what
he himself acknowledged to be central practices for "thinking," especially
in conjunction with computers: human writing and reading. And reading
took precedence, not least because in a future library the flow of information (as opposed to selection, comparison, and the like) would run
overwhelmingly from machine to human. The strategy of sidelining the
practice allowed his "future library" vision to forecast a dazzlingly intri-

cate informational world, but it left a void at the heart of its projected symbiosis. In that respect, it is interesting to note that Licklider's prophecy, for all its renown, was not only impractical at the time, but also distinctly unrepresentative of the technologies that were being advanced in order to design a future information culture. The projects that *did* shape the coming information age addressed issues of reading head-on, and they did so partly by repurposing the science that had addressed that practice in the previous decades—as so many other enterprises had done before them.

In fact, it is not an exaggeration to state that the issue of text-manipulation—and even in some cases the problem of teaching machines to read—became *the* task that possessed the imaginations of key protagonists in the shaping of the digital computer. This was true from the earliest days, when Herbert Simon could tout a 1955 programming initiative that permitted a machine "to learn to distinguish between figures representing the letter *O* and figures representing *A*."[43] Ten years later, leaders in the field were more ambitious. Take Douglas Engelbart, for example. Another veteran of military radar systems, Engelbart was now working on a "Framework for the Augmentation of Human Intellect" at the Stanford Research Institute. Like Licklider, he reckoned that the need to use typewriters and the need for linear reading were the two major logjams in human-computer collaboration. But he attacked those problems more directly. One reason for this was that Engelbart was more interested in text-manipulation itself. He was one of the handful of figures, along with Ted Nelson (and, to be sure, Licklider), whose experiments would eventually give rise to hypertext protocols. Like them, he had his own firmly held view of who would be practicing the techniques he was developing—not a "reader," as such, but a "user." Engelbart's "user" would have definite skills, however, which in a sense were the externalization of cognitive protocols that readers were assumed to exercise tacitly. And his team tended to design around that premise, the aim being, as one engineer put it, to be "in synch with the psychomotor capability of the human being." The system they created, the highly influential NLS (for oN-Line System), required practice to master. Someone faced with its texts could not even "read the goddamn stuff," confessed a member, without being "lessoned into the NLS culture," and Alan Kay reckoned that it took ten hours of "really exercising the thing" to get proficient. But the practice

had an end: working with it was then meant to become as second nature as thinking. And Engelbart worked on interfacing technologies that
might serve that end better than a keyboard. One early attempt was a
"chord keyset" device—a one-handed keyboard substitute similar to the
Microwriter that would briefly flourish in the 1980s.[44] Of greater lasting
significance was his determination to find some way of manipulating
symbols on a screen other than by typing. Thinking beyond the light pen
(and the RAND Corporation's contemporary tablet-plus-stylus technology known as GRAIL), Engelbart designed a small instrument that he
called a "bug." It was supposed to be used in coordination with the chord
keyset rather than a conventional keyboard, with one hand on each.[45] He
thought of it as a stopgap, to be adopted in lieu of a better solution that he
anticipated his group would soon invent. Renamed the mouse, it would
turn out to be much more than that.

We are so familiar with the mouse today that it can come as a surprise to realize that its utility was ever in doubt. But in the late 1960s
and early 1970s that was very much the case. The question of whether to
adopt the mouse as a pointing device was a difficult one. How to determine whether it was efficient, fast, or beneficial? Answering that question
turned out to depend on putting to use the theory of informational interaction that Paul Fitts had developed back in the 1950s, based on his use of
the science of reading to address issues in aviation cockpit design. But it
happened at Xerox PARC rather than at SRI. In 1970–71, Engelbart's project ran into terminal problems when an effort was made to scale up the
system. Newcomers found the learning curve to become his ideal users
too steep, and a system touted as working at the speed of thought proved
painfully slow when deployed on early networks. Engelbart's own eccen-
tricities of leadership made matters worse, until in 1971 many of the SRI
personnel decamped to Xerox. There they became associated with Alan
Kay's Dynabook project. The change of locale implied quite a profound
shift in the ideal reader for their designs. That reader was now not a programmer but a beginner. And because Xerox was a company dedicated
to documentation, the canonical practice of such a person became, not
programming or technical research, but text editing. It was here that for
the first time a user staring at a computer screen would look at a "desktop" with "documents" and even a "library" on it.

At PARC, the question of the mouse was posed to a group of experimenters led by a psychologist-in-training named Stuart Card. Card, cognitive theorist and AI pioneer Allen Newell, and others had launched an initiative of their own that they called the Applied Information-Processing Psychology Project (AIP). It was yet another bid to create a new research field around the issues of information and human creativity. In this case, Card and his colleagues envisaged a "science of the user," to be based in cognitive science and premised on treating the computer and the thinker as interrelated aspects of a system. Work with texts would be its prime focus. And it should have practical applications, in terms of leading to better designs for interfaces. Measuring the speed and efficiency of the mouse in text editing was therefore an ideal topic for this group. Card decided not only to test the device against the various possible alternatives, moreover, but to go further and develop a theory of pointing devices in general. To do that he adopted Fitts's technique of measuring how quickly and accurately someone could move a pointer to a new target location. This meant appropriating a well-understood but hitherto entirely discrete research tradition. Nobody since Fitts himself had thought to replace Fitts's generic targets with readable characters, but this was what Card's context required. In a sense, by making this substitution Card was undoing the shift away from the science of reading that Fitts had made a generation earlier.

His plan in place, Card recruited a group of Stanford students to put it into action. He measured their speed and accuracy in targeting character strings on a screen with four different pointing devices, the mouse being one, over hundreds of trials. The result was an emphatic victory. The mouse produced by far the fastest, most accurate targeting. But more than that, the mouse turned out to be what Card called a "Fitts's Law device"—that is, the graph of its results mirrored the predictions of Fitts's Law. (Two keyboard-based rivals, by contrast, were not "Fitts's Law devices" because their performance reflected the number of keystrokes needed.) And, even more significant, those results were extraordinarily close to the optimum predicted values based on Fitts's theory. "The limiting factor in moving the mouse," Card concluded, "is not in the mouse, but in the eye-hand coordinate system itself." What that meant was that the mouse was effectively a proxy for the human user's own abilities: it was as efficient as simple manual pointing. Barring some remarkable enhancement of human cognitive capacities, it was hard to imagine any future device performing much better, whatever its design might be.[46]

Card's decisive demonstration of the mouse's prowess was an immensely important moment in the designing of the information age. It was the pivotal event that persuaded a reluctant Xerox corporation to embrace the little device, and thence for other manufacturers to follow suit. But for him it was only one element in his hoped-for new discipline. In seeking to create their psychological science of human-computer interaction, he and his partners decided to treat the human as a computer-like entity, and Fitts's Law had a role here too. They modeled the user, now, not as an ideal reader or programmer, nor even as a child, but as a physical and cognitive "Model Human Processor." It had its own short- and long-term memory storage, processing capabilities, codes of operation, and so on, and a "cycle time" of the order of a tenth of a second (a value that was a relic of psychophysics, dating back to Cattell). This allowed for extended analyses of the tasks undertaken by a user together with a computer as if they were combined computing processes. But again, Xerox being the context for this work, specific tasks were at issue: an explicit aim was to compute predictions about how fast people could *read*. And in considering that question, Card found himself returning to eye-movement experiments dating back to Buswell in the 1920s. Based on those and other results, he now calculated that anyone claiming to read more than six hundred words per minute could not possibly be seeing every part of the text. "In other words, speed readers skim." (There is no record of what his boss, Kay, made of this, given his own claim to read at two thousand words per minute.) From there Card moved directly to Fitts's Law again. He reproduced the data from Fitts's original 1950s work at length, and then set his own reader an exercise based on the theory, involving the question of how best to design the keypad on a pocket calculator.[47] The point was that reading was now not the basis of human-computer interaction, but a part of user-computer cognition—and every one of those terms mattered. It must be taken into account in this way in design decisions.

Although only one element in what was a broadly ambitious disciplinary scheme, the model, theory, and testing protocol that Card and his group developed proved extraordinarily important in shaping the basic ways in which people and computers would interact. In the short term, they established a way of evaluating interaction technologies. They were hence at the heart not only of the decision to adopt the mouse at all, but also of the verdict on how many buttons it should have. (The answer turned out to be two, except in Cupertino.) How fast a cursor would move on the screen of a personal computer was also decided based on these criteria; so was the size of every icon. The results were built into Xerox's expensive and unlucky Star computer. And the combination was what Apple's Steve Jobs saw when he visited PARC in November 1979 to be given a demonstration of the Alto machine.

Not only did Apple end up creating an immeasurably more successful implementation than Xerox itself had, as has become notorious. At the same time, the man who had conducted the demonstration, Larry Tesler, departed from PARC and himself joined Apple. Tesler personified the text-centric position of the PARC group, partly because he had extensive experience in producing Palo Alto's Free University newsletter and in page-editing for a Xerox subsidiary called Ginn Publishing. He had pioneered word-processing and markup language. At Apple it was Tesler who insisted that a machine would have to have bitmapped graphics if it were to be usable to produce readable pages—and the Macintosh duly did. As Thierry Bardini concludes, after being relegated and reconstrued in so many ways, human reading abilities were once more at the forefront of information design. At Apple, the imagined reader who had been central to information design for decades was recast yet again, no longer as programmer, child, or algorithmic processor, but as everyone.[48]

Brainy Readers

To understand historically how we experience information, it turns out that we need to look at how reappropriations of the science of reading shaped that experience. A science that had started out with the ambition of attuning readers to texts became a way of attuning texts to readers—and now, as they were reconceived, users. It was the science of reading that bequeathed the techniques, devices, designs, and conceptions that shaped the everyday experience of information in the twenty-first century.[49]

The cognitive sciences—plural—of those users that Card, Licklider, and others envisaged did come into existence, and they continue to be exercised today. Technologies for amassing, sorting, sieving, and "read-ing" (by algorithms) information in enormous data sets are constituents of our everyday experiences. When we book travel or use Netflix, we are both receiving the outputs of such processes and contributing, willy-nilly, to their next iterations. The latter-day versions of Licklider's future librar-ies may more often be operating in concert with libraries than as outright replacements for them, but they have nevertheless become the essential tools with which we all do our thinking and acting. In that sense, the sci-ences traced in this and the preceding chapters do not simply describe our informational world. They have created it, and they continue to shape it today.

That certainly does not mean that reading has ceased to happen, of course, nor that it has been downgraded to the extent that Licklider seems to have desired. The kind of activity that he called "clerical" still cannot always be delegated safely to algorithms, much as software com-panies may like to claim that it can. Andrew Abbott's incisive account of the everyday life of research captures the extent to which successful thinkers have to retain responsibility for making judgments amid the al-gorithms. But at the same time reading is disaggregated into a complex array of practices, ranging from brutally fast scanning to engaged, intense perusal accompanied by lengthy note taking.[50] It is easy to conclude that, even though reading in these different forms remains essential, its frag-mentation is characteristic of the digital and networked world, and that something is being lost. It is all the more important, then, to remember what is really noteworthy. The fact is that reading practices were *always* various. Medieval schoolmen were well aware of that as they pored over Scripture to glean its different senses. What is new in the digital era is not the fact of variety but its constitution, and how we think about it. (It is interesting to note, however, that the first major text-processing proj-ect of the computer era was Jesuit priest Roberto Busa's effort to digitize the corpus of Thomas Aquinas, which was begun in the 1950s and gen-erated decades of introspection among medievalists.)[51] The elements of contemporary reading practices are not the same elements that Aquinas encountered, or for that matter Cattell, and we need to understand them with different tools and methods.

It is partly to meet that need that the science of reading has taken on new life in the twenty-first century. As our informational world has grown more complex, so the necessity for ways to understand how we "read" its representations has grown. The science that Stuart Card helped inaugurate has gone by various names over the years, but nowadays it is generally referred to by the shorthand moniker UX (for "user experience"). The acronym refers both to consumers' relatively passive engagements with advertising and to our more active interactions with operating systems, databases, library catalogs, and complex interactive systems like airline-ticketing sites. In today's UX, eye-tracking technologies have come back into favor. Or maybe they never really fell out of it: even in the 1960s, Licklider was working with human-factors engineer John Senders to bring eye-movement data from Fitts's cockpit research to bear on safety issues in automobiles. Such work continued without a break into the digital era, as we have seen (and Senders would also win an award later for inventing online journals).[52] But they are put to use nowadays rather more generally, not only to answer directly consequential questions to do with marketing or safety, but also to assess the consequences of different kinds of informational display. Eye-tracking shows, for example, that most people do not "read" a new website in the traditional sense. They seek out key words, initially reading across from left to right in much the way one would on a printed page, but then paying less and less attention to the rightward elements as their eyes move down the screen.[53] If some element of information is to be noticed, then, it should be placed to the left. The interactive quality of online information means that observations like this have taken on a new kind of importance—or perhaps that the positivistic dreams of the early scientists of reading are being revived in a new form.[54] And the applications extend from the imagery of icons on a domestic PC screen to the control of weapons, aircraft, and ships in militaries across the world.

How drastic the consequences of UX design could be in such environments was made clear in a number of very public disasters in the 1980s, 1990s, and early 2000s that were blamed on misconstruals of information displays, sometimes made in stressful and fast-changing conditions. Perhaps the foremost example was also the earliest: the case of the shooting-down of an Iranian airliner by the USS Vincennes in 1988. Initial accounts were murky, and the official report largely blamed the fog of war; the ship had been involved in confrontations with smaller Iranian vessels at the time, and in fact it was revealed later that the captain had crossed into

Iranian waters in a pursuit of them, the wisdom of which was contested. But a consensus soon developed that the information displays of the Vincennes's Aegis "combat display system" were at least partly to blame. When the ship's radar sensors detected signals, their multiple reports were received automatically in its Combat Information Center (CIC), but they were not integrated into one coherent display that a viewer could take in at a glance. The data corresponding to one aircraft would appear as numerical readouts on different screens, with one showing a reading on altitude, say, and another velocity (and the operator had to press keys to get even those to appear). Moreover, the displays did not show rates of change, so the operator also had to calculate those values himself by hand, at the same time as viewing successive readings. The cognitive and physical dexterity involved was formidable even in ideal conditions. At the moments when the decision to fire at the aircraft was being made on the Vincennes, conditions were not ideal. The ship was pitching into a hard turn in pursuit of the small Iranian boats, and the CIC was a chaotic mess of flying papers and other detritus.[55]

In hearings on the Vincennes disaster before the House Armed Services Committee, the social psychologist Richard Nisbett invoked Bruner's classic experiment in which he had had people read images of playing cards with a tachistoscope.[56] He implied that just as Bruner's students' readings, arrived at quickly, had rested on expectations as much as sensory inputs, so a system that demanded a confusing mixture of visual and physical coordinations had given rise on the Vincennes to a "reading" that matched trained expectations but not reality. The Aegis system was often touted as the world's most sophisticated, and it was. But it was sophisticated for a different purpose: fighting massive air-sea battles against formations of Soviet bombers. It turned out to be extraordinarily poorly designed for tracking a single airliner, especially when operated by stressed crewmen aboard a pitching ship inside hostile waters. (Nisbett's testimony in fact had a pointed double meaning, as his wording also implied that the navy's own report into the incident, which had exonerated all concerned, suffered from a similar information-processing failure.) Richard Pew, an ex-colleague of Licklider's and a veteran of human-computer interaction design from the 1960s and 1970s, and now a leading psychologist at Licklider's old company of BBN, added that the problem was but one example of "'glass cockpit' syndrome." Operators became overwhelmed by the displays of information on multiple CRT screens, he explained, adding that the phenomenon was well known from decades of research on the reading experiences of pilots. Their point does seem to have struck home. In the wake of the inquiry, the navy did institute a series of researches on decision-making and systems design, in which eye-tracking experiments played a significant role. It is impossible to document this definitively, but informants in today's UX industry have told me that the navy's initiative played a role in legitimating eye-movement research in that enterprise. At least, there exists a collective memory to this effect.[57]

Military, automotive, and aerospace uses of such research have been particularly striking, but in fact across the informational world research on reading is as vibrant and various as it has always been. Indeed, it has a disciplinary identity now that it did not possess until the end of the twentieth century: a professional society, the Society for the Scientific Study of Reading, was founded in 1993, and since 1997 it has published a regular journal of research, *Scientific Studies of Reading*.[58] But the field's place in the classification of knowledge has changed. For most of the twentieth century the science of reading was primarily a subfield of experimental psychology (or, to a lesser extent, of sociology). In the twenty-first century that affiliation remains very strong. As a rough indication, out of forty-four contributors to Snowling and Hulme's *Handbook* to the subject, thirty—nearly 70 percent—are based in psychology departments; the next most common home, a cognition or neuroscience unit, has only seven. But at the same time, the science of reading has also become one element in a self-consciously coherent cluster of scientific, educational, medical, corporate, and military enterprises embracing technology, mind, and body. The science of reading is certainly not alone in experiencing this change. It is a broad development that has taken effect across the disciplines, and in particular across those disciplines that relate to informatics and the life sciences.[59] But for the science of reading it has deeper roots than digitization and the Internet, as we have seen. It originated in the intellectual reconfigurations of the World War II era. If anything, the digital revolution rested on this hybrid discipline, rather than vice-versa.

One common element of this new disciplinary configuration about reading is its explicit focus on the brain. Market research that uses eye-tracking—even that based on AI pseudo-readers—is nowadays generally described as based on neuroscientific principles more than on those of cognitive science. And recent years have seen a small industry develop of trade books by scientists of reading explaining how fMRI and other imaging technologies can help us understand, if not quite what reading *is*, then at least the areas of the brain that are activated while it happens.[60] As with other developments in the field, it is possible to see this as a step toward realizing ambitions that are surprisingly old: Guy Buswell, as we have seen, was keen to use "brain-wave apparatus" before World War II in a bid to extend the experimental study of reading into the realms now claimed by neuroscientists. The technology and funding available for cognitive neuroscience promise a new wave of research interest, and with that a generation of proposals for new pedagogical strategies and remediation techniques for struggling readers. And it has been claimed that reading is the exemplary case of educational neuroscience in general, so more than literacy itself is potentially at stake. At present, however, how much the "neuro" component in proposals about reading really adds is not always clear. Neuroimaging technology still tends to lack an ability to make very fine distinctions, so claims based on it often do not diverge in practice very far from already familiar theories. At the same time, though, "neuromyths" abound, and are fomented in commercial marketing. Educationalists are understandably wary at the prospect of facing yet another round of demands to invest resources in hyped-up devices, this time said to solve problems of early reading by virtue of "brain-based learning." Moreover, even if in some cases neuroscience can help us judge between rival hypotheses—we may be able to decide whether a given reading disability is visual or linguistic in nature, for example—it rarely has much to suggest about how to teach. At the current moment, it seems best to adopt a reserved endorsement, pausing to listen to practicing teachers rather than hastily embracing neuro-determinism. The problem, as Cambridge University's Usha Goswami notes, is not so much neuroscience itself as what the "ed-tech" and publishing industries make of it. Many educators would surely agree. Perhaps, as Goswami has suggested, the neuroscientific moment may even offer an opportunity to reengineer an interaction between education and the commercialization of science that has long been dysfunctional.[61]

Neuroscientific approaches are often said to promise an answer to the most obvious question about reading in the digital age. Is there anything *different* about digital reading? When we read on a screen, is it essentially the same as reading on paper, or not? What if the screen is showing information that is not contained within the immediately present system, but networked worldwide? Subjectively, the answer seems to be that such reading is indeed different, at least for readers brought up on print and paper. They report that it is increasingly hard to concentrate. Or perhaps *concentrate* is not quite the right term—what is difficult to sustain is not intense concentration, because we still feel able to work, but the informally continuous sense of casualness or flow that used to characterize novel-reading. The old feeling that one could lose a sense of time while absorbed in *The Woman in White* seems to have disintegrated. It is partly that there is always some distracting possibility a click away. One becomes habituated to hopping quickly from location to location, gathering information in discrete snippets, to the extent that it requires conscious training—and the self-discipline of a recovering addict—to revert to reading sequentially and at length. But what evidence is there that reading itself is changing? The question is nowadays construed as essentially neuroscientific. Neuroscientists of reading look for alterations in the parts of readers' brains that "light up" in different reading contexts. They claim insights on that basis that will, by virtue of informing our behaviors, and thanks to the phenomenon of neuroplasticity, change the brains of future readers. As the science's most accessible champion, Maryanne Wolf, puts it, neuroscience is telling us "facts about reading and the reading brain" that "will lead to significant cognitive changes in you, the next generation, and possibly our species."[62]

It is appropriate to conclude the substantial story told in this book with the advent of the neuroscience of reading, because the contentions of neuroscience are recapitulating many of the arguments and claims of previous generations.[63] Cognitive neuroscientists are keen to point out, for example, that a good reader is constantly predicting what words are likely to come next, even as they enter the fringes of one's peripheral vision. They infer that an unknown array of past experiences are involved in any worthwhile act of reading. Even the much-maligned notion of "reading readiness" that was so much a theme in the conflicts over pedagogy that took place from the 1950s through the 1990s makes a comeback of sorts. And so, more significantly, does a convenient evolutionary mode of explanation, coupled with an appeal to a natural-historical account of ancient writing systems that Huey would surely have found congenial. In each case, however, these points now come with a reference to a neuronal circuit or a location in the brain, which is said to develop and change as a consequence of routine experiences. And that is a difference in knowledge. The doctrine of neuroplasticity does suggest possibilities for experimentation to trace its dauntingly complicated processes. But, as in the case of teaching, many of the recommendations presented as neuroscientific could have been arrived at without fMRI machines. By the time they reach the general public, such recommendations typically include guidance to take more time, turn off screens, and read deeply so as to rehabituate oneself to what Wolf nicely calls "cognitive patience." Such advice is well taken, of course. But similar strategies have been suggested in the past without access to such science—with less credibility, perhaps, but that suggests the commonplace thought that what imaging technologies really contribute, at least in their present state, is credibility itself. Ask further questions, about mechanisms for example, and one rapidly reaches the terrain of the "many unknowns" that lie at too small a scale, or in too complex a set of neuronal relationships, for current technology to cast much light on them. It becomes clear just how insensitive state-of-the-art devices still are in comparison to what we want them to tell us.[64]

A century and a half of history should temper any haughtiness one may be tempted to feel. Has there been any moment in the history of the science of reading, after all, when the field was *not* defined at least as much by its promises as by its achievements? And the tendency to confuse the two—to assume that what practitioners hoped to achieve had already been achieved, and that it constituted good grounds to change policy and practices—is similarly traditional. The gap between promise and accomplishment is large now, but it has always been large. To acknowledge this is not to denigrate the science, although obviously it does counsel caution about the more extravagant claims of its champions, perhaps especially in the nonspecialist media. On the contrary: it is to recognize that the science of reading is indeed worth doing. It is to accept that it is a *science*. That is, it is a disciplined practice of collective curiosity, dedicated to using all the techniques of the laboratory to explore natural phenomena that are themselves endlessly protean, fascinating, complex, and consequential. It is, as sciences should be, intrinsically open-ended.

It has already had huge effects—on how we think, how we learn, what we do, and who we are. And it will surely continue to have correspondingly great effects in its future neuroscientific guise. Some of its consequences have been good; others, to put it no more strongly, have been the subject of intense and sustained denunciation. And, finally, it has itself both shaped and been shaped by every current of modern American history, from the invention of the mass media to the advent of computing and modern life sciences. Any science worth its salt has to be emphatically historical in this way, through and through. The only way to grasp its character, achievements, and implications—for past, present, and future—is to take the trouble to appreciate it "warts and all," as the profoundly human enterprise it is.

Conclusion

Reading, Science, and History

We live today in a world shaped by the science of reading. The pages we look at, and the characters on them that we cannot help but interpret, bear the impress of this science. So do the screens we find ourselves gazing at for so much of the time, and even the tools we use to interact with the digital information we encounter when we do. In countless ways large and small, our everyday experiences reflect the development and impact of this scientific endeavor. The fact that its consequences are so ubiquitous is an obvious reason why the history of the science of reading matters. Among the sciences, that of reading may lack the self-evident grandeur of fields like cosmology and high-energy physics, which reveal the deep structure of the universe, and it cannot quite claim to tell us about the fundamentals of human nature in the way that genetics and neuroscience do. Whatever James McKeen Cattell and Samuel Renshaw may have thought, it generally cannot offer cures for diseases or means to extend human life. But how we all live and learn rests on the achievements of this endeavor. Understanding its history therefore helps us appreciate in new ways why it is that we think and act as we do.

The very ordinariness of reading is a major reason why the science that attends to it is worth doing, and in turn why that science is worth understanding historically. Reading is an activity that is quietly essential to modern life. We pay it little attention because it is simply so commonplace that in many cases we scarcely notice it happening at all. At the same time, as so many have observed, the practice is not natural,

417

and it should not be taken for granted. Learning how to read has always been a process that has required work, time, and resources, and for many people—arguably, for all people—reading is a skill never fully mastered. When the science of reading was invented, in the late nineteenth century, that discontinuity and those needs were not new, but they were newly visible. Darwinism in particular made the issue of reading's unnatural character consequential. The erosion of the boundary between human and animal lent special significance to this peculiar activity that only humans performed, and at the same time the advent of social Darwinism posed discomfiting questions about the acquisition of such a skilled practice and its relationship to grand issues of progress and degeneration. Such questions loomed large because of the intense presence of industrially printed media in late nineteenth-century society. Newspapers, billboards, signs, magazines, forms, and, of course, books were everywhere, to an extent that no previous society had known. Advanced industries required workers who could read and apply instructions, and do so in ever-changing contexts. In the era after the Civil War, this need for a skilled workforce combined with the common conviction that America should have an informed electorate to make the attainment of mass literacy into a national mission. Mass education came to be seen as central to the achievement of a modern democratic society. At the same time, the fate of individuals in such a society was increasingly believed to rest on the ability to be an expert, proficient reader. All this meant that the processes of reading and of learning to read had to be understood and managed scientifically. The first generation of psychophysics researchers recognized this need, and they responded to it by creating a new science.

The science of reading remains worthy of our attention today for similarly broad and deep reasons. The concept of reading itself may seem to be wider now, when we include within it our interactions with screens and various instruments. But it is no less true now than it was in Cattell's time that the act of reading itself is foundational to modern life. Meanwhile, it remains a difficult attainment for many people: in 2019 the US Department of Education estimated that 21 percent of American adults, some forty-three million people, lacked the ability to do basic literary tasks.[1] It seems that if we do not try to ascertain how texts "work"—and how we work with texts—then we will be admitting defeat in our efforts to understand and sustain what we may call civilization. It makes sense, then, that we are probably even more acutely concerned today than ever about how best to read, and especially about how best to become readers in the first place. As a society we devote enormous resources both to early learning and to efforts at remediation for those adult citizens for whom reading remains difficult. There is a consensus both that society at large requires that its members be proficient readers and that an inability to read well is a severe constriction on any particular individual's well-being. Questions of reading thus have a very pressing moral and practical edge. If you doubt that, just sit in on a meeting of any suburban elementary school's parent-teacher association. We want our society's efforts at teaching at all levels, especially the most fundamental, to be well-founded. One unavoidable reason why the science of reading will continue to be pursued intensively, then, is that the processes and institutions that shape us, our children, and virtually all our fellow citizens rely on that science. For us, no less than for Huey, Dodge, Thorndike, and Renshaw, to stop doing the science of reading would seem like an abandonment of the promise of America.

That highly charged context makes the science of reading unusually vulnerable to a paradox that affects all sciences, however. On the one hand, it claims to be producing reliable and robust—in a word, true—knowledge about its subject. But on the other, it produces that knowledge by virtue of being a continuing research enterprise, which means that its arguments are always provisional and subject to refinement or correction. When exposed to audiences accustomed to simplistic popular representations of science as a collection of irrefutable facts and theories, the conjunction is acutely unstable. If scientific views of reading alter over time, are they worthy of respect? Should my child be subjected to teaching methods based on arguments that we know, from past experience, will change? It is hardly surprising if patience wears thin. Lamenting that fewer than half of Americans were "proficient" readers, for example, in June 2021, *The Economist* warned that the nation remained "stuck" in a debate over teaching methods that had been grinding on for decades. The magazine lambasted what it called the "psycholinguistic guessing game" approach, and with it the "balanced" strategy inherited from the 1990s. Instead it hailed the state of Mississippi for embracing phonics wholesale. The way forward, it proclaimed, was to reject ill-founded prejudices and accept "the science of reading" lock, stock, and barrel.[2] One can sympathize to some extent with the opinion, but of course there was a deep irony to such a call. The very approaches that were being condemned were themselves products of the science of reading as it had been scarcely a generation before. It could be argued that the real problem here—for the magazine's writers as well as its readers—was not a failure to honor scientific claims, but an innocence about the historical constitution of such claims.

Whatever else may be said about the reading wars, they did, and do, make clear how great the stakes can be in any effort to understand reading and create more readers. They also make clear how great the stakes are in suggesting the need to historicize such understandings. All that makes it a little daunting to advance a proposal to view the science of reading historically, as a contingent body of knowledge, technology, and practice that has emerged from complex cultural processes. One runs the risk of being condemned for undermining the credibility of a particular method of instruction, and even of scientific expertise in general. In the 2020s the latter is not a risk to be disregarded lightly. Could it not be argued that we simply know more now than we used to, in which case the history is all moot? Experts may fear that if citizens are exposed to past scientific views of reading that are no longer embraced, they will wonder why they should trust what the science now tells us.

My own impression is that the focus of such anxiety would be misplaced. The blame for any erosion in public confidence should be placed not on scientists' knowing more, or on their knowing differently, but on such a changing enterprise being in conflict with an inaccurate and simplistic view of the nature of science in general—a view that is widely believed, in part, because popular scientific writers themselves have promulgated it in a misguided conviction that they are defending science amid a "crisis" of corrosive public skepticism.[3] If one has become used to assuming that science is unchanging, unquestioning, and stable, then of course any historical knowledge of its development is likely to be seen as a damaging revelation. Worse still, if the revelation is just about *one part* of science, then the reaction is liable to be an inference that this particular part was arrived at incompetently, or more probably (because it has passed as authoritative) corruptly. Somewhere, it will be assumed, there will exist another bit of science that will be as true as science in general is supposed to be, and such an alternative is, of course, putatively discoverable by lay "research." Such is the all-too-familiar structure of skepticism in contemporary America. The anxieties about expertise that such a reality gives rise to among scientists are real and reasonable. But the phenomenon emerges from too little knowledge—or the wrong kind of knowledge—rather than too much.[4]

My own impression is that knowing where the science of reading came from, how it has developed, and what its impact has been will have a positive social value. It should help undergird the kinds of conversations that a twenty-first century society needs to have, which necessarily cut across the distinction between expert communities and laity—or rather, which involve an acceptance that, especially in a field like this, "lay" people have substantial expertise in their own right, to which credentialed experts should be attentive. History teaches humility, on all sides, and that is generally no bad thing. After all, in training readers in the ways we do, society is making a choice. It is not a choice that can be shirked. We have to adopt some set of strategies, while alternatives to those strategies can and do exist—and have in fact existed in the past. And whatever choice we make does indeed have huge consequences, which may well be comprehended in terms of benefit and harm. For many of us, in our roles as parents, teachers, administrators, and politicians, those are often the appropriate terms to use. But at the same time—and it really should not be too hard to achieve this—we can also approach them in a spirit of curiosity, to see what may be learned. In our roles as thinking members of a reflective, intelligent culture, we can adopt a nonjudgmental stance, at least for the duration of the exercise. It is certainly the case that past knowledge about reading was not simply disastrous, because it did in fact generate real readers (and some great writers too). We may think about how such knowledge and methods have been decided upon in prior contexts, and consider them not only in terms of a one-dimensional scale of success and failure—a linear measure of ability at a defined skill—but also in those of the more sophisticated, multidimensional question of what *kinds* of reading and community they have given rise to. We can wonder how they have contributed to the formation and transformation of cultures. We can acknowledge that there are different kinds of reading, and try to evaluate the existence and consequences of such practices in even quite recent history. To understand the nature of twentieth-century literature, say, it can be helpful to accept the need to recover what practice of reading was applied to such literature. The history of the science of reading will help us do this. Understanding it is constructive, not corrosive. It is in that light that it can contribute to a proper approach to the learning and practice of reading today.

This still involves granting the point that what the scientists of reading did before our own time was not simply ill-conceived and futile. I have emerged from this experience with a conviction that the practice of that science was in truth immensely worthwhile, if only in the way that the dedicated pursuit of any sophisticated scientific tradition is worthwhile. Between psychophysics and cognitive science, the scientists and others who sought to understand the nature of reading developed concepts, questions, methods, instruments, analytical tools, and social networks that gave rise to knowledge. And that is always valuable. The knowledge that these scientists generated was real and consequential at the time, even if in later decades it would be sidelined, disregarded, or even dismissed (and occasionally revived, too). This is in fact typical of modern sciences. As Peter Galison has contended, a scientific discipline almost never transforms lock, stock, and barrel at one moment; instrumental, experimental, and theoretical traditions have their own continuities and discontinuities.[5] So it has been in the science of reading. Eye-movement cameras, say, show a continuous history back from today's digital technologies, through Buswell, to Dodge and Erdmann, even as the cognitive turn has transformed what such devices are thought to show. To understand that is to be equipped to suggest judicious and responsible stances with regard to today's claims. For example, one of the most important turning points in the reputation of this science came with the acknowledgment that childhood reading was not merely a crude version of mature, expert reading, but something different that warranted attention in its own right. In practical terms, that meant that in teaching one should not aim to inculcate the habits of expert readers directly—those would come later—but a distinct, somewhat mechanical practice. This marked the ascent of the "decoding" motif in the reading wars. But by the same token, it also implied that decoding was *not* so useful *after* a pupil had made those first steps. Recognizing this, one could point out that the notion of "balance" that became so prominent in those debates—the notion attacked in that *Economist* article—could still be valuable if interpreted in a diachronic sense. The balance could lie in stressing decoding techniques initially and other techniques later.

But there is much more to it than pedagogy. Beyond the concerns of contemporary citizens, teachers, and scientists—and, perhaps, underlying them too—there is a more profound reason why the science of reading is worth excavating historically. It forces us to confront a discomfiting double reality. On the one hand, reading, that foundation-stone of human knowledge in modernity, is understood only in imperfect, contingent, and partial ways. But on the other, our ideas about it are immensely and inescapably consequential. We cannot do without them, whether we may want to or not. It is largely because reading is such a difficult process to understand that beliefs about it have changed so radically over time. But that means that experiences of reading itself must also have changed radically, for the simple reason that everything about those experiences arises through education, training, and acculturation. Pages have been designed differently, circulated differently, and approached differently, all on the basis of changing accounts of the practice and effects of reading. We have seen, for example, how children's books have consequently taken on different forms and appearances at different stages in the history of the science of reading. The same is probably true of all genres, those for adults and professionals at least as much as those aimed at children and beginners. And readers themselves have been trained to perform reading in quite distinct ways. Even at the most intimate level, therefore—in the moment when a reader, apparently alone, silent, and unencumbered, comes face-to-face with the page of a book—what happens is profoundly shaped by educational and cultural processes. The philosopher Ian Hacking has written eloquently about how certain disciplines in the human sciences, psychology among them, produce classifications of people that then affect the people classified, producing a recursive "looping" effect that over time makes the phenomena themselves more real.[6] Something similar has happened historically with the science of reading. Because it has changed learning practices, this science has continuously and inevitably altered the very practice that it studies. (That may be one reason why the science itself has been so changeable, of course—unusually for a science, its very changes testify to its fidelity.) The implications of this for what we can know about ourselves, our culture, and the cultures of others are deep and serious.

Take the discipline of history, for example. History generally rests on the preservation and interpretation of written documents. When you read a dramatic account of the Battle of Actium or an interpretation of the effects of the Peace of Versailles, you are generally viewing a refined, interpreted version of evidence that has been accumulated in written or printed form. And just as historical facts rest on such traces—or perhaps *are* such traces—much of historical *explanation* is based more or less implicitly on assumptions about the powers of written and printed materials. This is especially true for those branches of history that focus primarily on culture and ideas: intellectual history, certainly, but areas like the history of science too. There is a largely tacit consensus fundamental to such fields, all the more important for being unacknowledged explicitly. It holds that the act of reading is at root the same now as it has always been, or at least that the differences are not radical. To be sure, good historians accept as a matter of course that context matters and that the original reading of, say, a sixteenth-century tract cannot be inferred from the text itself. As we like to say, the meaning of a work is in the hands of its users. But still we assume that we can understand how that meaning comes into being, and that we can do so by reading. At a fundamental level, grounded in the human frame itself, our experiences still seem to mirror those of our predecessors. True, the assumption is not simply naive. We assert that we can understand past meanings better by multiplying our own reading experiences, so we pore over many pamphlets rather than just one. But as that implies, context, in practice, means more text. That is the basis on which we remain confident that our reconstructions of, say, Reformation religious experience are at least well-founded.

Sometimes the things we most take for granted—the things we assume are fixed points in the cultural universe—are precisely the things the histories of which are most worth excavating. We calibrate our senses of self, constancy, change, and time by reference to them, so when we appreciate that they change we learn something especially significant about ourselves and our world. That is true of reading. Recall Raymond Dodge's disquiet in 1901 when he noticed that his habit of reading was constantly being activated automatically, whether he willed it or not, because he was always coming into contact with printed materials on the walls of buildings, the sides of tramcars, the advertisements of news stalls, and so on. He believed that this was one of the distinctive features of modern American society, and he was scared enough of it to call for the authorities to mandate redesigned carriages. A century later, we are now extremely familiar with such constant activation, and again we feel that we are undergoing something new and disruptive. Dodge contrasted the hectic experience of print in 1901 with a more stable and manageable prior world of information, and we in turn contrast our bedlam to an earlier information culture that we assume people managed comfortably. Of course, it need hardly be said that *our* lost halcyon days were *his* days of terrifying novelty and chaos. He believed that a radically new experience of reading demanded the creation of a science to understand it, and so do we; where he appealed to the latest psychophysics, we appeal to the latest neuroscience. The fate of a new generation of Americans was at stake, in his view, and that remains true for us today. And a corresponding set of perceptions may well arise in future generations. In other words, the conceit that reading is an automatic practice, based in human nature and accessible to scientific research, is itself historically specific, as well as the kind of science that is pursued on that premise. The skills that that science gives rise to vary accordingly: it is no more or less plausible to suppose that readers trained in the 1920s really did grasp words first than to assume that those trained in the 1700s or 1980s grasped characters—or that those of the 2020s do so now. Things were very different a mere century ago. Imagine how much more different they must have been five centuries ago, or a millennium. The way to an appreciation of such differences, I propose, is via an understanding of the knowledge and training through which skilled readers have been made. For a scientific age, that way lies through the science of reading.

That said, it is interesting to note that a good deal of current scientific opinion about reading actually supports the contention (here advanced on a historical footing) that it is a variegated and dynamic practice, not reducible to one basic and unchanging perceptual skill. In noting that, I do not mean to question the current consensus that a "decoding" model is the preferred basis for teaching early readers. That consensus is based on the experiences of countless teachers, after all, as well as on the researches of professional scientists. What I mean is that when we read in different ways, our brains really do change, at least according to the current doctrine of neuroplasticity. Changing habitual experiences actually correlate with different neural phenomena. Reading is contingent on cultural practice, all the way down to the level of neurons. This is the sense in which Maryanne Wolf can declare that the shift from print to screens not only will change how our brains operate, but has done already.[7] To be explicit, the neuroscientific claim is not that when our brains change in concert with differing reading practices those changes are inherited by our offspring. They are not genetically encoded. It is that new reading practices give rise to new neuronal pathways in the individual concerned, meaning that the resiliency of our reading habits does have a certain basis in the brain itself. We then create technologies, environments, and teaching practices that serve those habits and will tend to foster them in our followers.

That suggests a point about the interdependence of history and science. Much that the modern science of reading investigates, and everything that it claims to know about the practice, turns out to be cultural "all the way down." This is all the more apparent as the science and the history of this field converge on a shared understanding—an understanding that reading is indefinitely multiform and unsettled. It is shaped by cultural experience, by history itself. This is true even at the neuronal level. What might seem at first an ultra-reductive approach, neuroscientific imaging, here ends up forcing a broadening of its own perspective. It confirms that we do not learn everything about reading from its fMRI machines, any more than Judd and Buswell learned everything about it from their eye-camera instruments in the 1920s. Like those earlier machines, our instruments can and do confirm for us their own limits. We can know scientifically that science alone is not enough. Neuroscientists should not *become* historians, or vice versa—there are good reasons why the different disciplines exist, and the risks of trivialization in over-hasty claims to interdisciplinary insight are too great. But they will certainly need to heed them.

The implications of historicizing reading should be vertiginous for historians. They should feel that the epistemic ground they were standing on has suddenly lost its solidity. But it is worth suggesting that those implications should be no less unnerving for scientists too, for all that the latest neuroscientific claims are conveniently congruent with those advanced here. They pose the question of how it is that scientists themselves read: what particular properties such reading may possess, how the skill is learned, how it has developed over the centuries, and how it may be changing now. That question matters, not least, because of contemporary concerns over the ways in which we come to know whether science itself is trustworthy and productive—whether, indeed, it "works."

In the current generation, the churning out of scientific papers has reached absurd proportions. Estimates are that there are currently some

30,000 scientific journals in existence, many of them "predatory" organs, publishing about 2–2.5 million papers per year. Almost all of those papers are pointless as scientific contributions; they have value solely as entries in résumés and citation indexes. The publishing and measurement of research productivity have generated an arms race, pitting citation algorithms against entrepreneurial system-gaming enterprises that offer to research, write, and publish papers for ambitious careerists.[8] Whatever "scientific reading" may be, any practice conducted in the face of this mass of words, graphs, tables, and images surely cannot be readily identified with the act laypeople are familiar with. In the same period, meanwhile, warnings have multiplied of a "replication" or "reproducibility" crisis, especially but not only in psychology. All too many experiments published in reputable journals—over 50 percent in one much-touted study—turn out to fail replication checks. It even seems that this crisis may be magnified in its effects because nonreplicable papers are apparently far more likely to be cited than replicable ones, presumably because their claims tend to be more interesting or daring.[9] Such papers are not only passing review, then, but are being "read" more than other, presumably more reliable, pieces. And judgments of scientific value, credibility, or worth are evidently being arrived at and acted upon. The shape of scientific knowledge itself thus seems to reflect, at least to this extent, a distinctly odd kind of reading practice.

There is not the space or time to develop the thought here. But it does suggest that we should look much further and more deeply, not only at the science of reading, but at the reading of science—or rather, at the act of reading *in* science. Both historians and scientists themselves should look at how it is that scientists themselves learn their professional practices of "reading," to call it that. Those practices are likely to differ from discipline to discipline, and quite possibly from subfield to subfield, in ways that we have not yet even begun to map out. And with implications that will only become visible to us when we have done that work. One way to resolve the alleged crises of the scientific enterprise may lie in an understanding of those practices. We need to know what they are, how they developed, and whether they need to change.

My impression after delving into this subject is therefore that a radical historicization of reading does not need to be epistemologically defeatist. Quite the contrary, in fact: it opens up huge and potentially very consequential new prospects for research and enlightenment. I have only skimmed the surface here of a subject that merits much more attention, across a myriad of times and places. There are plenty of kinds of evidence with which we can work to build a new kind of understanding of how culture has been created, circulated, and maintained. The readers of eighteenth-century France, say, were poachers, and so were those of 1960s San Francisco, but they were different kinds of poachers, with different quarry, different tools and tactics, and different gamekeepers to steer clear of. They learned their poaching skills through experiences and sciences that can be brought back to light. All of those things have left traces and are investigable. The point of history, I think, is to explore the richness, variety, depth, and connectivity of human experiences, rather than to adjudicate between them, and attending seriously to the historical cultures of reading is a way to do that.

There is the possibility here for a new departure, or at least a new way of seeing other cultures, whether they be separated from ours by time or by space. The pioneer of information technologies J. C. R. Licklider put the matter nicely in 1960 when he was first conceiving of the kind of world we now inhabit. Licklider remembered back to the very earliest years treated in this book, invoking as he did so the mathematician and philosopher Henri Poincaré. An exact contemporary of the first scientists of reading, Poincaré invented a universal information system for mathematicians that was designed to make them into perfectly efficient disciplinary readers. In dreaming up the venture, he had come to realize that the very problem it might solve had not yet been properly identified. Perhaps our predicament is not so different. "The question is not, 'What is the answer,'" as Poincaré pointed out then, and as Licklider reiterated at a similar moment of potentiality. "The question is, 'What is the question?'"[10]

NOTES

Introduction

1   E. B. Huey, *The Psychology and Pedagogy of Reading* (New York: Macmillan, 1922 [1908]), 2, 5, 7–8.

2   For one example, the technology writer Nate Anderson draws on Nietzsche to suggest strategies for dealing with today's digital overload: *In Emergency, Break Glass: What Nietzsche Can Teach Us about Joyful Living in a Tech-Saturated World* (New York: W. W. Norton, 2022).

3   For some earlier thoughts on these lines, see A. Johns, "Science and the Book in Modern Cultural Historiography," *Studies in History and Philosophy of Science* 29 (1998): 167–94.

4   A large number of works could now be cited, but see, for example, A. Williams, *The Social Life of Books: Reading Together in the Eighteenth-Century Home* (New Haven: Yale University Press, 2017) and A. Blair, *Too Much to Know: Managing Scholarly Information before the Modern Age* (New Haven: Yale University Press, 2010). A key early manifesto for the field was R. Chartier, "Texts, Printings, Readings," in *The New Cultural History*, ed. L. Hunt (Berkeley: University of California Press, 1989), 154–75; other statements of that period include R. Chartier, "Culture as Appropriation: Popular Cultural Uses in Early Modern France," in *Understanding Popular Culture: Europe from the Middle Ages to the Nineteenth Century*, ed. S. L. Kaplan (Berlin: Mouton, 1984), 229–53, R. Chartier, "Du livre au lire," in *Pratiques de la lecture*, ed. R. Chartier (Paris: Editions Rivages, 1985), 62–88, and R. Darnton, "History of Reading," in *New Perspectives on Historical Writing*, ed. P. Burke (Cambridge: Polity, 1991), 140–67.

5   M. de Certeau, *The Practice of Everyday Life* (Berkeley: University of California Press, 1984), 165–76.

6   P. Bourdieu, *Distinction: A Social Critique of the Judgement of Taste*, trans. R. Nice (London: Routledge and Kegan Paul, 1984), 466–84; R. Hoggart, *The Uses of Literacy* (London: Chatto and Windus, 1957).

7   D. F. McKenzie, *Bibliography and the Sociology of Texts* (London: British Library, 1985).

8   See especially T. F. Gieryn, *Cultural Boundaries of Science: Credibility on the Line* (Chicago: University of Chicago Press, 1999).

9   See, for example, R. Smith, *Between Mind and Nature: A History of Psychology* (London: Reaktion, 2013); W. J. M. Levelt, *A History of Psycholinguistics: The Pre-Chomskyan Era* (Oxford: Oxford University Press, 2013); and, classically, E. G. Boring, *A History of Experimental Psychology* (New York: Century, 1929; 2nd ed., New York: Appleton-Century-Crofts, 1950). More general is R. Smith, *The Norton History of the Human Sciences* (New York: Norton, 1997).

10   J. W. Cortada, *All the Facts: A History of Information in the United States Since 1870* (Oxford: Oxford University Press, 2016), 28–47, 92–235; P. Starr, *The Creation of the Media: Political Origins of Modern Communications* (New York: Basic Books, 2004), 233–66, esp. 240–50.

11   On this debate, see J. Auerbach, *Weapons of Democracy: Propaganda, Progressivism, and American Public Opinion* (Baltimore: Johns Hopkins University Press, 2015), and S. Herbst, *A Troubled Birth: The 1930s and American Public Opinion* (Chicago: University of Chicago Press, 2021).

12   For more detailed accounts of this subject, see A. Johns, *The Nature of the Book: Print and Knowledge in the Making* (Chicago: University of Chicago Press, 1998), 380–443, and A. Johns, "The Physiology of Reading," in *Books and the Sciences in History*, ed. N. Jardine and M. Frasca-Spada (Cambridge: Cambridge University Press, 2000), 291–314.

13   See in general E. Shagan, *The Rule of Moderation: Violence, Religion, and the Politics of Restraint in Early Modern England* (Cambridge: Cambridge University Press, 2011).

14   A. Winter, *Mesmerized: Powers of Mind in Victorian Britain* (Chicago: University of Chicago Press, 1998), 306–43.

15   D. Hartley, *Various Conjectures on the Perception, Motion, and Generation of Ideas (1746)*, trans. and ed. R. E. A. Palmer and M. Kallich (Los Angeles: William Andrews Clark Memorial Library, University of California, 1959), 53–55; D. Hartley, *Observations on Man, His Frame, His Duty, and his Expectations*, 2 vols. (London: printed by S. Richardson for J. Leake and W. Frederick, 1749), I, sig. a3r, 268–323; II, 255. More generally, see Johns, "Physiology of Reading."

16   A. Warwick, *Masters of Theory: Cambridge and the Rise of Mathematical Physics* (Chicago: University of Chicago Press, 2003).

17   Johns, *Nature of the Book*, chap. 6; R. Darnton, "Readers Respond to Rousseau: The Fabrication of Romantic Sensitivity," in *Great Cat Massacre* (New York: Basic Books, 1994), 215–56, esp. 242–43; G. J. Barker-Benfield, *The Culture of Sensibility: Sex and Society in Eighteenth-Century Britain* (Chicago: University of Chicago Press, 1992), chap. 2.

18   Quoted in Johns, "Physiology of Reading," 308.

19   J. S. Mill, "Attack on Literature" (1831), in *The Collected Works of John Stuart Mill*, vol. 22, *Newspaper Writings: Vol. A*, ed. A. P. Robson and J. M. Robson (Toronto: University of Toronto Press, 1986), 318–27; Mill, "Comparison of the Tendencies of French and English Intellect" (1833), in *The Collected Works of John Stuart Mill*, vol. 23, *Newspaper Writings August 1831-October 1834*, ed. A. P. Robson and J. M. Robson (Toronto: University of Toronto Press, 1986), 442–47, esp. 445–46; Mill, "Civiliza-

tion" (1836), in *The Collected Works of John Stuart Mill*, vol. 18, *Essays on Politics and Society*, ed. J. M. Robson (Toronto: University of Toronto Press, 1977), 117–47, esp. 133–35; Mill, "Austin's Lectures on Jurisprudence" (1832), in *The Collected Works of John Stuart Mill*, vol. 21, *Essays on Equality, Law, and Education*, ed. J. M. Robson (Toronto: University of Toronto Press, 1984), 51–60, esp. 54–55; Mill, "The Present State of Literature" (1827), in *The Collected Works of John Stuart Mill*, vol. 26, *Journals and Debating Speeches*, ed. J. M. Robson (Toronto: University of Toronto Press, 1988), 409–17, esp. 410–13.

20   Mill to Florence May, before November 1868, in *The Collected Works of John Stuart Mill*, vol. 16, *The Later Letters of John Stuart Mill 1849–1873*, ed. F. E. Mineka and D. N. Lindley (Toronto: University of Toronto Press, 1972), 1472–75.

21   J. S. Mill, "Autobiography" (1873), in *The Collected Works of John Stuart Mill*, vol. 1, *Autobiography and Literary Essays*, ed. J. M. Robson and J. Stillinger (Toronto: University of Toronto Press, 1981), 3–290, esp. 137–92; Mill, "A System of Logic," in *The Collected Works of John Stuart Mill*, vol. 8, *A System of Logic . . . Books IV–VI and Appendices*, ed. J. M. Robson (Toronto: University of Toronto Press, 1974), 861–70, 875–78, 951–52; Mill, "The Education Bill (2)," and "Election to the School Boards (2)," in *The Collected Works of John Stuart Mill*, vol. 29, *Public and Parliamentary Speeches July 1869–March 1873*, ed. J. M. Robson and B. L. Kinzer (Toronto: University of Toronto Press, 1988), 391–96 and 398–401; M. Poovey, *A History of the Modern Fact: Problems of Knowledge in the Sciences of Wealth and Society* (Chicago: University of Chicago Press, 1998), 323–25.

22   K. J. Mays, "The Disease of Reading and Victorian Periodicals," in *Literature in the Marketplace: Nineteenth-Century British Publishing and Reading Practices*, ed. J. O. Jordan and R. L. Patten (Cambridge: Cambridge University Press, 1995), 165–94.

23   L. E. Javal, "Essai sur la physiologie de la lecture," *Annales d'oculistique* 80 (1878): 240–74, and *Annales d'oculistique* 82 (1879): 242–53; N. J. Wade and B. W. Tatler, "Origins and Applications of Eye Movement Research," in *The Oxford Handbook of Eye Movements*, ed. S. P. Liversedge, I. D. Gilchrist, and S. Everling (Oxford: Oxford University Press, 2011), 17–43.

24   N. J. Wade and B. W. Tatler, "Did Javal Measure Eye Movements During Reading?" *Journal of Eye Movement Research* 2, no. 5 (2008): 1–7; [?] Lamare, "Des mouvements des yeux dans la lecture," *Bulletins et mémoires de la Société Francaise d'Ophtalmologie* 10 (1892): 354–64, esp. 357, 361; E. Landolt, "Nouvelles recherches sur la physiologie des mouvements des yeux," *Archives d'ophtalmologie* 11 (1891): 385–95, esp. 390–95; Landolt, "Un instrument destiné à contrôler la direction et les mouvements des yeux pendant le travail," *Bulletins et mémoires de la Société Francaise d'Ophtalmologie* 10 (1892): 253. Lamare's initial is sometimes given as M., but as far as I can tell in the original papers this always represented "Monsieur." Interestingly, in the discussion following his paper one participant drew attention to the American practice of teaching to read by syllabus and words, "fluently," rather than by letters.

25   E. Hering, "Über Muskelgeräusche des Auges," *Sitzberichte des Kaiserlichen Akademie der Wissenschaften in Wien. Mathematisch-naturwissenschaftliche Klasse* 79 (1879): 137–54; Wade and Tatler, "Origins and Applications," 24; N. J. Wade, "Scanning the Seen: Vision and the Origins of Eye-Movement Research," in *Eye Movements: A Window on Mind and Brain*, ed. R. P. G. van Gompel, M. H. Fischer, W. S. Murray, and R. L. Hill (London: Elsevier, 2007), 32–63, esp. 33–53.

26   D. Phillips, *Acolytes of Nature: Defining Natural Science in Germany, 1770–1850* (Chicago: University of Chicago Press, 2012).

27   A. Fyfe, *Steam-Powered Knowledge: William Chambers and the Business of Publishing, 1820–1860* (Chicago: University of Chicago Press, 2012), 1–25, 34–36; J. A. Secord, *Visions of Science: Books and Readers at the Dawn of the Victorian Age* (Oxford: Oxford University Press, 2014), 236–46; A. Johns, "The Identity Engine: Printing and Publishing at the Beginning of the Knowledge Economy," in *The Mindful Hand: Inquiry and Invention from the Late Renaissance to Early Industrialisation*, ed. L. Roberts, S. J. Schaffer, and P. Dear (Amsterdam: Koninklijke Nederlandse Akademie van Wetenschappen, 2007), 403–28.

28   This was most famously the case in France, but similar ideologies played out in the United Kingdom and the United States too. See F. Furet, *Reading and Writing: Literacy in France from Calvin to Jules Ferry* (Cambridge: Cambridge University Press, 1982), D. Vincent, *Literacy and Popular Culture: England, 1750–1914* (Cambridge: Cambridge University Press, 1989), and D. Vincent, *The Rise of Mass Literacy: Reading and Writing in Modern Europe* (Cambridge: Polity, 2000).

29   Excellent synoptic accounts of the relationship between print and nation in this period may be found in volumes 3 and 4 of D. D. Hall, ed., *A History of the Book in America*, 5 vols. (Chapel Hill: University of North Carolina Press/American Antiquarian Society, 2000–2009). See also A. Johns, *Piracy: The Intellectual Property Wars from Gutenberg to Gates* (Chicago: University of Chicago Press, 2009), 291–326; R. D. Brown, *The Strength of a People: The Idea of an Informed Citizenry in America 1650–1870* (Chapel Hill: University of North Carolina Press, 1996), 154–95; M. Schudson, *The Good Citizen: A History of American Civic Life* (New York: Free Press, 1998), 144–87; A. D. Chandler and J. W. Cortada, eds., *A Nation Transformed by Information: How Information has Shaped the United States from Colonial Times to the Present* (Oxford: Oxford University Press, 2000), 39–136; J. Yates, *Control Through Communication: The Rise of System in American Management* (Baltimore: Johns Hopkins University Press, 1989); A. D. Chandler, *The Visible Hand: The Managerial Revolution in American Business* (Cambridge, MA: Harvard University Press, 1977), 145–87.

Chapter One

1   M. M. Sokal, "The Education and Psychological Career of James McKeen Cattell, 1860–1904" (PhD diss., Case Western Reserve University, 1972), 56.

2   J. Radway, "Learned and Literary Print Cultures in an Age of Professionalization and Diversification," in *A History of the Book in America*, general ed. D. D. Hall, vol. 4, *Print in Motion: The Expansion of Publishing and Reading in the United States, 1880–1940*, ed. C. F. Kaestle and J. A. Radway (Chapel Hill: University of North Carolina Press, 2009), 197–233.

3   There is a large literature on these developments in the sciences. See, for example, D. Phillips, *Acolytes of Nature: Defining Natural Science in Germany, 1770–1850* (Chicago: University of Chicago Press, 2012); D. Cahan, ed., *From Natural Philosophy to the Sciences: Writing the History of Nineteenth-Century Science* (Chicago: University of Chicago Press, 2003); N. Wise, ed., *The Values of Precision* (Princeton: Princeton University Press, 1995).

4   G. Finkelstein, *Emil Du Bois-Reymond: Neuroscience, Self, and Society in Nineteenth-Century Germany* (Cambridge, MA: MIT Press, 2013), 123.

5   There is a large literature on psychophysics, but for this context see especially J. Canales, *A Tenth of a Second: A History* (Chicago: University of Chicago Press, 2009), 21–86; A. Hui, *The Psychophysical Ear: Musical Experiments, Experimental Sounds, 1840–1910* (Cambridge, MA: MIT Press, 2013), 1–21, and R. M. Brain, *The Pulse of Modernism: Physiological Aesthetics in Fin-de-Siècle Europe* (Seattle: University of Washington Press, 2015), 5–36.

6   R. Benschop and D. Draaisma, "In Pursuit of Precision: The Calibration of Minds and Machines in Late Nineteenth-Century Psychology," *Annals of Science* 57, no. 1 (2000): 1–25.

7   Brain, *Pulse of Modernism*, 11–23.

8   J. M. Cattell, "The School and the Family," *Popular Science Monthly* 74 (1909): 84–95, esp. 95.

9   M. M. Sokal, "The Unpublished Autobiography of James McKeen Cattell," *American Psychologist* 26, no. 7 (July 1, 1971): 626–35.

10 J. M. Cattell, "On the Time Required for Recognizing and Naming Letters and Words, Pictures and Colors," translated from German in J. M. Cattell, *James McKeen Cattell: Man of Science*, 2 vols. (Lancaster: The Science Press, 1947), 1:13–25. In point of fact, this paper included both the Baltimore experiments and the Leipzig ones, and it is not always clear which took place where.

11 Cattell, "On the Time Required," 23, 25.

12 M. M. Sokal, ed., *An Education in Psychology: James McKeen Cattell's Journal and Letters from Germany and England, 1880–1888* (Cambridge, MA: MIT Press, 1981), 127.

13 Sokal, "Education and Psychological Career," 196–97.

14 Sokal, "Education and Psychological Career," 174–75.

15 Sokal, "Education and Psychological Career," 324.

16 Sokal, "Education and Psychological Career," 176, 180, 186–88; J. M. Cattell, "Über die Trägheit der Netzhaut und des Sehcentrums," *Philosophische Studien* 3 (1885): 94–127; Cattell, "The Inertia of the Eye and Brain," *Brain* 8 (1885): 295–312, esp. 304–5, 311–12, my italics.

17 Cattell, "Inertia of the Eye and Brain," 306–9.

18 Sokal, "Education and Psychological Career," 165. An abridgment in English was published the following year as "The Time It Takes to See and Name Objects," *Mind* 11, no. 41 (January 1886): 63–65.

19 Sokal, "Education and Psychological Career," 264.

20 Sokal, "Education and Psychological Career," 266.

21 J. M. Cattell, "Darwin, and After Darwin," *Psychological Review* 3, no. 4 (July 1, 1896): 437–43, esp. 447.

22 J. M. Cattell and L. Farrand, "Physical and Mental Measurements of the Students of Columbia University," *Psychological Review* 3, no. 6 (November 1, 1896): 618–48.

23 For a brief account of its influence, see J. Carson, *The Measure of Merit: Talents, Intelligence, and Inequality in the French and American Republics, 1750–1940* (Princeton: Princeton University Press, 2007), 172–77.

24 Cattell and Farrand, "Physical and Mental Measurements," 641–42, 648. It made Cattell proud when physicists borrowed equipment from his psychology laboratory to measure the rate of fall of bodies in liquids and found it less constant than the psychological values he was measuring: J. M. Cattell, "Address of the President before the American Psychological Association, 1895," *Psychological Review* 3, no. 2 (March 1, 1896): 134–48, esp. 140, 142.

25 Sokal, "Education and Psychological Career," 569.

26 Radway, "Learned and Literary Print Cultures," 197–233, esp. 212–14 for *The Atlantic* in these years.

27 Sokal, "Education and Psychological Career," 586–88.

28 For Comtean schemes of this kind, which originated in pre-1848 Paris but survived into the later part of the century, see J. Tresch, *The Romantic Machine: Utopian Science and Technology After Napoleon* (Chicago: University of Chicago Press, 2012).

29 J. Jastrow, "Experimental Psychology in Leipzig," *Science* 8, no. 198 (November 19, 1886), Supplement, 459–62, esp. 461. The idea of extending lifespans may have been inspired by a similar thought expressed in Percy Bysshe Shelley's notes to *Queen Mab*.

30 "Labor May Approve Psychological Test," *New York Times*, December 29, 1925, 14.

31 J. M. Cattell, "The Perception of Light," in *System of Diseases of the Eye*, 4 vols., ed. W. F. Norris and C. A. Oliver (Philadelphia: J. B. Lippincott, 1896–1900), 1:505–38, esp. 506.

32 J. M. Cattell, "On Relations of Time and Space in Vision," *Psychological Review* 7, no. 4 (July 1900): 325–43, esp. 340.

33 J. M. Cattell, "The Visual Perception of Space," *Science*, n.s., 14, no. 346 (August 1901): 263–66, esp. 265–66; Cattell, "On Relations of Time and Space in Vision," 325, 328–29n3, 330, 343.

34 Cattell, "Vision with the Moving Eye," *Psychological Review* 1 (1894): 507–8.

35 J. M. Cattell, *The University and Business Methods* (Garrison, NY: n.p., 1905) (reprinted from *The Independent* December 28, 1905); Cattell, *The Carnegie Foundation for the Advancement of Teaching* (Lancaster, PA: New Era, 1916).

36 J. M. Cattell, *University Control* (New York and Garrison, NY: The Science Press, 1913), 5–6, 25, 31.

37 J. M. Cattell, "Columbia University and Professor Cattell," *Science*, n.s., 46, no. 1189 (October 12, 1917): 363–64; *Science*, n.s., 46, no. 1191 (October 26, 1917): 411–13; G. Jonçich, *The Sane Positivist: A Biography of Edward L. Thorndike* (Middletown, CT: Wesleyan University Press, 1968), 377.

38 J. M. Cattell, "Education and Democracy," *Science*, n.s., 39, no. 996 (January 30, 1914): 154–64; Cattell, "The School and the Family." See also J. M. Cattell, "More Successful Life is Psychology's Promise," *New York Times*, January 20, 1924, 3; Cattell, "Address of the President," 144; Sokal, "Education and Psychological Career," 497.

39 J. M. Cattell, "The Journal *Science* and the American Association for the Advancement of Science," *Science*, n.s., 64, no. 1658 (October 8, 1926), 342–47; J. M. Cattell and others, "Correspondence in Regard to the Censorship of Scientific Journals," *Science*, n.s., 96, no. 2488 (September 4, 1942): 216–21; Brubaker, "The Cattell Family and Early-Twentieth-Century Printing Here," *Intelligencer Journal*, February 12, 2010, A11.

Chapter Two

1 For the statistic on school-leavers see R. L. Venezky and C. F. Kaestle, "From McGuffey to Dick and Jane: Reading Textbooks," in *A History of the Book in America*, general ed. D. D. Hall, vol. 4, *Print in Motion: The Expansion of Publishing and Reading in the United States, 1880–1940*, ed. C. F. Kaestle and J. A. Radway (Chapel Hill: University of North Carolina Press, 2009), 415–30, esp. 418.

2 E. B. Huey, *The Psychology and Pedagogy of Reading* (New York: Macmillan, 1908), vii, 8, 385–418, 387; J. Cohen-Cole, "The Politics of Psycholinguistics," *Journal of the History of the Behavioral Sciences* 51 (2015): 54–77.

3   See, in general, N. J. Wade and B. W. Tatler, "Origins and Applications of Eye Movement Research," in *The Oxford Handbook of Eye Movements*, ed. S. P. Liversedge, I. D. Gilchrist, and S. Everling (Oxford: Oxford University Press, 2011), 17–43; N. J. Wade and B. W. Tatler, *The Moving Tablet of the Eye: The Origins of Modern Eye Movement Research* (Oxford: Oxford University Press, 2005).

4   See, in general, C. R. Acland, "The Swift View: Tachistoscopes and the Residual Modern," in *Residual Media*, ed. C. R. Acland (Minneapolis: University of Minnesota Press, 2007), 361–84; D. S. Coombs, "An Untrained Eye: The Tachistoscope and Photographic Vision in Early Experimental Psychology," *History and Technology* 28, no. 1 (March 2012): 107–17; and R. Benschop, "What is a Tachistoscope? Historical Explorations of an Instrument," *Science in Context* 11, no. 1 (1998): 23–50. The media theorist Friedrich Kittler focused on the tachistoscope in his highly influential *Discourse Networks 1800/1900* (Stanford: Stanford University Press, 1990), likening it to the typewriter; as will become obvious, I do not agree with his interpretation.

5   R. Dodge, "The Psychology of Reading," *Psychological Review* 8, no. 1 (January 1901): 56–60.

6   W. F. Dearborn, *The Psychology of Reading: An Experimental Study of the Reading Pauses and Movements of the Eye* (New York: The Science Press, 1906); W. J. M. Levelt, *A History of Psycholinguistics: The Pre-Chomskyan Era* (Oxford: Oxford University Press, 2013), 469.

7   Wade and Tatler, "Origins and Applications of Eye Movement Research," 30; Wade and Tatler, *Moving Tablet of the Eye*, 115–60; C. H. Judd and G. T. Buswell, *Silent Reading: A Study of the Various Types* (Chicago: University of Chicago Press, 1921); Levelt, *History of Psycholinguistics*, 470; Huey, *Psychology and Pedagogy of Reading*.

8   R. Dodge, "Beschreibung eines Neuen Chronographen," *Zeitschrift für Psychologie und Physiologie der Sinnesorgane* 10 (1896): 414–20; B. Erdmann and R. Dodge, *Psychologische Untersuchungen Über das Lesen auf Experimenteller Grundlage* (Halle: M. Niemeyer, 1898).

9   R. Dodge and T. S. Cline, "The Angle Velocity of Eye Movements," *Psychological Review* 8, no. 2 (March 1901): 145–57, esp. 151.

10   Dodge and Cline, "Angle Velocity."

11   R. Dodge, "The Act of Vision," *Harper's Monthly* 104 (December 1, 1901): 937–41.

12   For biographical details see Judd's interesting account in C. H. Judd, "Charles H. Judd," *A History of Psychology in Autobiography*, vol. 2, ed. C. Murchison (Worcester, MA: Clark University Press, 1932), 207–35.

13   A. C. Armstrong ("with the assistance of Mr. C. H. Judd"), "The Imagery of American Students," *Psychological Review* 1, no. 5 (September 1894): 496–505.

14   C. H. Judd, "Uber Raumwahrnehmungen im Gebiete des Tastsinnes," *Philosophische Studien* 12 (1896): 409–63 (Stratton's role as a subject is noted on 417, 420, 426–27, 433, 436, and 448), which received a dismissive review ("not weighty—save in bulk") by Harvard's Herbert Nichols in *Psychological Review* 3 (1896): 577–78; G. Stratton, "Ueber die Wahrnehmung von Druckänderungen bei verschiedenen Geschwindigkeiten," *Philosophische Studien* 12 (1896): 525–86 (especially 527, 547, 552–53, 567, 575, and 578-85 for Judd's involvement).

15   W. Wundt, *Outlines of Psychology*, trans. C. H. Judd (Leipzig: W. Englemann; New York: G. E. Stechert, 1897).

16   Parts of it were published as C. H. Judd, "Movement and Consciousness," *Psychological Review Monograph Supplements* 7, no. 1 (29) (March 1905): 199–226, and it was

clearly related to work in Judd, "An Experimental Study of Writing Movements," *Philosophische Studien* 19 (1902): 243–59.

17 "The Physiology of Art," *Examiner* 3623 (July 7, 1877): 846–48; R. M. Brain, *The Pulse of Modernism: Physiological Aesthetics in Fin-de-Siècle Europe* (Seattle: University of Washington Press, 2015), xiii, 95–149 and passim.

18 G. M. Stratton, "Eye-Movements and the Aesthetics of Visual Form," *Philosophische Studien* 20 (1902): 336–59; Stratton, "Symmetry, Linear Illusions, and the Movements of the Eye," *Psychological Review* 13, no. 2 (March 1906): 82–96; V. Lee and C. Anstruther-Thomson, "Beauty and Ugliness," *Contemporary Review* 72 (July 1, 1897): 544–69 and 669–88.

19 C. H. Judd, "Experimental [review of Stratton]," *Psychological Review* 10, no. 3 (May 1903): 336–37.

20 C. H. Judd, C. N. McAllister, and W. M. Steele, "General Introduction to a Series of Studies of Eye Movements by Means of Kinetoscopic Photographs," *Psychological Review Monograph Supplements* 7, no. 1 (March 1905) (Yale Psychological Studies, n.s., 1, no. 1): 1–16.

21 C. H. Judd, "The Müller-Lyer Illusion," *Psychological Review Monograph Supplements* 7, no. 1 (March 1905): 55–81, esp. 72.

22 E. H. Cameron and W. M. Steele, "The Poggendorff Illusion," *Psychological Review Monograph Supplements* 7, no. 1 (March 1905): 83–111, esp. 90–91; C. H. Judd and H. C. Courten, "The Zöllner Illusion," *Psychological Review Monograph Supplements* 7, no. 1 (March 1905): 112–39, esp. 123, 139.

23 C. H. Judd, "Movement and Consciousness," 216–17; C. H. Judd, "Practice without Knowledge of Results," *Psychological Review Monograph Supplements* 7, no. 1 (March 1905): 185–98, esp. 198.

24 C. H. Judd, "Evolution and Consciousness," *Psychological Review* 17, no. 2 (March 1910): 77–97.

25 E. B. Huey, "Hygienic Requirements in the Printing of Books and Papers," *Popular Science Monthly* 70 (1907): 542–48.

26 D. K. Hartman and D. H. Davis, "Edmund Burke Huey: The Formative Years of a Scholar and Field," *57th Yearbook of the Literacy Research Association* (Oak Creek, WI: Literacy Research Association, 2008), 41–55, esp. 46.

27 Hartman and Davis, "Edmund Burke Huey," 47–49.

28 Hartman and Davis, "Edmund Burke Huey," 51.

29 Huey, *Psychology and Pedagogy of Reading*, 33.

30 Huey, *Psychology and Pedagogy of Reading*, 41–43.

31 Huey, *Psychology and Pedagogy of Reading*, 52–53.

32 Huey, *Psychology and Pedagogy of Reading*, 61, 63, 72–74, 110–11.

33 Huey, *Psychology and Pedagogy of Reading*, 67–68.

34 Huey, *Psychology and Pedagogy of Reading*, 102, 103.

35 Huey, *Psychology and Pedagogy of Reading*, 104–8.

36 Huey, *Psychology and Pedagogy of Reading*, 178–80.

37 A. B. Smuts, *Science in the Service of Children 1893–1935* (New Haven: Yale University Press, 2006), 31–48, esp. 36.

38 Huey, *Psychology and Pedagogy of Reading*, 200, 206, 220–21.

39 Huey, *Psychology and Pedagogy of Reading*, 126, 224, 225.

40 Huey, *Psychology and Pedagogy of Reading*, 239.

41 Huey, *Psychology and Pedagogy of Reading*, 276.

42 Huey, *Psychology and Pedagogy of Reading*, 244.

43 Huey, *Psychology and Pedagogy of Reading*, 278–79, 284, 287.

44 Huey, *Psychology and Pedagogy of Reading*, 45–46.

45 Huey, *Psychology and Pedagogy of Reading*, 300.

46 Huey, *Psychology and Pedagogy of Reading*, 301–3.

47 Huey, *Psychology and Pedagogy of Reading*, 125, 382.

48 Huey, *Psychology and Pedagogy of Reading*, 308–9, 319, 330, 337, 339, 348–49.

49 Huey, *Psychology and Pedagogy of Reading*, 365–68.

50 Huey, *Psychology and Pedagogy of Reading*, 402, 404–5.

51 A. Rabinbach, *The Human Motor: Energy, Fatigue, and the Origins of Modernity* (New York: Basic Books, 1990), 146–78.

52 Huey, *Psychology and Pedagogy of Reading*, 390–92.

53 Huey, *Psychology and Pedagogy of Reading*, 394, 398. See also the exhaustive discussion in H. Cohn, *The Hygiene of the Eye in Schools*, trans. W. P. Turnbull (London: Simpkin, Marshall, 1886), 94–219.

54 Huey, *Psychology and Pedagogy of Reading*, 395–96.

55 R. A. Meckel, *Classrooms and Clinics: Urban Schools and the Protection and Promotion of Child Health, 1870–1930* (New Brunswick: Rutgers University Press, 2013), 11, 29, 91. For examples, see: F. B. Dresslar, *School Hygiene* (New York: Macmillan, 1913), 82; Cohn, *Hygiene of the Eye in Schools*; A. Newsholme, *School Hygiene: The Laws of Health in Relation to School Life* (London: Swan Sonnenschein, 1898), 12–16; E. R. Shaw, *School Hygiene* (London: Macmillan, 1902), 135–62; G. G. Groff, *School Hygiene* (New York and Chicago: E. L. Kellogg, 1889), 5–8.

56 Huey, *Psychology and Pedagogy of Reading*, 424–26. In point of fact, reading researchers did occasionally voice the need to address scripts in Chinese, Hebrew, Arabic, and Japanese, all of which involved movements other than the English-language standard of left to right along a horizontal line, as well as with Braille. But their efforts were overwhelmingly concentrated on printed alphanumerics in English. Hansun Hsiung has recently been exploring this topic with respect to Chinese.

57 Huey, *Psychology and Pedagogy of Reading*, 429.

58 Huey, *Psychology and Pedagogy of Reading*, 413–18, 431.

59 J. B. Carroll, foreword to E. B. Huey, *The Psychology and Pedagogy of Reading* (repr. Cambridge: MIT Press, 1968), vii–xii; Huey, *Psychology and Pedagogy of Reading*, 397–98; J. B. Reed and R. J. Meyer, "Edmund Burke Huey (1870–1913): A Brief Life with an Enduring Legacy," in *Shaping the Reading Field: The Impact of Early Reading Pioneers, Scientific Research, and Progressive Ideas*, ed. S. E. Israel and E. J. Monaghan (Newark, DE: International Reading Association, 2007), 159–75, esp. 160–64; E. B. Huey, *Backward and Feeble-Minded Children* (Baltimore: Warwick and York, 1912), 3.

Chapter Three

1 Chicago's primacy was widely acknowledged across the field. See, for example: M. A. Tinker, "Eye Movements in Reading," *Journal of Educational Research* 30, no. 4 (December 1936): 241–77, esp. 241; J. A. O'Brien, *Silent Reading* (New York: Macmillan, 1922), 13–16; M. D. Vernon, *The Experimental Study of Reading* (Cambridge: Cambridge University Press, 1931), xiii.

2   C. H. Judd, *Genetic Psychology for Teachers* (New York: D. Appleton, 1906 [1903]), 236–64; Judd, *Psychology* (New York: Charles Scribner's Sons, 1907), 371–75.

3   N. B. Smith, "Reading: Seventy-Five Years of Progress," in *Reading: Seventy-Five Years of Progress*, ed. H. A. Robinson, Supplementary Educational Monographs 98 (Proceedings of the University of Chicago Annual Conference on Reading, vol. 28. Chicago: University of Chicago Press, 1966), 3–12.

4   C. H. Judd and G. T. Buswell, *Silent Reading: A Study of the Various Types* (Chicago: University of Chicago, 1922), 21–22.

5   Smith, "Reading," 6. For the importance and measurement of silent reading, see also the Russell Sage Foundation's effort to establish standards and tests based on Gray and others: M. A. Burgess, *The Measurement of Silent Reading* (New York: Russell Sage Foundation, 1921), 14–15, 20. For the trade in public reading skills (of which Marshall McLuhan's mother was a practitioner), see P. McDowell, "Elsie McLuhan's Vocal Science," *Publications of the Modern Language Association* 135, no. 2 (March 2020): 378–86. A comprehensive account of the laboratory science of reading at the moment of its greatest prestige is M. D. Vernon, *The Experimental Study of Reading* (Cambridge: Cambridge University Press, 1931). Magdalen Vernon's account was written in the distant setting of Cambridge, and under the eye of the always-interesting F. C. Bartlett; it is the gateway to a distinct version of the science of reading that was fascinating in its own right but deserves a study to itself.

6   W. S. Gray, "The Diagnostic Study of an Individual Case in Reading," *Elementary School Journal* 21, no. 8 (April 1921): 577–94; W. S. Gray, *Remedial Cases in Reading: Their Diagnosis and Treatment*, Supplementary Educational Monographs 22 (Chicago: University of Chicago, 1922), 6–7, 13, 188–204; W. S. Gray and R. Munroe, *The Reading Interests and Habits of Adults* (New York: Macmillan, 1930); W. S. Gray and B. Rogers, *Maturity in Reading* (Chicago: University of Chicago Press, 1956).

7   J. M. Robertson, "Guy Thomas Buswell (1891–1994)," *American Psychologist* 51, no. 2 (February 1996): 152.

8   Judd and Buswell, *Silent Reading*, 58–61, 155–56. See also C. H. Judd, *Reading: Its Nature and Development* (Chicago: University of Chicago Press, 1918), 62–65.

9   Judd and Buswell, *Silent Reading*, 26.

10   Judd and Buswell, *Silent Reading*, 154–55.

11   Judd and Buswell, *Silent Reading*, 4.

12   G. T. Buswell, *How Adults Read*, Supplementary Educational Monographs 45 (Chicago: University of Chicago, 1937), 1.

13   Buswell, *How Adults Read*, 2.

14   Buswell, *How Adults Read*, 42–53.

15   Buswell, *How Adults Read*, 37, 42, 139–41.

16   Buswell, *How Adults Read*, 142–43.

17   Buswell, *How Adults Read*, 108–116.

18   G. T. Buswell, *Remedial Reading at the College and Adult Levels: An Experimental Study*, Supplementary Educational Monographs 50 (Chicago: University of Chicago, 1939), 63.

19   P. Otlet, *Traité de documentation: Le livre sur le livre* (Brussels: Editiones de Mundaneum, 1934), 320.

20   E. A. Taylor, *Controlled Reading: A Correlation of Diagnostic, Teaching, and Corrective Techniques* (Chicago: University of Chicago Press, 1937), viii, 52–105, 152. See also

S. E. Taylor, *Eye-Movement Photography with the Reading Eye* (Huntington, NY: Educational Developmental Laboratories, 1960), 4–9.

21 Taylor, *Controlled Reading*, ix–xv, 4–6.

22 Buswell, *Remedial Reading*, 19–23, 29–30, 67.

23 Buswell, *Remedial Reading*, 1.

24 Buswell, *Remedial Reading*, 1, 66.

25 Buswell, *Remedial Reading*, 69–70.

26 See, e.g., M. A. Tinker, "Eye Movements in Reading," *Journal of Educational Research* 30, no. 4 (December 1936): 241–77, esp. 255–56; M. A. Tinker, "Use and Limitations of Eye-Movement Measures of Reading," *Psychological Review* 40 (1933): 381–87.

27 Tinker, "Use and Limitations of Eye-Movement Measures of Reading"; M. A. Tinker and A. Frandsen, "Evaluation of Photographic Measures of Reading," *Journal of Educational Psychology* 25 (1934): 96–100; M. A. Tinker, "Reliability and Validity of Eye-Movement Measures of Reading," *Journal of Experimental Psychology* 19 (1936): 732–46; D. G. Paterson and M. A. Tinker, *How to Make Type Readable: A Manual for Typographers, Printers and Advertisers* (New York: Harper, 1940); P. Witty and D. Kopel, *Reading and the Educative Process* (Boston: Ginn, 1939), 208–10.

28 G. T. Buswell, *How People Look at Pictures: A Study of the Psychology of Perception in Art* (Chicago: University of Chicago Press, 1935), 4–9, 17.

29 Buswell, *How People Look at Pictures*, 9, 10–16.

30 Buswell, *How People Look at Pictures*, 35–36.

31 Buswell, *How People Look at Pictures*, 41, 45.

32 Buswell, *How People Look at Pictures*, 59.

33 Buswell, *How People Look at Pictures*, 80–81. Buswell does not name the cathedral in his photograph, but it is attributed to a New York photographer and it is clearly St. Patrick's.

34 Buswell, *How People Look at Pictures*, 85–92.

35 Buswell, *How People Look at Pictures*, 115–16.

36 Buswell, *How People Look at Pictures*, 117–18, 134–35.

37 Buswell, *How People Look at Pictures*, 131–33.

38 Buswell, *How People Look at Pictures*, 136–41.

39 R. L. Venezky and C. F. Kaestle, "From McGuffey to Dick and Jane: Reading Textbooks," in *A History of the Book in America*, general ed. D. D. Hall, vol. 4, *Print in Motion: The Expansion of Publishing and Reading in the United States, 1880–1940*, ed. C. F. Kaestle and J. A. Radway (Chapel Hill: University of North Carolina Press / American Antiquarian Society, 2009), 415–30.

40 For the consequences in the particularly fraught area of human biology, see A. Shapiro, *Trying Biology: The Scopes Trial, Textbooks, and the Antievolution Movement in American Schools* (Chicago: University of Chicago Press, 2013), 14–38 and 78–86.

41 U. Sinclair, *The Goslings: A Study of the American Schools* (Pasadena: Published by the author, 1924), 185.

42 E. L. Thorndike, "Reading as Reasoning: A Study of Mistakes in Paragraph Reading," *Journal of Educational Psychology* 8, no. 6 (June 1917): 323–32; also published separately as Thorndike, *Reading as Reasoning: A Study of Mistakes in Paragraph Reading* (Baltimore: Warwick and York, 1917).

43 Thorndike, "Reading as Reasoning," 329; Venezky and Kaestle, "From McGuffey to Dick and Jane," 419.

44 G. T. Buswell and W. H. Wheeler, *The Silent Reading Hour: First Reader* (Chicago: Wheeler Publishing Company, 1924), iii–vi.

45 C. Kismaric and M. Heiferman, *Growing Up With Dick and Jane: Learning and Living the American Dream* (San Francisco: Collins, 1996), 10–14, 21, 25, 30; C. Lauritzen, "William Scott Gray (1885–1960): Mr. Reading," in Israel and Monaghan, eds., *Shaping the Reading Field*, 307–26. The subject of Sharp and Gray's collaboration on the Dick and Jane readers needs more research, but there are some interesting reminiscences highlighting the context of competition in the publishing industry in H. A Foreman, *These Things I Remember* (Chicago: Scott, Foreman, 1949), esp. 65–73.

46 J. Hinshelwood, *Congenital Word-Blindness* (London: H. K. Lewis, 1917), 1–39.

47 A. I. Gates, *The Improvement of Reading: A Program of Diagnostic and Remedial Methods* (New York: Macmillan, 1927), 272–73 (quoting Burt).

48 C. Schmitt, "Developmental Alexia: Congenital Word-Blindness, or Inability to Read," *Elementary School Journal* 18, no. 9 (May 1918): 680–700 and 18, no. 10 (June 1918): 757–69, esp. 765–68 and 767n1; Judd, *Reading*, 119–34; W. S. Gray, *Remedial Cases in Reading: Their Diagnosis and Treatment*, Supplementary Educational Monographs 22 (Chicago: University of Chicago, 1922), 6–7, 13–14.

49 W. F. Dearborn, "The Etiology of Congenital Word-Blindness," in E. E. Lord, L. Carmichael, and W. F. Dearborn, *Special Disabilities in Learning to Read and Write*, Harvard Monographs in Education I, 2, 1 (Cambridge, MA: Harvard University, 1925), 50–76, esp. 68–70.

50 F. D. Brooks, *The Applied Psychology of Reading* (New York: D. Appleton, 1926), 6–7, 176–77.

51 For Orton's biography and major papers, see S. T. Orton (ed. J. L. Orton), *"'Word-Blindness' in School Children" and Other Papers on Strephosymbolia (Specific Language Disability—Dyslexia) 1925–1946* (Pomfret, CT: Orton Society, 1966).

52 J. F. Lyday, "The Greene County Mental Clinic," *Mental Hygiene* 10 (1926): 759–86.

53 S. T. Orton, "'Word-Blindness' in School Children," *Archives of Neurology and Psychiatry* 14, no. 5 (November 1925): 581–615; Lyday, "Greene County," 769, 786.

54 Orton, "'Word-Blindness' in School Children," 598–99.

55 Orton, "'Word-Blindness' in School Children," 610.

56 M. Monroe, "Methods for Diagnosis and Treatment of Cases of Reading Disability," *Genetic Psychology Monographs* 4, nos. 4–5 (October–November 1928), 335–456, esp. 335–39, 425–26; G. M. Fernald and H. Keller, "The Effect of Kinaesthetic Factors in the Development of Word Recognition in the Case of Non-Readers," *Journal of Educational Research* 4, no. 5 (December 1921): 355–77; A. I. Gates and C. C. Bennett, *Reversal Tendencies in Reading: Causes, Diagnosis, Prevention and Correction* (New York: Teachers College, Columbia University, 1933), 32–33.

57 Orton, "'Word-Blindness' in School Children," 608–9.

58 Orton, "'Word-Blindness' in School Children," 611.

59 S. T. Orton, "The 'Sight Reading' Method of Teaching Reading, as a Source of Reading Disability," *Journal of Educational Psychology* 20, no. 2 (February 1929): 135–43.

60 M. M. Mathews, *Teaching to Read, Historically Considered* (Chicago: University of Chicago Press, 1966), 139–40. I do not trace the subsequent history of dyslexia research and treatment in this book, because the field became a large and somewhat autonomous one; but good summaries may be found in parts 5 and 7 of Snowling and Hulme's *Science of Reading*.

Chapter Four

1   For an introduction, see M. Bulmer, *The Chicago School of Sociology: Institutionalization, Diversity, and the Rise of Sociological Research* (Chicago: University of Chicago Press, 1984).

2   D. Waples, *People and Print: Social Aspects of Reading in the Depression* (Chicago: University of Chicago Press, [1938]), vii.

3   L. Heide, *Punched-Card Systems and the Early Information Explosion 1880–1945* (Baltimore: Johns Hopkins University Press, 2009), 68–104; L. Gitelman, *Paper Knowledge: Toward a Media History of Documents* (Durham: Duke University Press, 2014); A. Csiszar, *The Scientific Journal: Authorship and the Politics of Knowledge in the Nineteenth Century* (Chicago: University of Chicago Press, 2018), 241–79.

4   P. Otlet, *Traité de documentation: Le livre sur le livre* (Brussels: Editiones Mundaneum, 1934); M. Buckland, "On the Cultural and Intellectual Context of European Documentation in the Early Twentieth Century," in *European Modernism and the Information Society: Informing the Past, Understanding the Present*, ed. W. B. Rayward (Aldershot: Ashgate, 2008), 45–57, esp. 48–53.

5   F. P. Keppel, "The Carnegie Corporation and the Graduate Library School: A Historical Outline," *Library Quarterly* 1, no. 1 (January 1931): 22–25; W. S. Learned, *The American Public Library and the Diffusion of Knowledge* (New York: Harcourt, Brace, 1924), 12.

6   D. Waples, "An Approach to the Synthetic Study of Interest in Education," *Journal of Educational Psychology* 11 (1920): 301–15, 369–84, 445–58, 502–16. For Waples's work, see S. Karetzky, *Reading Research and Librarianship: A History and Analysis* (Westport, CT: Greenwood, 1982), 93–230. For his biography, see D. Waples, *On the March: A Short Autobiography for Friends and Family* (privately printed, n.d.) and G. Kamberelis and M. K. Albert, "Douglas Waples (1893–1978): Crafting the Well-Read Public," in *Shaping the Reading Field*, ed. S. E. Israel and E. J. Monaghan (Newark, DE: IRA, 2007), 247–77.

7   D. Waples, "The Graduate School at Chicago," *Library Quarterly* 1, no. 1 (January 1931): 26–36; J. Dewey, *The Sources of a Science of Education* (New York: Horace Liveright, 1929); S. Norman, "*The Library Quarterly* in the 1930s: A Journal of Discussion's Early Years," *Library Quarterly* 58, no. 4 (October 1988): 327–51, esp. 336–37. Waples's later synthetic account was published as *Investigating Library Problems* (Chicago: University of Chicago Press, 1939).

8   C. C. Williamson, "The Place of Research in Library Service," *Library Quarterly* 1, no. 1 (January 1931): 1–17; C. S. Thompson, "Do We Want a Library Science?" *Library Journal* 56 (1931): 581–87; J. Richardson Jr., *The Spirit of Inquiry: The Graduate Library School at Chicago, 1921–51* (Chicago: American Library Association, 1982), 90–93. For the rumor about Judd and the presidency, see University of Chicago, Judd Papers, Box 1, folder 14.

9   P. Butler, *An Introduction to Library Science* (Chicago: University of Chicago Press, 1933), 29. For Butler and his work, see J. V. Richardson, ed., *The Gospel of Scholarship: Pierce Butler and a Critique of American Librarianship* (Metuchen, NJ: Scarecrow, 1992).

10   Butler, *Introduction*, 33–34, 47, 52–53, 73–74.

11   Butler, *Introduction*, 83–84, 93–95, 99, 106, 111–13, 114–15.

12 D. M. Kennedy, *Freedom from Fear: The American People in Depression and War, 1929–1945* (Oxford: Oxford University Press, 1999), 255–57.

13 Waples, *People and Print*, 229; see also the foreword by the Committee on Studies in Social Aspects of the Depression, vii–ix.

14 J. T. Sparrow, *Warfare State: World War II Americans and the Age of Big Government* (Oxford: Oxford University Press, 2011), 29–41.

15 L. R. Wilson and E. A. Wright, *County Library Service in the South: A Study of the Rosenwald County Library Demonstration* (Chicago: University of Chicago Press, 1935), v–vii, 4 (italics in original), 16, 87.

16 Wilson and Wight, *Library Service*, 81.

17 Wilson and Wight, *Library Service*, 150.

18 Wilson and Wight, *Library Service*, 207–8.

19 Wilson and Wight, *Library Service*, 116–17, 123, 126–27, 142.

20 Wilson and Wight, *Library Service*, 202, 204–5, 210–12, 222–31; *School Money in Black & White* (Chicago: Julius Rosenwald Fund, 1935), 24.

21 L. R. Wilson, *The Geography of Reading: A Study of the Distribution and Status of Libraries in the United States* (Chicago: American Library Association / University of Chicago Press, 1938), vii–viii, 1–2.

22 Wilson, *Geography of Reading*, 44–45, 71.

23 Wilson, *Geography of Reading*, 71, 76, 188, 191–92.

24 Wilson, *Geography of Reading*, 4, 44–46, 58, 64–66, 87, 182–86, 191–93, 434.

25 Wilson, *Geography of Reading*, 14–22, 29–31, 32–34.

26 Wilson, *Geography of Reading*, 79–80, 109, 437–38.

27 Wilson, *Geography of Reading*, 325–44, 352, 354, 385–88, 421–23.

28 Wilson, *Geography of Reading*, 125–28, 166–67.

29 Wilson and Wight, *Library Service*, 83.

30 Wilson and Wight, *Library Service*, 222, 229.

31 Wilson, *Geography of Reading*, 38, 95, 106–8, 115, 168–69, 193, 436–37.

32 L. Terman, "Genius and Stupidity: A Study of Some of the Intellectual Processes of Seven 'Bright' and Seven 'Stupid' Boys," *Pedagogical Seminary* 13 (1906): 307–73, esp. 312, 338–41.

33 D. J. Kevles, "Testing the Army's Intelligence: Psychologists and the Military in World War I," *Journal of American History* 55, no. 3 (December 1968): 565–81, esp. 566, 569. For the adoption of intelligence testing by eugenics proponents in this era, see in general D. J. Kevles, *In the Name of Eugenics: Genetics and the Uses of Human Heredity* (Cambridge, MA: Harvard University Press, 1995 [1985]), 76–84.

34 E. L. Thorndike, "The Measurement of Ability in Reading," *Teachers College Record* 15, no. 4 (September 1914): 207–77.

35 R. M. Yerkes, ed., *Psychological Examining in the United States Army*, Memoirs of the National Academy of Sciences, vol. 15 (US Government Printing Office, 1921), 279, 347 (for Thorndike's literacy test, adapted by Terman). For the conduct and consequences of the Army tests, see Stephen Jay Gould's unequaled demolition in *The Mismeasure of Man* (New York: W. W. Norton, 1981), 188–233.

36 G. Jonçich, *The Sane Positivist: A Biography of Edward L. Thorndike* (Middletown: Wesleyan University Press, 1968), 361–75.

37 L. Stoddard, *The Revolt Against Civilization: The Menace of the Under Man* (New York: Charles Scribner's Sons, 1922), 56–74.

38 W. Lippmann, "The Mental Age of Americans," *New Republic* October 25, 1922, 213–15, and succeeding articles in the series.

39 S. L. Pressey, *Introduction to the Use of Standard Tests* (Yonkers, NY: World Book Company, 1923), 1, 108–20; Stoddard, *Revolt*, 57. Thorndike later used the opportunity of his being awarded an honorary degree at the University of Chicago to attack proposals for equalization in education and support selective schools: Jonçich, *Sane Positivist*, 486.

40 J. S. Chall, "This Business of Readability," *Educational Research Bulletin* 26, no. 1 (January 1947): 1–13, esp. 1–2.

41 P. A. Jones, "Elizabeth Cleveland Morris (1877–1960): Leader of the Literacy and Adult Elementary Education Movement in North Carolina," *Information and Culture* 52, no. 2 (2017): 186–206.

42 D. Waples, B. Berelson, and F. R. Bradshaw, *What Reading Does to People: A Summary of Evidence on the Social Effects of Reading and a Statement of Problems for Research* (Chicago: University of Chicago Press, 1940), 21–22; M. J. Adler, *How to Read a Book: The Art of Getting a Liberal Education* (New York: Simon and Schuster, 1940).

43 Waples, *Investigating Library Problems*, 50.

44 W. S. Gray and B. E. Leary, *What Makes a Book Readable* (Chicago: University of Chicago Press, 1935), 1–5.

45 Wilson, *Geography of Reading*, 439, 442.

46 M. Wilkinson et al., *The Right Book for the Right Child* (New York: John Day, 1933 and subsequent editions); B. A. Lively and S. L. Pressey, "A Method for Measuring the 'Vocabulary Burden' of Textbooks," *Educational Administration and Supervision* 9 (1923): 389–98. For the connection between Waples's work and the science of readability, see G. R. Klare and B. Buck, *Know Your Reader: The Scientific Approach to Readability* (New York: Hermitage House, 1954), 54–55.

47 L. Bryson, "Testing Books for their Readability," *New York Times*, February 23, 1936, Education, 2.

48 Wilson, *Geography of Reading*, 224, 290.

49 Klare and Buck, *Know Your Reader*, 109; Chall, "This Business of Readability," 9, 11–12; Chall, "This Business of Readability: A Second Look," *Educational Research Bulletin* 35, no. 4 (April 1956): 89–99 and 111–12.

50 H. D. Lasswell, *Democracy Through Public Opinion* (Menasha, MI: George Banta Publishing Company, 1941); American Library Association, *The Equal Chance: Books Help To Make It* (Chicago: ALA, 1936), 5.

51 W. S. Gray and R. Munroe, *The Reading Interests and Habits of Adults* (New York: Macmillan, 1929), 198–205.

52 Waples, Berelson, and Bradshaw, *What Reading Does to People*, 27–31.

53 J. D. Peters, *Speaking Into the Air: A History of the Idea of Communication* (Chicago: University of Chicago Press, 1999), 10–22.

54 H. Lasswell, *Propaganda Technique in the World War* (New York: A. A. Knopf, 1927); D. Waples, "Propaganda and Leisure Reading," *Journal of Higher Education* 1, no. 2 (February 1930): 73–77; D. Waples and H. Lasswell, *National Libraries and Foreign Scholarship* (Chicago: University of Chicago Press, 1936), 4.

55 D. Waples, ed., *Print, Radio, and Film in a Democracy* (Chicago: University of Chicago Press, 1942).

56 G. D. Jaworski, "Goffman's Interest in Spies and Espionage: The University of Chicago Context," *Symbolic Interaction* 44, no. 2 (2020): 1–20, esp. 9–12.

57 K. Wahl-Jorgensen, "How Not to Found a Field: New Evidence on the Origins of Mass Communication Research," *Journal of Communication* 54, no. 3 (September 2004): 547–64; Waples, *On the March*, 5–8.

58 D. J. Czitrom, *Media and the American Mind from Morse to McLuhan* (Chapel Hill: University of North Carolina Press, 1982), 91–133. For the development of the field of communication studies in general, see E. M. Rogers, *A History of Communication Study* (New York: Free Press, 1994), esp. 137–243 for Chicago and Lasswell and 244–315 for Lazarsfeld.

Chapter Five

1 E. C. Morriss, M. V. Morse, and E. Phillips, *An Experimental Reading Study in the Joint Library-Adult Elementary Education Field* (New York: Bureau of Publications, Teachers College, Columbia University, 1935), esp. iii, 1–3, 8, 11–12, 13–15 (italics in original).

2 J. S. Chall, "This Business of Readability," *Educational Research Bulletin* 26, no. 1 (January 1947): 1–13, esp. 2.

3 E. Dale and R. W. Tyler, "A Study of the Factors Influencing the Difficulty of Reading Materials for Adults of Limited Reading Ability," *The Library Quarterly* 4, no. 3 (July 1934): 384–412.

4 L. Boissoneault, "Literacy Tests and Asian Exclusion Were the Hallmarks of the 1917 Immigration Act," *Smithsonian Magazine*, February 6, 2017.

5 J. R. Givens, *Fugitive Pedagogy: Carter G. Woodson and the Art of Black Teaching* (Cambridge, MA: Harvard University Press, 2021), 3–4, 7, 12–13, and passim.

6 V. P. Franklin, "Black Social Scientists and the Mental Testing Movement, 1920–1940," in *Black Psychology*, ed. R. L. Jones, 3rd ed. (Berkeley: Cobb and Henry, 1991), 207–24, esp. 212–13.

7 For Bond's biography see W. J. Urban, *Black Scholar: Horace Mann Bond, 1904–1972* (Athens, GA: University of Georgia Press, 1992) and M. Fultz, "A 'Quintessential American': Horace Mann Bond, 1924–1939," *Harvard Educational Review* 55, no. 4 (November 1985): 416–42.

8 Franklin, "Black Social Scientists," 214–15.

9 H. M. Bond, *Black American Scholars: A Study of their Beginnings* (Detroit: Balamp, 1972), 88–89.

10 C. S. Johnson and H. M. Bond, "The Investigation of Racial Differences Prior to 1910," *Journal of Negro Education* 3, no. 3 (July 1934): 328–39; H. M. Bond, "Intelligence Tests and Propaganda," *The Crisis* 28, no. 2 (June 1924): 61–64, esp. 61; W. J. Urban, "The Black Scholar and Intelligence Testing: The Case of Horace Mann Bond," *Journal of the History of the Behavioral Sciences* 25 (October 1989): 323–34. For a comprehensive list of Bond's publications on intelligence tests, see Fultz, "'Quintessential American,'" 427n56.

11 H. M. Bond, "What the Army 'Intelligence' Tests Measured," *Opportunity* 2, no. 19 (July 1924): 197–202, esp. 198; W. Lippmann, "The Mental Age of Americans," *New Republic* 32, no. 412 (October 25, 1922): 213–15. (This was the first of a series of articles by Lippmann against Stoddard and his allies.)

12 Bond, "Intelligence Tests and Propaganda," 64; Bond, "What the 'Army Intelligence' Tests Measured," 198–202. See also a study that Bond cited: H. B. Alexander, "A Comparison of the Ranks of American States in Army Alpha and in Social-Economic Status," *School and Society* 16, no. 405 (September 30, 1922): 388–92.

13 H. M. Bond, "An Investigation of the Nonintellectual Traits of a Group of Negro Adults," *Journal of Abnormal Psychology* 21 (October 1926): 267–76, esp. 267, 268, 276.

14 C. Foreman, *Environmental Factors in Negro Elementary Education* (New York: W. W. Norton, for the Julius Rosenwald Fund, 1932), 7, 13, and passim; Fultz, "'Quintessential American,'" 426; P. Irons, *Jim Crow's Children: The Broken Promise of the* Brown *Decision* (New York: Viking, 2002), 33. For Buswell and Judd see their brief exchange in University of Chicago Special Collections, Judd papers, Box 2, folder 3.

15 H. M. Bond, *The Education of the Negro in the American Social Order* (New York: Octagon Books, 1966 [1934]), 10.

16 Bond, *Education of the Negro*, 172–87, esp. 186–87. See also H. M. Bond, "The Negro Elementary School and the Cultural Pattern," *Journal of Educational Sociology* 13, no. 8 (April 1940): 479–89, and Fultz, "'Quintessential American,'" 419–20, 429.

17 A. Fairclough, "'Forty Acres and a Mule': Horace Mann Bond and the Lynching of Jerome Wilson," *Journal of American Studies* 31, no. 1 (April 1997), 1–17.

18 H. M. Bond and J. W. Bond, *The Star Creek Papers: Washington Parish and the Lynching of Jerome Wilson*, ed. A. Fairclough (Athens, GA: University of Georgia Press, 1997).

19 H. M. Bond, "Self-Respect as a Factor in Racial Advancement," *Annals of the American Academy of Political and Social Science* 140 (November 1928): 21–25, esp. 24–25.

20 H. M. Bond, "Social and Economic Influences on the Public Education of Negroes in Alabama, 1865–1930" (PhD diss., University of Chicago, 1936), 1–2.

21 Bond, "Social and Economic Influences," 30–37, 492.

22 Bond, "Social and Economic Influences," 50–51, 176–77, 195–202, 232–42, 517–18; R. C. Morris, *Reading, 'Riting, and Reconstruction: The Education of Freedmen in the South, 1861–1870* (Chicago: University of Chicago Press, 1981 [1976]), 174–77.

23 Bond, "Social and Economic Influences," 299–300, 326, 332–34, 351–52, 358–59, 362–73, 383–94, 404–11.

24 Bond, "Social and Economic Influences," 472–88. For the Rosenwald Fund's efforts, see: A. M. Johnson-Jones, *The African American Struggle for Library Equality: The Untold Story of the Julius Rosenwald Fund Library Program* (Lanham: Rowman and Littlefield, 2019), 15–18 and passim; P. T. Graham, *A Right to Read: Segregation and Civil Rights in Alabama's Public Libraries, 1900–1965* (Tuscaloosa: University of Alabama Press, 2002), 26–32; W. A. Wiegand and S. A. Wiegand, *The Desegregation of Public Libraries in the Jim Crow South: Civil Rights and Local Activism* (Baton Rouge: Louisiana State University Press, 2018), 35–39; C. Knott, *Not Free, Not for All: Public Libraries in the Age of Jim Crow* (Amherst: University of Massachusetts Press, 2015), 105–19; D. M. Battles, *The History of Public Library Access for African Americans in the South: Or, Leaving Behind the Plow* (Lanham: Scarecrow Press, 2009), 67.

25 C. B. Goodlett and A. H. Calloway, *The Reading Abilities of the Negro Elementary School Child in Kanawha County West Virginia* (Institute, WV: West Virginia State College, 1940), 9, 43.

26 H. M. Bond, "The Educational and Other Social Implications of the Impact of the Present Crisis Upon Racial Minorities," *Journal of Negro Education* 10, no. 3 (July 1941): 617–22.

27 Knott, *Not Free*, 137–39.

28 E. A. Gleason, *The Southern Negro and the Public Library* (Chicago: University of Chicago Press, 1941), 88–109, esp. 108; Battles, *Public Library Access*, 81, 107.

29 Gleason, *The Southern Negro and the Public Library*, 85.

30   Gleason, *The Southern Negro and the Public Library*, 142–44; Knott, *Not Free*, 137.

31   Gleason, *The Southern Negro and the Public Library*, 30–68, esp. 66–68.

32   Gleason, *The Southern Negro and the Public Library*, 189–91. On the importance of recruiting African American librarians for providing suitable books (and not just hand-me-downs from white libraries), see Knott, *Not Free*, 169–237, and Wiegand and Wiegand, *Desegregation*, 39–43.

33   See the remarkable little pamphlet the school issued to attract students: *Libraries, Librarians, and the Negro* (Atlanta: Atlanta University, 1944).

34   Recent accounts of public-library access for African Americans in this era include D. Hanbury, *The Development of Southern Public Libraries and the African American Quest for Library Access, 1898–1963* (Lanham, MD: Lexington, 2020); Wiegand and Wiegand, *Desegregation*; Knott, *Not Free*; and Battles, *Public Library Access*.

35   L. D. Reddick, "Where Can a Southern Negro Read a Book?" *New South* 9, no. 1 (January 1954): 5–11.

36   Wiegand and Wiegand, *Desegregation*, 15, 48–50; Battles, *Public Library Access*, 82–83.

37   Knott, *Not Free*, 140–42.

38   Hanbury, *Development of Southern Public Libraries*, 45–46.

39   Knott, *Not Free*, 238–62; Wiegand and Wiegand, *Desegregation*, 46.

40   J. P. Jackson, "'Racially Stuffed Shirts and Other Enemies of Mankind': Horace Mann Bond's Parody of Segregationist Psychology in the 1950s," in *Defining Difference: Race and Racism in the History of Psychology*, ed. A. S. Winston (Washington, DC: American Psychological Association, 2004), 261–83; Fultz, "'Quintessential American,'" 423–34; H. M. Bond, "Cat on a Hot Tin Roof," *Journal of Negro Education* 27, no. 4 (Autumn 1958): 519–25. For the role of social scientists in general, see J. P. Jackson, *Social Scientists for Social Justice: Making the Case Against Segregation* (New York: NYU Press, 2001).

41   R. Kluger, *Simple Justice: The History of* Brown v. Board of Education *and Black America's Struggle for Equality* (New York: Vintage, 2004 [1975]), 618–29, 639–53.

42   On the call for equalization, see also H. M. Bond, "Redefining the Relationship of the Federal Government to the Education of Racial and Other Minority Groups," *Journal of Negro Education* 7, no. 3 (July 1938): 454–59, esp. 455–56, 459.

43   Bond, *The Education of the Negro in the American Social Order*, vi–vii, x–xii, 475–77.

Chapter Six

1   "He Sees Everything in a Flash," *Ohio State University Monthly*, June 1954.

2   *The Lantern*, Ohio State University, May 28, 1925, 1.

3   *The Lantern*, May 27, 1935; May 29, 1935; May 31, 1935. The convict, Peter Treadway, was executed the day after Renshaw's report of his experience with the instrument appeared.

4   J. M. Larsen Jr., "Samuel Renshaw (1892–1981)," *American Psychologist* 38, no. 2 (February 1983): 226; A. P. Weiss, *A Theoretical Basis of Human Behavior*, 2nd ed. (Columbus, OH: R. G. Adams, 1929).

5   *The Lantern*, September 4, 1925, 3; September 28, 1925, 7; *Ohio State University Monthly*, February 1926, 27.

6   *The Lantern*, April 12, 1934; May 9, 1934; J. D. Weinland and W. S. Schlauch, "An Examination of the Computing Ability of Mr Salo Finkelstein," *Journal of Experimental Psychology* 21, no. 4 (October 1937): 382–402; Weiss, *Theoretical Basis*, 331–57.

Another stimulus for Renshaw's interest may have been a study he was commissioned to do in the late 1920s on the effects of movies on children's sleep patterns, one section of which involved psychophysical research on "flickering" light: S. Renshaw, V. L. Miller, and D. P. Marquis, *Children's Sleep* (New York: Macmillan, 1933), 209–22, and OSU Archives, Renshaw Papers, RG.40.117, folder 45.

7   The introduction of the tachistoscope to the United States was associated with Dodge: see, for example, R. Dodge, "The Reaction Time of the Eye," *Psychological Review* 6 (1899): 477–83. For tachistoscopes in general, see R. Benschop, "What is a Tachistoscope? Historical Explorations of an Instrument," *Science in Context* 11 (1998): 23–50.

8   *The Lantern*, April 21, 1948.

9   S. Renshaw, "Tachistoscopic Studies on Visual Perception," *Psychological Optics* 1, no. 8 (May 1940): 1–4.

10 D. G. Wittels, "You're Not as Smart as You Could Be: Part 2," *Saturday Evening Post* April 24, 1948, 30–31, 142–147, esp. 146–47; S. Rensaw, "The Visual Perception and Reproduction of Forms by Tachistoscopic Methods," *Journal of Psychology* 20 (1945): 217–32, esp. 217–22; *The Lantern*, May 16 1945; April 21, 1948; "He Sees Everything in a Flash," *Ohio State University Monthly*, June 1954.

11 D. G. Wittels, "You're Not as Smart as You Could Be: Part 1," *Saturday Evening Post*, April 17, 1948, 47.

12 Renshaw, "Visual Perception and Reproduction of Forms"; Wittels, "You're Not as Smart as You Could Be: Part 1," 49; Wittels, "You're Not as Smart as You Could Be: Part 2," 142–43.

13 Wittels, "You're Not as Smart as You Could Be: Part 1," 20–21 and 44–52, esp. 20; *The Lantern*, April 14, 1948.

14 F. W. Hebbard, "A Brief Biography of Glenn A. Fry," *American Journal of Optometry and Physiological Optics* 53, no. 7 (July 1976): 339.

15 *The Lantern*, October 22, 1937.

16 *Psychological Optics* (Duncan, OK: Optometric Extension Program, 1939–1962?). A number of volumes are accessible at Archive.org, but the most complete sequence is held at Ohio State University.

17 S. Renshaw, "Some New Contributions to the Problem of Human and Animal Learning," *Ohio State Monthly*, May 1935.

18 This is part of the history of the policing of information that I mean to trace in a future book.

19 *The Lantern*, January 23, 1946.

20 D. G. Wittels, "You're Not as Smart as You Could Be: Part 3," *Saturday Evening Post*, May 1, 1948, 30, 115–19.

21 This example was used in C. R. Carpenter, "The War and Vision," *Visual Digest* 7, no. 4 (Spring 1944): 36–39 (reprinted from the Optometric Extension Program).

22 *Ohio State University Monthly*, November 1942; P. Fountain, "Eyes That See," *Flying* (August 1945): 30–31, 132, 136–38, esp. 30–31.

23 Fountain, "Eyes That See."

24 "Recognizing Planes," *New York Times*, November 21, 1943, E13. There are copies of the slides in the OSU archives: Renshaw Papers, RG 40/98. Interestingly, the images of Axis aircraft are often more blurry than those of Allied.

25 Renshaw, "Visual Perception," 227–30; "He Sees Everything in a Flash"; A. C. Vicory, *A Brief History of Aircraft Identification Training* (Washington, DC: George Wash-

ington University HumRRO, Professional Paper 27–68, August 1968), 2–3. Chicago journalist Jack Cooper later implied that the Grable story referred to a class of aircraft mechanics: J. Cooper, "Secret of Speed Reading is Seeing Fast, Accurately," *Chicago Daily Tribune*, April 1, 1951, A1.

26 *The Lantern*, October 7, 1942; October 15, 1942; October 19, 1942; May 12, 1943; March 6, 1944; *Ohio State Monthly*, November 1942.

27 Fountain, "Eyes That See"; *The Lantern*, December 5, 1944; *Ohio State University Monthly*, March 1945; *The Lantern*, April 27, 1943.

28 C. R. Carpenter, "The War and Vision," *Visual Digest* 7, no. 4 (Spring 1944): 36–39, esp. 38.

29 H. L. Sherman (with the collaboration of R. L. Mooney and G. A. Fry), *Drawing By Seeing: A New Development in the Teaching of the Visual Arts Through the Training of Perception* (New York: Hinds, Hayden and Eldredge, 1947), 4.

30 R. Mooney to P. C. Galpin, May 18, 1948: OSU Archives, Sherman Papers, UA.RG40.142.0002.40.142, folder 2. See also OSU Archives, Sherman Papers, e.g., UA.RG40.142.0001.40.142, Box 1, folder 2; *The Lantern*, February 19, 1941; E. W. Rannells, "A Critique of Drawing By Seeing," *College Art Journal* 12, no. 2 (Winter 1953): 140–47, esp. 140–41. Rannells explicitly attributed credit to Renshaw, albeit in the context of a highly critical account of Sherman's aesthetic claims.

31 G. Bader, *Hall of Mirrors: Roy Lichtenstein and the Face of Painting in the 1960s* (Cambridge, MA: MIT Press, 2010), 1–41, esp. 7.

32 Sherman, *Drawing By Seeing*, esp. 5–25; H. Sherman, "The Eye in the Arts," *Educational Research Bulletin* 33, no. 1 (January 19, 1944): 1–6; E. C. Okerbloom, "Hoyt Sherman's Experimental Work in the Field of Visual Form," *College Art Journal* 3, no. 4 (May 1944): 143–47; M. Torlen, "Hit With a Brick: The Teachings of Hoyt L. Sherman," *Visual Inquiry* 2 (2013): 313–26. For dentistry, see OSU Archives, Sherman Papers, UA.RG40.0001.40.142, Box 1, folder 2. For the flash helmet, see OSU Archives, Sherman Papers, UA.RG.40142.0003.40.142, folders 3, 4.

33 Sherman, *Drawing by Seeing*, 56.

34 Army Air Forces *Statistical Digest: World War II* (Office of Statistical Control, December 1945), 309.

35 P. M. Fitts and R. E. Jones, "Analysis of Factors Contributing to 460 'Pilot Error' Experiences in Operating Aircraft Controls" (orig. 1947); reprinted in H. W. Sinaiko, ed., *Selected Papers on Human Factors in the Design and Use of Control Systems* (New York: Dover, 1961), 332–58, esp. 333.

36 Fitts and Jones, "Analysis," 338–40. There is a brief and fluent, but not wholly reliable, account of Fitts's work in C. Kuang with R. Rabricant, *User Friendly: How the Hidden Rules of Design are Changing the Way we Live, Work, and Play* (New York: MCD / Farrar, Straus and Giroux, 2019), 75–87.

37 P. M. Fitts and R. E. Jones, "Psychological Aspects of Instrument Display. I: Analysis of 270 'Pilot-Error' Experiences in Reading and Interpreting Aircraft Instruments" (orig. 1947); reprinted in Sinaiko, *Selected Papers*, 359–96, esp. 360–61, 370–71, 374–75, 377, 379, 390–91.

38 Fitts and Jones, "Psychological Aspects," 364, 395; P. M. Fitts, "Psychological Research on Equipment Design in the AAF," *The American Psychologist* 2 (1947): 93–98, esp. 93.

39 F. C. Bartlett, "Instrument Controls and Display: Efficient Human Manipulation" (FPRC Report 565, December 1943).

40   R. B. Loucks, "Legibility of Aircraft Instrument Dials" (AAF School of Aviation Medicine, Randolph Field, Texas, Project 265, Reports 1–2, 1944).

41   Fitts, "Psychological Research on Equipment Design," 93.

42   E.g., P. M. Fitts, R. E. Jones, and J. L. Milton, "Eye Fixations of Aircraft Pilots: III" (USAF Air Materiel Command, AF Technical Report 5967, 1949), 17.

43   P. M. Fitts, R. E. Jones, and J. L. Milton, "Eye Movements of Aircraft Pilots During Instrument-Landing Approaches," *Aeronautical Engineering Review* 9, no. 2 (February 1950): 24–29.

44   P. M. Fitts and C. W Simon, "The Arrangement of Instruments, the Distance between Instruments, and the Position of Instrument Pointers as Determinants of Performance in an Eye-Hand Coordination Task" (AAF Technical Report 5832, WADC, February 1952).

45   P. M. Fitts, "Psychological Requirements in Aviation Equipment Design," *Journal of Aviation Medicine* 17 (1946): 270–75, esp. 273–74.

46   Fitts, "Psychological Research on Equipment Design"; P. M. Fitts, "Introduction to Psychological Research on Equipment Design," in Fitts, ed., *Psychological Research on Equipment Design* (Washington, DC: US Government Printing Office. AAF Aviation Psychology Program Research Report 19, 1947), 1–19; R. W. Pew, "Paul Morris Fitts, 1912–1965," in *Division 21 Members Who Made Distinguished Contributions to Engineering Psychology*, ed. H. L. Taylor (Washington, DC: American Psychological Association, 1994), 23–44, esp. 23–27. For some of the results of Fitts's program at Dayton, see the series of Wright Air Development Center (WADC) reports entitled *Eye Fixations of Aircraft Pilots*, issued through the late 1940s and 1950s.

47   S. D. King and R. C. Oldfield, "The Movements of the Eyes in Searching for an Inconspicuous Object" (Flying Personnel Research Committee FPRC 678, 1946); R. C. Oldfield, "Visual Coverage Afforded in a Single Fixation" (FPRC 679, 1946).

48   "AFROTC 'Wars,'" *The Lantern*, November 28, 1955; J. Kangilaski, "Psychology Lab Practices Radar Wars," *The Lantern*, February 24, 1958; M. E. Holmes, "Performing Proficiency: Applied Experimental Psychology and the Human Engineering of Air Defense, 1940–1965" (PhD diss., University of Chicago, 2014), 165, 182. For Renshaw's radar project, see OSU Archives, UA.1996.057.003.RG38.0, Project 258.

49   Fitts, Jones, and Milton, "Eye Movements of Aircraft Pilots"; R. B. Loucks, "Project Report," Project 265, 2, October 27, 1944, AAF School of Aviation Medicine, Randolph Field, TX; B. Hartman and P. M. Fitts, "The Development of Techniques and Procedures for the Study of Alertness in Aviation Personnel," Ohio State University Research Foundation, November 1950; P. M. Fitts, "Psychological Research on Equipment Design," Army Air Forces Aviation Psychology Program Research Report No. 19 (Washington, DC: US Government Printing Office, 1947); P. M. Fitts, "The Information Capacity of the Human Motor System in Controlling the Amplitude of Movement," *Journal of Experimental Psychology* 47, no. 6 (June 1954): 381–91. See also Fitts, "Psychological Requirements," 270, 272.

50   E. L. Milam, *Creatures of Cain: The Hunt for Human Nature in Cold War America* (Princeton: Princeton University Press, 2019), 79–88.

51   *The Lantern*, May 16, 1945.

52   Fountain, "Eyes That See," 138.

53   Wittels, "You're Not as Smart as You Could Be: Part 1," 49; Wittels, "You're Not as Smart as You Could Be: Part 2," 142–43.

54 *The Lantern*, February 8, 1946; April 17, 1946; April 14, 1948; January 11, 1951; November 6, 1951.

55 *The Lantern*, November 18, 1953.

56 *The Lantern*, November 9, 1951. For the Trainer, see S. Renshaw Papers, Cummings Center for the History of Psychology, Akron, OH, M1226 (February 23, 1949).

57 Wittels, "You're Not as Smart as You Could Be"; R. A. Heinlein, "Gulf," *Astounding Science Fiction* 44, no. 3 (November 1949): 53–90 and 44, no. 4 (December 1949): 54–79; R. A. Heinlein, *Citizen of the Galaxy* (New York: Charles Scribner's Sons, 1957); R. A. Heinlein, *Stranger in a Strange Land* (New York: G. P. Putnam's Sons, 1961).

58 *Brooklyn Eagle*, April 14, 1949, 4.

59 "Getting Through the Pile," *Newsweek* 43, no. 2 (January 11, 1954): 48.

60 A. Heilman, "Rapid Reading: Uses and Abuses," *Journal of Developmental Reading* 5, no. 3 (Spring 1962): 157–63, esp. 159. There is a survey of some of the tachistoscopic devices on offer in C. R. Acland, *Swift Viewing: The Popular Life of Subliminal Influence* (Durham, NC: Duke University Press, 2012), 80–85.

61 "Getting Through the Pile."

62 K. S. Bartlett, "Lollypops As a Cure for Lip Movers," *Boston Globe*, November 2, 1952, A51.

63 H. E. Dickhuth, "Reading Speed Stepped Up By New Machines," *New York Herald Tribune*, August 8, 1950, 29.

64 R. Freedburg, "Are You a Slow Reader? Try New Course at Home," *Chicago Daily Tribune*, May 11, 1952, S7.

65 "Better Reading Improves Learning, Say Educators," *The DePaulia*, April 6, 1951, 5; J. Cooper, "Secret of Speed Reading is Seeing Fast, Accurately," *Chicago Daily Tribune*, April 1, 1951, A1.

66 J. Cooper, "New Device Aids Military to See Enemy Planes," *Chicago Daily Tribune*, July 1, 1951, W4; J. Cooper, "Improve Device Used to Speed Reading Time," *Chicago Daily Tribune*, July 1, 1951, W5.

67 W. J. Pauk, "The Cornell University Reading Improvement Program," in *Techniques and Procedures in College and Adult Reading Programs*, ed. O. S. Causey (Fort Worth: Texas Christian University Press, 1957), 16–22, esp. 20–21.

68 A. E. Texler, "Value of Controlled Reading: Summary of Opinion and Research," *Journal of Experimental Education* 11, no. 4 (June 1943): 280–92; M. L. Yoder, "Use of the Harvard Reading Films with Preparatory Class Students at Gallaudet College," *American Annals of the Deaf* 87, no. 5 (November 1942): 395–407.

69 L. L. Miller, "Evaluation of Workbooks for College Reading Programs," in *Techniques and Procedures*, ed. Causey, 75–85, esp. 75, 77.

70 W. C. Schwarzbek, "Some Factors Which Influence the Impression and Immediate Reproduction of Digits" (PhD diss., Ohio State University, 1935). W. C. Schwarzbek Papers, Cummings Center for the History of Psychology, Akron, OH, passim.

71 H. O. Patterson, "A Reading Improvement Program in Industry," in *Techniques and Procedures*, ed. Causey, 102–13; H. O. Patterson, "Reading Improvement Programs in General Motors," in *Reading and Revolution: The Role of Reading in Today's Society*, ed. D. M. Dietrich and V. H. Mathews (Newark, DE: International Reading Association, 1970), 31–37. See also E. A. Sullivan Jr., "Some Experiences with Reading Improvement at General Motors," *Journal of Developmental Reading* 2, no. 2 (Winter 1959): 12–16.

72 H. O. Patterson, "A Survey of Reading Improvement Programs in Industry," in *Techniques and Procedures*, ed. Causey, 121–33, esp. 121.

73 For the ambiguities of this role, see S. Shapin, *The Scientific Life: A Moral History of a Late Modern Vocation* (Chicago: University of Chicago Press, 2008), 127–64.

74 Patterson, "Survey of Reading Improvement Programs in Industry."

75 Air University, *The Air University Reading Improvement Program* (Montgomery, AL: Air University, 1948); CIA Reading Improvement Course Orientation Document, CIA-RDP78–03362A001800170001–1 (and other files declassified from the CIA's "Reading Techniques 1951–1956" files); M. L. Miller, "Devices and Instruments for Use in High School Reading Instruction," *The High School Journal* 39, no. 4 (January 1956): 227–32; C. L. Rosen, "Mechanical Devices for Increasing Speed of Reading," *Journal of Reading* 10, no. 8 (May 1967): 569–76; B. Schmidt, "Mechanical Devices and Reading Instruction," *Journal of Developmental Reading* 7, no. 4 (Summer 1964): 221–22; E. D. Weed Jr., "A Reading Improvement Program in Industry," *Journal of Developmental Reading* 1, no. 2 (Winter 1958): 27–32; M. A. Tinker, "Devices to Improve the Speed of Reading," *The Reading Teacher* 20, no. 7 (April 1967): 605–9; R. M. Williams, "Some Misconceptions about Teaching Reading Improvement to Businessmen," *Journal of Developmental Reading* 1, no. 3 (Spring 1958): 3–10; G. E. Wilson and U. W. Leavell, "An Experiment with Accelerator Training," *Peabody Journal of Education* 34, no. 1 (July 1956): 9–18; C. L. John Legere and William R. Tracey, "Reading Improvement in An Army Service School," *Journal of Developmental Reading* 4, no. 1 (Autumn, 1960): 41–46, esp. 42, 46.

76 R. L. Dowell, "Techniques and Procedures in Government Reading Programs," in *Techniques and Procedures*, ed. Causey, 91–101.

77 Compare S. Igo, *The Known Citizen: A History of Privacy in Modern America* (Cambridge, MA: Harvard University Press, 2018), 102–8.

Chapter Seven

1 The actual origin of the phrase is unclear: McLuhan much later dated it to "a radio conference" in July 1957, but it seems that there is no record to confirm this, and it may be that he was misremembering the date.

2 "Radio Gets Top Billing at Vancouver May Meet of U. of Brit. Columbia," *Variety*, April 29, 1958, 47; UBC President's Report, 1959: both quoted in C. McEwen, "Communications Program at UBC," McLuhan's New Sciences (blog), April 24, 2019, https://mcluhansnewsciences.com/mcluhan/2019/04/communications-programme-at-ubc/.

3 *Radio in the Future of Canada: A National Conference, Vancouver, May 5–9, 1958* (Vancouver: UBC, 1958), quoted in C. McEwen, "The Medium is the Message in 1958," McLuhan's New Sciences (blog), April 23, 2019, https://mcluhansnewsciences.com/mcluhan/2019/04/the-medium-is-the-message/#fn-50840-1. McLuhan later recalled that he had come up with the phrase while on the plane to Vancouver: W. T. Gordon, *Marshall McLuhan: Escape into Understanding* (New York: Basic Books, 1997), 397n49.

4 M. McLuhan, "The Role of Mass Communications in Meeting Today's Problems," *Proceedings of Conference on Educational Television* (US Department of Health, Education and Welfare: Office of Education, Circular 574), 14–17, esp. 14.

5   M. McLuhan. "Grammars of the Media," 1959, https://archive.org/details/naeb
     -b066-f09/page/n85/mode/1up.

6   M. McLuhan, "Myth and Mass Media," *Daedalus* 88, no. 2 (Spring 1959): 339–48,
     esp. 343.

7   M. McLuhan, "Electronic Revolution: Revolutionary Effects of New Media" (1959),
     in *Understanding Me: Lectures and Interviews*, ed. S. McLuhan and D. Staines (Cam-
     bridge, MA: MIT Press, 2003), 1–11, esp. 2–3, 5–7.

8   A. Johns, "Watching Readers Reading," *Textual Practice* 35, no. 9 (2021): 1429–52,
     esp. 1443–44.

9   R. L. Mooney, "Creation in the Classroom Setting," *Explorations in Creativity*, ed.
     R. L. Mooney and T. A. Razik (New York: Harper and Row, 1967), 206–15.

10   M. McLuhan, *Report on Project in Understanding New Media* (NAEB, 1960), 3–4;
     M. McLuhan to B. Muller-Thym, May 5, 1960, in M. Molinaro, C. McLuhan, and
     W. Toye, eds., *Letters of Marshall McLuhan* (Oxford: Oxford University Press, 1987),
     270–72.

11   M. McLuhan, *The Gutenberg Galaxy: The Making of Typographic Man* (London:
     Routledge and Kegan Paul, 1962), e.g., 24–28.

12   J. Miller, *Marshall McLuhan* (New York: Viking, 1971), 85–86.

13   McLuhan, *Gutenberg Galaxy*, 4–6; J. Z. Young, *Doubt and Certainty in Science: A
     Biologist's Reflections on the Brain* (Oxford: Clarendon Press, 1951), 67–68, 111; J. Z.
     Young, *A Model of the Brain* (Oxford: Clarendon Press, 1964), e.g., 94–111.

14   M. McLuhan, "The Effect of the Printed Book on Language in the 16th Century,"
     Explorations 7 (1954): 96–105, esp. 96–97, 100; [McLuhan], "Electronics as ESP," *Ex-
     plorations* 8 (1954): 7–8; "Verbi-Voco-Visual," *Explorations* 8 (1954): 22–24.

15   McLuhan, *Gutenberg Galaxy*, 124.

16   S. Currell, "Streamlining the Eye: Speed Reading and the Revolution of Words,
     1870–1940," in *Residual Media*, ed. C. R. Acland (Minneapolis: University of Minne-
     sota Press, 2007), 344–60, esp. 354–55.

17   McLuhan, *Gutenberg Galaxy*, 43, 47–49.

18   *Varsity*, University of Toronto, March 4, 1966, Review, 4, 9; A. Szende, "Second Read-
     ing Course to be Sponsored by SAC," *Varsity*, University of Toronto, 84, no. 29 (De-
     cember 2, 1964): 1; see also G. Koffman, "Students are Victims of Experiment," *Varsity*,
     University of Toronto, 84, no. 29 (December 2, 1964): 7, for the claim that the program
     was needed to remedy poor elementary-school training due to "look-see" methods.

19   T. Wolfe, "Suppose He is What He Sounds Like," in *McLuhan Hot & Cool*, ed. G. E.
     Stearn (Harmondsworth: Penguin, 1968 [1967]), 37–56, esp. 37; G. P. Elliott, "McLu-
     han's Teaching is Radical," *McLuhan Hot & Cool*, 87–95, esp. 87.

20   Several historians have described this moment. See especially: J. Cohen-Cole, *The
     Open Mind: Cold War Politics and the Sciences of Human Nature* (Chicago: Uni-
     versity of Chicago Press, 2014), 35–62; J. Isaac, *Working Knowledge: Making the
     Human Sciences from Parsons to Kuhn* (Cambridge, MA: Harvard University Press,
     2012), 158–90; A. Pickering, *The Cybernetic Brain: Sketches of Another Future* (Chi-
     cago: University of Chicago Press, 2010); and F. Turner, *The Democratic Surround:
     Multimedia and American Liberalism from World War II to the Psychedelic Sixties*
     (Chicago: University of Chicago Press, 2013), 151–80. David Kaiser's *How the Hippies
     Saved Physics: Science, Counterculture, and the Quantum Revival* (New York: W. W.
     Norton, 2012) deals with a slightly later period, but to some extent the perspectives
     of its characters reflected this process too.

21 On the subject of teaching machines and programmed learning, see: A. Watters, *Teaching Machines: The History of Personalized Learning* (Cambridge, MA: MIT Press, 2021); L. Cuban, *Teachers and Machines: The Classroom Use of Technology since 1920* (New York: Teachers College, 1986); and P. Saettler, *A History of Instructional Technology* (New York: McGraw-Hill, 1968), 49–75, 250–67. A. Watters, *The Monsters of Education Technology* (n.p.: Createspace, 2014) is a comprehensive discrediting of the field.

22 E. L. Thorndike and A. I. Gates, *Elementary Principles of Education* (New York: Macmillan, 1929), 239–44, esp. 242; G. Jonçich, *The Sane Positivist: A Biography of Edward L. Thorndike* (Middletown: Wesleyan University Press, 1968), 86–88, 126–31, 225–28, 239–45, 268–77, 282–83, 286, 317, 386, 394–96.

23 S. L. Pressey and F. P. Robinson, *Psychology and the New Education*, rev. ed. (New York: Harper and Brothers, 1944 [1933]), 367–71; W. L. Bryan and N. Harter, "Studies in the Telegraphic Language: The Acquisition of a Hierarchy of Habits," in *Casebook of Research in Educational Psychology*, ed. S. L. Pressey and J. E. Janney (New York: Harper and Brothers, 1937 [1897]), 221–28. See also E. J. Swift, "Studies in the Psychology and Physiology of Learning," *American Journal of Psychology* 14 (1903): 201–51, esp. 224–30.

24 C. Gere, *Pain, Pleasure, and the Greater Good: From the Panopticon to the Skinner Box and Beyond* (Chicago: University of Chicago Press, 2017), 166–69; B. F. Skinner, *The Technology of Teaching* (Englewood Cliffs, NJ: Prentice-Hall, 1968), 22–28. See in general A. Rutherford, *Beyond the Box: B. F. Skinner's Technology of Behavior from Laboratory to Life, 1950s–1970s* (Toronto: University of Toronto Press, 2009), 27–33 and passim.

25 W. C. Schwarzbek Papers, Archives, Cummings Center for the History of Psychology, Akron, OH, boxes M1655, M1656, M1659 (folders 2, 3).

26 S. Ramo, "A New Technique of Education," in *Teaching Machines and Programmed Learning: A Source Book*, ed. A. A. Lumsdaine and R. Glaser (Washington, DC: National Education Association, 1960), 367–81 (orig. in *Engineering and Science Monthly*, October 21, 1957); R. Glaser, "Christmas Past, Present, and Future," in *Teaching Machines*, 23–31, esp. 27–28.

27 Lumsdaine and Glaser, eds., *Teaching Machines*.

28 S. L. Pressey, F. P. Robinson, and J. E. Horrocks, *Psychology in Education* (New York: Harper and Brothers, 1959 [1933]), 372, 388n5, 389.

29 S. L. Pressey, "Autoinstruction: Perspectives, Problems, Potentials," in *Theories of Learning and Instruction*, ed. E. R. Hilgard (Chicago: University of Chicago Press / National Society for the Study of Education, 1964), 354–70.

30 G. Steiner, J. Miller, and A. Forge, "As for Blake," in *McLuhan Hot and Cool*, 268–74, esp. 269; Wolfe, "Suppose He Is What He Sounds Like," 50.

31 D. A. Hollinger, "The Defense of Democracy and Robert K. Merton's Formulation of the Scientific Ethos," in *Current Perspectives on the History of the Social Sciences*, ed. R. A. Jones and H. Kuklick (Greenwich, CT: JAI Press. *Knowledge and Society* 4, 1983), 1–15.

32 Cohen-Cole, *Open Mind*, 1–4.

33 S. Ghamari-Tabrizi, "Cognitive and Perceptual Training in the Cold War Man-Machine System," in *Uncertain Empire: American History and the Idea of the Cold War*, ed. J. Isaac and D. Bell (Oxford: Oxford University Press, 2012), 267–93.

34 Cohen-Cole, *Open Mind*, 65–81, 100, 124–25, 141.

35 Cohen-Cole, *Open Mind*, 24–27.

36 J. S Bruner, *Mandate From the People* (New York: Duell, Sloan and Pearce, 1944).

37 J. Isaac, *Working Knowledge: Making the Human Sciences from Parsons to Kuhn* (Cambridge, MA: Harvard University Press, 2012), 159–60, 167–79, 179–90; Cohen-Cole, *Open Mind*, 82–84, 87, 143.

38 J. Cohen-Cole, "Instituting the Science of Mind: Intellectual Economies and Disciplinary Exchange at Harvard's Center for Cognitive Studies," *British Journal for the History of Science* 40 (2007), 567–97.

39 Cohen-Cole, *Open Mind*, 187. For Miller's perspective, see G. A. Miller, "The Cognitive Revolution: A Historical Perspective," *Trends in Cognitive Science* 7, no. 3 (2003): 141–44.

40 Cohen-Cole, *Open Mind*, 153, 159–62.

41 J. S. Bruner and L. Postman, "An Approach to Social Perception," in *Current Trends in Social Psychology*, ed. W. Dennis (Pittsburgh: University of Pittsburgh Press, 1948), 71–118. For the centrality of vision to cybernetics in this and a slightly later period, see the treatment in O. Halpern's generally very pertinent *Beautiful Data: A History of Vision and Reason since 1945* (Durham, NC: Duke University Press, 2014), 204–15.

42 L. Postman and J. S. Bruner, "Perception Under Stress," *Psychological Review* 55 (1948): 314–23.

43 Bruner and Postman, "An Approach to Social Perception," 72, 85.

44 J. S. Bruner, "Personality Dynamics and the Process of Perceiving," in *Perception: An Approach to Personality*, ed. R. R. Blake and G. V. Ramsey (New York: Ronald Press, 1951), 121–47.

45 Bruner and L. Postman, "An Approach to Social Perception."

46 R. L. Solomon and L. Postman, "Frequency of Usage as a Determinant of Recognition Thresholds for Words," *Journal of Experimental Psychology* 43 (1952): 195–201.

47 J. S. Bruner and L. Postman, "Emotional Selectivity in Perception and Reaction," *Journal of Personality* 16, no. 1 (September 1947): 69–77.

48 J. S. Bruner and L. Postman, "Perception, Cognition, and Behavior," *Journal of Personality* 18 (1949): 28–29; E. McGinnies, "Emotionality and Perceptual Defense," *Psychological Review* 56 (1949): 244–51; J. McCleary and R. Lazarus, "Autonomic Discrimination Without Awareness: An Interim Report," *Journal of Personality* 18 (1949): 171–79.

49 L. Postman and J. S. Bruner, "Perception Under Stress," *Psychological Review* 55 (1948): 314–23.

50 Bruner, "Personality Dynamics and the Process of Perceiving," 123.

51 L. Postman, J. S. Bruner, and R. D. Walk, "The Perception of Error," *British Journal of Psychology* 42, nos. 1–2 (March-May 1951): 1–10, esp. 7–8.

52 J. S. Bruner and L. Postman, "On the Perception of Incongruity: A Paradigm," *Journal of Personality* 18, no. 2 (December 1949): 206–23. In a later version, the tachistoscope projected a romantic scene in which the figure of the beloved was replaced by "an ugly old hag," and it produced the same reactions of dominance and compromise. The hag was perceived as a young woman, or both figures were seen as middle-aged. In general, perceptual mechanisms tried to select aspects of the stimulus habitat that tended to suit the prevailing state of the organism. The tachistoscope could also be used to identify the "response thresholds" passed through on

the way to a correct perception. For example, someone shown a horse in repeated exposures might identify it as first a smudge, then a thing, then an animal, then a horse. Bruner and Postman, "Perception, Cognition, and Behavior," 14–31.

53 Compare D. Kaiser, "Thomas Kuhn and the Psychology of Scientific Revolutions," in *Kuhn's 'Structure of Scientific Revolutions' at Fifty: Reflections on a Science Classic*, ed. R. J. Richards and L. Daston (Chicago: University of Chicago Press, 2016), 71–95.

54 T. S. Kuhn, *The Structure of Scientific Revolutions*, 3rd ed. (Chicago: University of Chicago Press, 1996), 62–65.

55 J. S. Bruner, "On Perceptual Readiness," *Psychological Review* 64, no. 2 (1957): 123–52, esp. 123.

56 Cohen-Cole, *Open Mind*, 173–75, 179.

57 Bruner, "On Perceptual Readiness," 127.

58 McCleary and Lazarus, "Autonomic Discrimination Without Awareness: An Interim Report"; McCleary and Lazarus, "Autonomic Discrimination Without Awareness: A Study of Subception," *Psychological Review* 58 (March 1951): 113–22. For the phenomenon in general, see C. R. Acland, *Swift Viewing: The Popular Life of Subliminal Influence* (Durham, NC: Duke University Press, 2012).

59 P. D. Bricker and A. Chapanis, "Do Incorrectly Perceived Tachistoscopic Stimuli Convey Some Information?" *Psychological Review* 60, no. 3 (1953): 181–88.

60 G. A. Miller, *Language and Communication* (New York: McGraw-Hill, 1951), 200–204, 207–22.

61 A. C. Hoffman, B. Wellman, and L. Carmichael, "A Quantitative Comparison of the Electrical and Photographic Techniques of Eye-Movement Recording," *Journal of Experimental Psychology* 24, no. 1 (January 1939): 40–53.

62 L. Carmichael, "'Comprehension Time,' Cybernetics, and Regressive Eye Movements in Reading," *Acta Psychologica* 15, no. 1 (January 1959): 126–27; L. Carmichael, "Reading and Visual Fatigue," *Proceedings of the American Philosophical Society* 92, no. 1 (March 8, 1948): 41–41; L. Carmichael and W. F. Dearborn, *Reading and Visual Fatigue* (Boston: Houghton Mifflin, 1947), e.g., 179–82, 219–28, 266–300.

63 "La possibilité d'appliquer les concepts de la théorie des communications à des problèmes non électrotechniques constitue le principal postulat de la cybernétique de Norbert Wiener": B. Mandelbrot, "Structure formelle des textes et communication," *Word* 10, no. 1 (April 1954): 1–27, esp. 26.

64 A. Chapanis, [review of] "The Mathematical Theory of Communication," *Quarterly Review of Biology* 26, no. 3 (1951): 321.

65 C. E. Shannon, "A Mathematical Theory of Communication," *Bell System Technical Journal* 27, no. 3 (July 1948): 379–423, and 27, no. 4 (October 1948): 623–56, esp. 384–89, 398–99; C. E. Shannon, "Prediction and Entropy of Printed English," *Bell System Technical Journal* 30, no. 1 (January 1951): 50–64, esp. 50–51. See also C. E. Shannon, "The Redundancy of English," in *Cybernetics: The Macy Conferences, 1946–1953*, vol. I, *Transactions*, ed. C. Pias (Berlin: Diaphanes, 2003), 248–72.

66 N. G. Burton and J. C. R. Licklider, "Long-Range Constraints in the Statistical Structure of Printed English," *American Journal of Psychology* 68, no. 4 (1955): 650–53.

67 G. A. Miller and E. A. Friedman, "The Reconstruction of Mutilated English Texts," *Information and Control* 1 (1957): 38–55.

68 J. C. R. Licklider, K. N. Stevens, and J. R. M. Hayes, "Studies in Speech, Hearing, and Communications," Technical Report, MIT Acoustics Laboratory, September 30,

1954: cited in J. R. Pierce and J. E. Karlin, "Reading Rates and the Information Rate of a Human Channel," *Bell System Technical Journal* 36, no. 2 (March 1957): 497–516, esp. 497–98.

69   Pierce and Karlin, "Reading Rates and the Information Rate of a Human Channel," 498; J. R. Pierce and J. E. Karlin, "Maximum Information Rate through a Human Channel in Reading," *Science* 122, no. 3175 (November 4, 1955): 879.

70   W. L. Taylor, "'Cloze Procedure': A New Tool for Measuring Readability," *Journalism Quarterly* 30, no. 3 (fall 1953): 415–33. For Taylor and Schramm, see E. M. Rogers, *A History of Communication Study* (New York: Free Press, 1994), 440, 445–95.

71   D. H. Howes and R. L. Solomon, "Visual Duration Threshold as a Function of Word-Probability," *Journal of Experimental Psychology* 41 (1951): 401–10; R. L. Solomon and L. Postman, "Frequency of Usage as a Determinant of Recognition Thresholds for Words," *Journal of Experimental Psychology* 43 (1952): 195–201.

72   J. B. Carroll, "Communication Theory, Linguistics, and Psycholinguistics," *Review of Educational Research* 28, no. 2 (April 1958): 79–88.

73   For the connections, see J. S. Bruner, "Going Beyond the Information Given," in *Contemporary Approaches to Cognition*, ed. Bruner et al. (Cambridge, MA: Harvard University Press, 1957), 41–69, esp. (for reading/tachistoscopic discussions) 43–49, 55–56, 58–60.

74   C. H. Greenewalt, *The Uncommon Man: The Individual in the Organization* (New York: McGraw-Hill, 1959), 50–51; Cohen-Cole, *Open Mind*, 55–56.

75   See S. Shapin, *The Scientific Life: A Moral History of a Late Modern Vocation* (Chicago: University of Chicago Press, 2008), 127–64.

76   F. Turner, *The Democratic Surround: Multimedia and American Liberalism From World War II to the Psychedelic Sixties* (Chicago: University of Chicago Press, 2013), 115–48, 151–80, 220–58. See also Halpern, *Beautiful Data*, 95–102, 122–44.

77   For the British counterpart to this, see A. Johns, *Death of a Pirate: British Radio and the Making of the Information Age* (New York: W. W. Norton, 2011), 149.

78   J. B. Carroll, foreword to, and P. A. Kolers, introduction to, E. B. Huey, *The Psychology and Pedagogy of Reading* (Cambridge, MA: MIT Press, 1968 [1908]), vii–xii and xiii–xxxix, esp. x–xii, xiii–xiv, xviii–xx, xxii–xxiii, xxv, xxxiv–xxxix.

79   G. Swanson, *Swanson on Swanson: An Autobiography* (New York: Random House, 1980), 299–300, 344, 467–84.

80   S. M. Shearer, *Gloria Swanson: The Ultimate Star* (New York: Thomas Dunne, 2013), 282–89; T. Welsch, *Gloria Swanson: Ready for Her Close-Up* (Jackson, MS: University Press of Mississippi, 2013), 299–301, 303.

81   The best account I have read of the Talking Typewriter is W. Lockett, "Cybernetic Child Psychology: A Genealogy of the User" (PhD diss., NYU, 2019), 114–69.

82   W. Lockett, "Cybernetic Child Psychology," 61–62, 86–113, 120, 140–47.

83   Cohen-Cole, *Open Mind*, 197–98.

84   For the controversies surrounding the MACOS project, see E. L. Milam, *Creatures of Cain: The Hunt for Human Nature in Cold War America* (Princeton: Princeton University Press, 2019), 60–78, 169–75. For Bruner and the new math, see C. J. Phillips, *The New Math: A Political History* (Chicago: University of Chicago Press, 2015), 86–95.

85   B. A. Sanderson and D. W. Kratochvil, "The Edison Responsive Environment Learning System, or the Talking Typewriter" (Washington, DC: DHEW/OE Report AIR-21900-1-72-TR-12, January 1972), 15.

86 O. K. Moore, "Autotelic Response Environments and Exceptional Children," in *Experience, Structure, and Adaptability*, ed. O. J. Harvey (New York: Springer, 1966), 169–216.

87 Sanderson and Kratochvil, "Edison Responsive Environment," 20–21; Swanson, *Swanson on Swanson*, 512. For the installation and use of Talking Typewriters down to 1970, see J. Hill, "The Edison Responsive Environment: Its Development and Use," *Programmed Learning and Educational Technology* 7, no. 1 (1970): 29–42.

88 Lockett, "Cybernetic Child Psychology," 165–67.

89 R. Kobler, "The Talking Typewriter—and the Learning of Reading in a Disadvantaged Community," *Computers and Automation* 16, no. 11 (November 1967): 37–40.

90 For the Sullivan system, see L. J. Thompson, "The Sullivan Reading Program" (Washington, DC: DHEW/OE Report AIR-21900-11-71-TR-5, November, 1971), esp. 9–13.

91 Sanderson and Kratochvil, "The Edison Responsive Environment," 3–4.

92 Sanderson and Kratochvil, "Edison Responsive Environment," 11–12; "Voice Mirror . . . It Gives Instant Backtalk," *Popular Science*, April 1973, 99.

93 "Company Names McLuhan," *New York Times*, December 2, 1967, 63.

94 "Knowledge Decentralized," *New Yorker*, May 11, 1968, 33–34; Hill, "Edison Responsive Environment," 38.

95 Sanderson and Kratochvil, "Edison Responsive Environment," 7–8, 19.

96 Lockett, "Cybernetic Child Psychology," 110–11, 121.

97 For Kay's role in the mythology of PARC, see M. A. Hiltzik, *Dealers of Lightning: Xerox PARC and the Dawn of the Computer Age* (New York: Harper, 1999), 80–96.

98 Kay was referring here to contemporary arguments about sharing and piracy; see A. Johns, *Piracy: The Intellectual Property Wars from Gutenberg to Gates* (Chicago: University of Chicago Press, 2009), 474–84.

99 A. C. Kay, "A Personal Computer for Children of All Ages" (Palo Alto: Xerox PARC, 1972), [10].

100 A. Kay, "Computer Applications: A Dynamic Medium for Creative Thought," presentation at NCTE, University of Minnesota, 1972.

101 For a contemporary survey of this field, see D. Hansen, "Computer-Assisted Instruction and the Individualization Process," *Journal of School Psychology* 6, no. 3 (Spring 1968): 177–85.

102 Kay, "Personal Computer"; A. C. Kay, "The Early History of Smalltalk" (1993); L. Berlin, *Troublemakers: Silicon Valley's Coming of Age* (New York: Simon & Schuster, 2017), 104–05.

103 See especially F. Turner, *From Counterculture to Cyberculture: Stewart Brand, the Whole Earth Network, and the Rise of Digital Utopianism* (Chicago: University of Chicago Press, 2006).

104 R. Kobler, "The Talking Typewriter and the Learning of Reading in a Disadvantaged Community," reprinted in *Computers and People* 31, nos. 1–2 (January–February 1981): 7–10, esp. Berkeley's editorial comment on p. 10.

105 J. L. Rankin, *A People's History of Computing in the United States* (Cambridge, MA: Harvard University Press, 2018), 166–227.

Chapter Eight

1 W. C. Booth, "*Lady Chatterley's Lover* and the Tachistoscope," *Carleton Miscellany* (Spring 1960): 41–47; reprinted in *Journal of Developmental Reading* 3, no. 4 (Summer 1960): 232–37.

2 M. Biederman, *Scan Artist: How Evelyn Wood Convinced the World that Speed-Reading Worked* (Chicago: Chicago Review Press, 2019). William Schwarzbek took the trouble to meet Wood in person and emerged at best noncommittal: Schwarzbek Papers, Cummings Center for the History of Psychology, Akron, OH, M1651, folder 3.

3 R. D. McFadden, "Reading Machine is Cited in Hoax," *New York Times*, January 24, 1972, 23; G. Lichtenstein, "Parents Win Refund on a Reading Aid," *New York Times*, November 1, 1972, 24.

4 *Federal Trade Commission Decisions* 53 (1956): 1347–48.

5 B. Schmidt, "Mechanical Devices and Reading Instruction," *Journal of Developmental Reading* 7, no. 4 (Summer 1964): 221; W. C. Booth, "*Lady Chatterley's Lover* and the Tachistoscope," *Journal of Developmental Reading* 3, no. 4 (Summer 1960): 232–37. For Orne, see A. Winter, *Memory: Fragments of a Modern History* (Chicago: University of Chicago Press, 2012), 148–55.

6 L. L. Miller, "Evaluation of Workbooks for College Reading Programs," in *Techniques and Procedures in College and Adult Reading Programs*, ed. O. S. Causey (Fort Worth: Texas Christian University Press, 1957), 75–85, esp. 75, 77; H. O. Patterson, "A Survey of Reading Improvement Programs in Industry," in *Techniques and Procedures*, ed. Causey, 121–33.

7 Biederman, *Scan Artist*, 87–99.

8 E.g., N. G. Burton and J. C. R. Licklider, "Long-Range Constraints in the Statistical Structure of Printed English," *American Journal of Psychology* 68, no. 4 (1955): 650–53, esp. 651. For McLuhan, see H. M. McLuhan, *The Mechanical Bride: Folklore of Industrial Man* (New York: Vanguard, 1951), 40–42.

9 R. Flesch, *Why Johnny Can't Read—And What You Can Do About It* (New York: Harper and Brothers, 1955), 53–55.

10 Flesch, *Why Johnny Can't Read*, 3, 6–7.

11 L. Bloomfield, "Linguistics and Reading," *Elementary English Review* 19, no. 4 (April 1942): 125–30 and 19, no. 5 (May 1942), 183–86.

12 W. S. Gray, *On Their Own in Reading: How to Give Children Independence in Attacking New Words* (Chicago: Scott, Foresman, 1948), 29–33; R. Wardhaugh, "The Teaching of Phonics and Comprehension: A Linguistic Evaluation," in *Psycholinguistics and the Teaching of Reading*, ed. K. S. Goodman and J. T. Fleming (Newark, DE: International Reading Association, 1969), 79–90, esp. 84–85.

13 Flesch, *Why Johnny Can't Read*, 42.

14 Flesch, *Why Johnny Can't Read*, 117–18.

15 Flesch, *Why Johnny Can't Read*, 60–78, 109; A. I. Gates, G. L. Bond, and D. H. Russell, *Methods of Determining Reading Readiness* (New York: Teachers College, Columbia University, 1939), 3 and passim.

16 Flesch, *Why Johnny Can't Read*, 99, 124–34.

17 J. and N. Morgan, *Dr. Seuss and Mr. Geisel: A Biography* (New York: Random House, 1995), 153–56. For Spaulding senior's use of rhyme, see F. E. Spaulding and C. T. Bryce, *Learning to Read: A Manual for Teachers using the Aldine Readers* (New York: Newson and Company, 1918 [1907]), 8, and for his advice on texts, see F. E. Spaulding, *Measuring Textbooks* (New York: Newson and Company, 1923).

18 "Soviet Claiming Lead in Science," *New York Times*, October 5, 1957, 2; A. Trace, *What Ivan Knows that Johnny Doesn't* (New York: Random House, 1961), 6, 9–29. Trace's book focused primarily on learning to read and thus on the humanities, but its implications for science and technology were clear and explicit.

19   W. J. Urban, *More than Science and Sputnik: The National Defense Education Act of 1958* (Tuscaloosa: University of Alabama Press, 2010), esp. 32, 76–80, 172–77.

20   For the perspectives of both Chall and Dale on reading, see their contributions to I. K. Tyler and C. M. Williams, eds., *Educational Communication in a Revolutionary Age* (Worthington, OH: C. A. Jones, 1973) (Dale's talked at length about the notion of technologies as extensions of man).

21   M. M. Mathews, *Teaching to Read: Historically Considered* (Chicago: University of Chicago Press, 1966), 196–208.

22   J. S. Chall, *Learning to Read: The Great Debate*, updated edition (New York: McGraw-Hill, 1983), 190, 313–14.

23   Chall, *Learning to Read*, 7, 74, 136–39, 297–300, 307–8.

24   Chall, *Learning to Read*, 101, 104n4, 313, and 313n2. This was the same period that also saw the advent of major systems for the indexing and counting of scientific publications: A. Csiszar, "Provincializing Impact: From Imperial Anxiety to Algorithmic Universalism," in A. Johns and J. Evans, eds., *Beyond Craft and Code: Human and Algorithmic Cultures, Past and Present* (*Osiris* 38, forthcoming).

25   Chall, *Learning to Read*, 87, 89n1, 93, 94–98, 279.

26   Chall, *Learning to Read*, 291.

27   Chall, *Learning to Read*, 39–41, 56, 59, 61–64, 268.

28   For connections between Chall, the cognitive science movement, and Licklider, see G. A. Miller, ed., *Communication, Language, and Meaning: Psychological Perspectives* (New York: Basic Books, 1973), which includes contributions from all three.

29   G. A. Miller, "Psychology as a Means of Promoting Human Welfare," *American Psychologist* 24, no. 12 (1969): 1063–75.

30   K. S. Goodman, "Reading: A Psycholinguistic Guessing Game," *Journal of the Reading Specialist* 6, no. 4 (1967): 126–35.

31   K. S. Goodman, "A Linguistic Study of Cues and Miscues in Reading," *Elementary English* 42, no. 6 (October 1965): 639–43.

32   K. S. Goodman, *On Reading: A Common-Sense Look at the Nature of Language and the Science of Reading* (Portsmouth, NH: Heinemann, 1996), 3–10.

33   Kolers, "Reading is Only Incidentally Visual," in Goodman and Fleming, *Psycholinguistics*, 8–16.

34   R. B. Buddell, "Psycholinguistics Implications for a Systems of Communication Model," in Goodman and Fleming, *Psycholinguistics*, 61–78, esp. 72–73.

35   H. A. Robinson, foreword to Goodman and Fleming, *Psycholinguistics*, iii.

36   F. Smith, *Understanding Reading: A Psycholinguistic Analysis of Reading and Learning to Read* (New York: Holt, Rinehart and Winston, 1971), ix, 228–29.

37   F. Smith, preface to F. Smith, ed., *Psycholinguistics and Reading* (New York: Holt, Rinehart and Winston, 1973), v–viii, esp. vii.

38   D. E. Rosenbaum, "Mrs. Nixon Honorary Chairman of U. S. Right-to-Read Project," *New York Times*, September 24, 1969, 19.

39   D. E. Rosenbaum, "Dr. Allen Says Partisan Restraints Interfered with his Duties," *New York Times*, June 12, 1970, 21; Rosenbaum, "Federal Project in Reading Gains," *New York Times*, August 18, 1970, 25; R. L. Holloway, "Beyond the Ringing Phrase," *The Reading Teacher* 25, no. 2 (November 1971): 118–28.

40   G. Gent, "TV Workshop Will Dramatize Reading Lessons for Children," *New York Times*, June 23, 1971, 91.

41   M. Mayer, "The Electric Company: Easy Reader and a lot of Other Hip Teachers," *New York Times*, January 28, 1973, 15–26.

42   Chall, *Learning to Read*, 4–5.

43   "Illiteracy Fight Lacks Momentum," *New York Times*, September 12, 1971, 57.

44   R. Rheingold, "Johnny Still Can't Read Very Well," *New York Times*, June 16, 1974, 195.

45   "U.S. Office of Education Forming Bureau to Combine 23 Programs," *New York Times*, March 4, 1979, 50.

46   R. C. Anderson et al., *Becoming a Nation of Readers: The Report of the Commission on Reading* (Washington, DC: National Academy of Education, 1985), 105; *What Works: Research about Teaching and Learning* (Washington, DC: US Department of Education, 1986). For a contemporary critique, see J. L. Davidson, ed., *Counterpoint and Beyond: A Response to 'Becoming a Nation of Readers'* (Urbana, IL: National Council of Teachers of English, 1988).

47   M. Lee, "Study to Urge More Parental Involvement in Education," *LA Times*, February 28, 1986, 8.

48   B. Staples, "How California Betrayed Its Schools," *New York Times*, February 10, 1997, A14.

49   R. L. Colvin, "Phonics Method Endorsed as the Best Tool for Reading," *LA Times*, May 1996, 1. Compare Rothstein, "Consensus," B7.

50   S. Porter, "Mounting Illiteracy is a Disgrace to the Nation," *San Francisco Chronicle*, September 21, 1985, 51; P. Holt, "The Reading Crisis in America," *San Francisco Chronicle*, January 11, 1987, 1.

51   *Becoming a Nation of Readers*, 38.

52   T. Armstrong, "Bringing Back Dick and Jane," *San Francisco Chronicle*, May 10, 1991, B3; D. G. Savage, "School Failure a Teaching Trial—the Unread Tale," *LA Times*, September 6, 1985, 1; K. E. Perkins, "Teachers Don't Want Human Computers Who Can't," *Sacramento Bee*, July 22, 1984, A17.

53   D. Kollars, "Turning Children on to Reading," *Sacramento Bee*, December 21, 1986, B1.

54   N. Marsden, "'Real Books' for California Classrooms," *LA Times*, August 27, 1989, E3.

55   "Court: Get Public Input on Textbooks," *Sacramento Bee*, December 28, 1991, B1.

56   P. Schrag, "Politics and CLAS Bias," *Sacramento Bee*, September 30, 1994, B8; J. A. Fleming, "Commentary on Education," *LA Times*, May 7, 1995, 13; D. Anderluh, "State's Kids Read Poorly, Test Reveals," *Sacramento Bee*, April 28, 1995, A1; R. L. Colvin, "Reading Skills Lagging in State and Across U.S.," *LA Times*, April 28, 1995, A1; "Putting the State to the Test," *LA Times*, March 27, 1995, WVB10.

57   R. E. Colvin, "State's Reading, Math Reforms Under Review as Scores Fall," *LA Times*, March 23, 1995, A1; Colvin, "Teachers Speak Out in Favor of Reading Aloud Education," *LA Times*, May 5, 1995, 3.

58   R. E. Colvin, "Teachers Speak Out in Favor of Reading Aloud Education," *LA Times*, May 5, 1995, 3.

59   M. Ko, "Some Parents Angered by Teacher's Letter," *LA Times*, June 16, 1995, 7.

60   S. Engelberg, "A New Breed of Hired Hands Cultivates Grass-Roots Anger," *New York Times*, March 17, 1993, A1.

61   R. Rivenburg, "Gale-Force Phyllis," *LA Times*, July 3, 1996, 1.

62   "'Hooked on Phonics' Agrees to Tone Down its Ads," *San Francisco Chronicle*, December 15, 1994, A10; *LA Times*, December 14, 1994, 6; December 15, 1994, 6.

63   D. J. Saunders, "Another Skirmish in America's Education War," *San Francisco Chronicle*, June 19, 1995, A19; D. Margolis, "Beleaguered O.C. Marketer of Reading Kit

Plans Comeback," *LA Times*, October 31, 1995, 1; K. Pender, "Little Company, Huge Ad Budget," *San Francisco Chronicle*, October 17, 1991, C1; "Phonics Settles with FTC," *San Francisco Chronicle*, June 3, 1995, D2.

64   C. Saillant, "Researchers Have Eye on Students in Linking Vision to Reading Success," *LA Times*, August 8, 1995, B1; D. Jordan, "Educational Reforms Review," *LA Times*, April 3, 1995, 4; D. J. Saunders, "Californian Flunks Again," *San Francisco Chronicle*, March 31, 1995, A25.

65   R. L. Colvin, "State Poised to Lead U.S. in Reading Reform," *LA Times*, May 9, 1996, 3.

66   N. Lemann, "The Reading Wars," *The Atlantic* 280, no. 5 (November 1997): 128–34; R. L. Colvin, "Her Best Subject," *LA Times*, November 19, 1995, 1.

67   "The ABC Bill," *Sacramento Bee*, July 6, 1995, B6; K. Chavez, D. Anderluh, "Wilson Signs Back-to-Basics Education Bills," *Sacramento Bee*, October 12, 1995, A3; "Governor Signs Bills on Basic Spelling, Math," *San Francisco Chronicle*, October 12, 1995, A17.

68   *Every Child a Reader* (Sacramento: California Department of Education, 1995); D. A. Bee, "We Goofed, Says Eastin," *Sacramento Bee*, September 14, 1995, A1; R. L. Colvin, "State Report Urges Return to Basics in Teaching Reading," *LA Times*, September 13, 1995, 1; Colvin, "Eastin Vows Quick Action on School Reform," *LA Times*, September 14, 1995, 3; "Progress by Looking to the Past," *LA Times*, September 14, 1995, 8.

69   R. L. Colvin, "Teacher Training Lags as State Shifts its Strategy on Reading," *LA Times*, September 26, 1995, 3; Colvin, "Schools Still Waiting for New Reading Plan," *LA Times*, January 22, 1996, 3; Colvin, "State Board OKs Reading Policy Based on Phonics," *LA Times*, May 11, 1996, 20; Colvin, "Wilson to Propose Cutting Class Size in 1st, 2nd Grades," *LA Times*, May 19, 1996, 1; C. Alcala, "Reading at Top of Wish Lists," *Sacramento Bee*, May 28, 1996, B1; Colvin, "School Goals Spelled Out in 'ABC Bill,'" *LA Times*, July 5, 1995, 3.

70   N. Asimov, "Novel Plan for Teaching Reading," *San Francisco Chronicle*, May 9, 1996, A1; N. Asimov, "State Approves Return of Phonics to Schools," *San Francisco Chronicle*, May 11, 1996, A1.

71   No Child Left Behind Act of 2001, 20 USC 6368.

72   D. C. Simmons and E. J. Kame'enui, *A Consumer's Guide to Evaluating a Core Reading Program Grades K-3* (University of Oregon, 2003).

73   T. F. Gieryn, *Cultural Boundaries of Science: Credibility on the Line* (Chicago: University of Chicago Press, 1999).

74   A. Winter, *Memory: Fragments of a Modern History* (Chicago: University of Chicago Press), 225–55.

75   See, e.g., R. Rothstein, "Consensus in Reading War If Sides Would Only Look," *New York Times*, September 5, 2001, B7; D. Goldstein, "A Return to Phonics to Mold Better Readers," *New York Times*, February 15, 2020, A1, A23; J. Winter, "The Rise and Fall of Vibes-Based Literacy," *New Yorker*, September 1, 2022.

76   D. J. Schemo, "Federal-Local Clash in War Over Teaching Reading," *New York Times*, March 9, 2007, A16.

77   D. Goldstein, "A Return to Phonics to Mold Better Readers," *New York Times*, February 15, 2020, A1, A23.

78   D. Kipen, "Fighting to Reopen a Closed Book," *New York Times*, December 29, 2016, C6; J. Reyhner, "The Reading Wars: Phonics versus Whole Language," revised January 3, 2020, https://jan.ucc.nau.edu/jar/Reading_Wars.html.

79   J. Barzun, introduction to C. C. Walcutt, ed., *Tomorrow's Illiterates: The State of Reading Instruction Today* (Boston: Little, Brown and Company, 1961), xi–xvi.

80   N. Oreskes and E.M. Conway, *Merchants of Doubt: How a Handful of Scientists Obscured the Truth on Issues from Tobacco Smoke to Global Warming* (New York: Bloomsbury, 2010), 14–34, 240–242.

81   See, e.g., J. S. Bruner, *Mandate of the People* (New York: Duell, Sloan and Pearce, 1944), 159–69 (on the complex public response to a Beveridge-style plan for a welfare state in America), H. D. Lasswell, *Democracy Through Public Opinion* (Menasha, WI: George Banta, 1941), 35–44, and Lasswell, *Propaganda Technique in World War I* (Cambridge, MA: MIT Press, 1971 [1927]), 185–94, 208–13, 220–22.

82   For regression, see Winter, *Memory*, 103–23. For Velikovsky, see M. Gordin, *The Pseudoscience Wars: Immanuel Velikovsky and the Birth of the Modern Fringe* (Chicago: University of Chicago Press, 2012).

83   L. Buder, "'aa' as in Apple," *New York Times*, November 25, 1956, 310.

84   A. Weinberg, "The Impact of Large-Scale Science on the United States," *Science* 134 (1961): 161–64.

85   Chall, *Learning to Read*, 305.

Chapter Nine

1   S. Schwarzkopf, "In Search of the Consumer: The History of Market Research from 1890 to 1960," in *The Routledge Companion to Marketing History*, ed. D. G. B. Jones and M. Tadajewski (London: Routledge, 2016), 61–83; D. B. Ward, *A New Brand of Business: Charles Coolidge Parlin, Curtis Publishing Company, and the Origins of Market Research* (Philadelphia: Temple University Press, 2010), 117–19.

2   D. Starch, *Principles of Advertising* (Chicago: A. W. Shaw, 1923), 657–69. For his work on reading (done contemporaneously with that on advertising), see D. Starch, "The Measurements of Handwriting," *Journal of Educational Psychology* 4, no. 8 (October 1913): 445–64; Starch, "The Measurement of Efficiency in Reading," *Journal of Educational Psychology* 6, no. 1 (January 1915): 1–24; Starch, "The Measurement of Efficiency in Writing," *Journal of Educational Psychology* 6, no. 2 (February 1915): 106–14; Starch, "The Measurement of Efficiency in Spelling, and the Overlapping of Grades in Combined Measurements of Reading, Writing and Spelling," *Journal of Educational Psychology* 6, no. 3 (March 1915): 167–86.

3   T. M. Shepard, "The Starch Application of the Recognition Technique," *Journal of Marketing* 6, no. 4 (April 1942): 118–24, esp. 119, 122–24.

4   S. Craig, "Daniel Starch's 1928 Survey: A First Glimpse of the U.S. Radio Audience," *Journal of Radio and Audio Media* 17, no. 2 (2010): 182–94, esp. 182, 185–87.

5   F. N. Stanton and H. E. Burtt, "The Influence of Surface and Tint of Paper on the Speed of Reading," *Journal of Applied Psychology* 19 (1935): 683–93.

6   F. N. Stanton, "Memory for Advertising Copy Presented Visually vs. Orally," *Journal of Applied Psychology* 18 (1934): 45–64.

7   For the general problem, see A. Johns, *Death of a Pirate: British Radio and the Making of the Information Age* (New York: W. W. Norton, 2011), 72–74.

8   J. J. Karol, "Measuring Radio Audiences," *Printers' Ink* 177, no. 8 (November, 1936): 44–56 began as one of these pamphlets.

9   M. J. Socolow, "The Behaviorist in the Boardroom: The Research of Frank Stanton, Ph.D.," *Journal of Broadcasting and Electronic Media* 52, no. 4 (2008): 526–43, esp. 529–34; H. M. McLuhan, *The Mechanical Bride: Folklore of Industrial Man* (New York: Vanguard, 1951), 48–49.

10   J. Auerbach, *Weapons of Democracy: Propaganda, Progressivism, and American Public Opinion* (Baltimore: Johns Hopkins University Press, 2015), 16–48.

11   H. Cantril and G. W. Allport, *The Psychology of Radio* (New York: Harper Brothers, 1935), 178, 254.

12   For Lazarsfeld's fascinating account of the story, see P. F. Lazarsfeld, "An Episode in the History of Social Research: A Memoir," in *The Intellectual Migration: Europe and America, 1930–1960*, ed. D. Fleming and B. Bailyn (Cambridge, MA: Belknap Press, 1969), 270–337.

13   P. F. Lazarsfeld, *Radio and the Printed Page: An Introduction to the Study of Radio and Its Role in the Communication of Ideas* (New York: Duell, Sloan and Pearce, 1940), 38.

14   Lazarsfeld, *Radio and the Printed Page*, xiv, 45–47, 94–132.

15   Lazarsfeld, *Radio and the Printed Page*, 139–45, 179.

16   Lazarsfeld, *Radio and the Printed Page*, 169–70n26, 199, 262–63, 272–74, 290; Cantril and Allport, *Psychology of Radio*, 159–63.

17   For the Fugger newsletters, see, most recently, A. Pettegree, *The Invention of News: How the World Came to Know about Itself* (New Haven: Yale University Press, 2014), 113–16.

18   Lazarsfeld, *Radio and the Printed Page*, 333. Lazarsfeld used this historical parallel many times, starting with a report to the Rockefeller Corporation in 1939 and then continuing through years of lecturing on mass media and society, at least into the mid-1950s. See Lazarsfeld, "Episode in the History of Social Research," 320–21.

19   Lazarsfeld. "Episode in the History of Social Research," 327; P. F. Lazarsfeld, "Communication Research and the Social Psychologist," in *Current Trends in Social Psychology*, ed. W. Dennis (Pittsburgh: University of Pittsburgh Press, 1948), 218–73, esp. 234–35.

20   For "frame of reference," see Lazarsfeld, *Radio and the Printed Page*, 105.

21   R. L. Venezky, "Research on Reading Processes: A Historical Perspective," *American Psychologist* (May 1977): 339–45, esp. 344.

22   A. L. Yarbus, *Eye Movements and Vision*, trans. B. Haigh (New York: Plenum Press, 1967). For the context in the USSR, see S. Gerovitch, *From Newspeak to Cyberspeak: A History of Soviet Cybernetics* (Cambridge, MA: MIT Press, 2002), e.g., 108–12, 222–24.

23   For examples, see R. M. Johnson, "Measuring Advertising Effectiveness," in *Handbook of Marketing Research*, ed. R. Ferber (New York: McGraw-Hill, 1974), part C, 4, nos. 151–64. Other authors in Ferber's handbook provide more examples of eye-movement tracking. See also W. Meyers, "The Eyes Have it," *New York Times*, December 5, 1982, A29.

24   A. Pirisi, "Eye-Catching Advertisements," *Psychology Today*, February 1, 1997.

25   N. J. Wade and B. W. Tatler, "Origins and Applications of Eye Movement Research," in *The Oxford Handbook of Eye Movements*, ed. S. P. Liversedge, I. D. Gilchrist, and S. Everling (Oxford: Oxford University Press, 2011), 17–43, esp. 32–34; J. Calderone, "Eye Tracking in Google Glass: A Window into the Soul?" *Scientific American*, January 1, 2015.

26   T. Wu, *The Attention Merchants: The Epic Scramble to Get Inside Our Heads* (New York: Knopf, 2016).

27   T. B. Lee, "Consumer Reports: Tesla Autopilot a 'Distant Second' to GM Super Cruise," *Ars Technica*, October 28, 2020, https://arstechnica.com/cars/2020/10/report-tesla-autopilot-has-best-performance-gm-super-cruise-is-safest/.

28 J. G. Proudfoot, J. L. Jenkins, J. K. Burgoon, and J. F. Nunamaker Jr., "More than Meets the Eye: How Oculometric Behaviors Evolve Over the Course of Automated Deception Detection Interactions," *Journal of Management Information Systems* 33, no. 2 (2016): 332–60, esp. 351–52.

29 "Eye Tracking Market Share, Size, Insights," *Market Watch*, March 29, 2020, https:// www.marketwatch.com/press-release/industry-report-eye-tracking-market-size -predicted-to-reach-us-2142-million-by-2026-2020-07-14. For an earlier estimate on the same scale see Rahul Kumar, "Eye Tracking Market Expected to Reach $1,818.1 Million, Globally, by 2024," Allied Market Research, April 2018, https://www .alliedmarketresearch.com/press-release/eye-tracking-market.html.

30 See, canonically, S. Zuboff, *The Age of Surveillance Capitalism: The Fight for a Human Future at the New Frontier of Power* (New York: Public Affairs, 2019), but the concern predated Zuboff's blockbuster.

31 For an introduction, see O. Etzioni, M. Banko, and M. J. Cafarella, "Machine Reading," in Y. Gil and R. J. Mooney, chairs, *Proceedings of the 21st Annual AAAI Conference on Artificial Intelligence* (Menlo Park: AAAI Press, 2006), 1517–19.

32 "Machine Reading Comprehension," AI Lab Projects, Microsoft, accessed April 9, 2022, https://www.microsoft.com/en-us/ai/ai-lab-machine-reading.

33 Feng-GUI, Predictive Visual Analytics, accessed April 9, 2022, https://feng-gui.com.

34 M. M. Waldrop, *The Dream Machine: J. C. R. Licklider and the Revolution that Made Computing Personal* (San Francisco: Stripe, 2018 [2001]), 79, 105, 115–17.

35 J. C. R. Licklider, "Man-Computer Symbiosis," *Transactions of Human Factors in Electronics* (March 1960): 4–11, esp. 6–7.

36 Licklider, "Man-Computer Symbiosis," 4; J. C. R. Licklider and W. E. Clark, "On-Line Man-Computer Communication," *AIEE-IRE Proceedings, May 1–3, 1962, Joint Computer Conference* (San Francisco), 113–28, esp. 120. Boulton Paul was known for having built the Defiant, a fighter plane that could only shoot backwards.

37 J. C. R. Licklider, *Libraries of the Future* (Cambridge, MA: MIT Press, 1965), 28.

38 Later he defined "mulling over" as searching for undetected relations within a body of knowledge in the precognitive system. Licklider, *Libraries of the Future*, 19–20, 33, 117–19.

39 Licklider, *Libraries of the Future*, 146–47, 177; D. G. Bobrow, R. Y. Kain, B. Raphael, and J. C. R. Licklider, "A Computer-Program System to Facilitate the Study of Technical Documents," *American Documentation* 17, no. 4 (October 1966), 186–89.

40 I. E. Sutherland, "Sketchpad, A Man-Machine Graphical Communication System" (PhD diss., MIT, 1963), republished online at www.cl.cam.ac.uk/techreports/UCAM -CL-TR-574.pdf.

41 I. E. Sutherland, "Sketchpad: A Man-Machine Graphical Communication System," AFIPS Spring Joint Computer Conference, May 21–23 1963, 329–46, esp. 329, 332, 344; T. Bardini, *Bootstrapping: Douglas Engelbart, Coevolution, and the Origins of Personal Computing* (Stanford: Stanford University Press, 2000), 81–95.

42 Licklider and Clark, "On-Line Man-Computer Communication," 122, 123; Licklider, *Libraries of the Future*, 126–27; Licklider, "Man-Computer Symbiosis," 9–10.

43 G. P. Dinneen, "Programming Pattern Recognition," *Proceedings of the March 1–3, 1955, Western Joint Computer Conference* (Association for Computing Machinery, 1955), 94–100; H. Simon and A. Newell, "Heuristic Problem-Solving: The Next Advance in Operations Research," *Operations Research* 6, no. 1 (January–February 1958): 1–10, esp. 6–7. See also J. Evans, T. Reigeluth, and A. Johns, "The Craft and

Code Binary: Before, During, and After," in A. Johns and J. Evans, eds., *Beyond Craft and Code: Human and Algorithmic Cultures, Past and Present* (*Osiris* 38, forthcoming).

44 For those too young to remember the Microwriter, see "Microwriter," Buxton Collection, Microsoft, updated April 2011, https://www.microsoft.com /buxtoncollection/detail.aspx?id=5.

45 Bardini, *Bootstrapping*, esp. 138–42 for the famous "mother of all demos."

46 S. K. Card, W. K. English, and B. J. Burr, "Evaluation of Mouse, Rate-Controlled Isometric Joystick, Step Keys, and Text Keys for Text Selection on a CRT," *Ergonomics* 21, no. 8 (1978): 601–13, esp. 601–2, 605–6, 608–9, 612–13; S. K. Card and T. P. Moran, "User Technology: From Pointing to Pondering," *Proceedings of the ACM Conference on the History of Personal Workstations* (1986), 183–98, esp. 185. For Card's general theory, see his doctoral dissertation: S. K. Card, "Studies in the Psychology of Computer Text Editing Systems" (PhD diss., Carnegie Mellon University, 1978).

47 S. K. Card, T. P. Moran, and A. Newell, *The Psychology of Human-Computer Interaction* (Hillsdale, NJ: Lawrence Erlbaum Associates, 1983), 44–57; Card and Moran, "User Technology," 185.

48 Bardini, *Bootstrapping*, 145–49, 164–66, 169–72, 175–77; Card and Moran, "User Technology," 185.

49 See also C. Kuang and R. Fabricant, *User Friendly: How the Hidden Rules of Design Are Changing the Way We Live, Work, and Play* (New York: Farrar, Straus and Giroux, 2019), 75–96.

50 A. Abbott, *Digital Paper: A Manual for Research and Writing with Library and Internet Materials* (Chicago: University of Chicago Press, 2014), e.g., 110–28, 129–48.

51 M.-G. Verbergt, "Borgesian Dreams and Epistemic Nightmares: The Effects of Early Computer-Use on French Medievalists (1970–1995)," *Storia della Storiografia* 75, no. 1 (2019): 83–104.

52 S. Baron, "Control Systems R&D at BBN," in *A Culture of Innovation: Insider Accounts of Computing and Life at BBN*, ed. D. Walden and R. Nickerson (East Sandwich, MA: Waterside, 2011), 189–218, esp. 191.

53 N. S. Baron, *Words Onscreen: The Fate of Reading in a Digital World* (Oxford: Oxford University Press, 2015), 42–43, 169.

54 For examples, see J. R. Bergstrom and A. J. Schall, *Eye Tracking in User Experience Design* (Waltham, MA: Morgan Kaufmann, 2014) and J. Nielsen and K. Pernice, *Eyetracking Web Usability* (Berkeley: New Riders, 2010).

55 E. J. Lerner, "Lessons of Flight 655," *Aerospace America* April 1989, 18–26; J. Jacky, "Aegis User Interface Changes Planned," in *Software Engineering Notes* 14, no. 1 (January 1989): 6–9; S. Squires, "Peril for Pilots," *Los Angeles Times*, December 5, 1988, 3; W. M. Fogarty, *Investigation Report: Formal Investigation into the Circumstances Surrounding the Downing of Iran Air Flight 655 on 3 July 1988* (US Department of Defense: August 18, 1988), 48.

56 *Iran Air Flight 655 Compensation: Hearings before the Defense Policy Panel of the Committee on Armed Forces*, 100th Cong., 2nd session, 1988 (Washington, DC: USGPO, 1989), 190–91.

57 For examples see L. E. Sibert and R. J. K. Jacob, "Evaluation of Eye Gaze Interaction," *CHI '00: Proceedings of the SIGCHI Conference on Human Factors in Computing* (April 2000): 281–88; and P. Hearn, "MSU Eye-Tracking Technology Helping Navy Assess Thinking Skills" (on psychologist Stephanie Doane), Mississippi State Uni-

versity, August 5, 2004, https://www.newsarchive.msstate.edu/newsroom/article/2004/08/msu-eye-tracking-technology-helping-navy-assess-thinking-skills.

58 Society for the Scientific Study of Reading (SSSR), accessed April 9, 2022, https://www.triplesr.org/.

59 See, in general, J. Chandler and A. I. Davidson, eds., "The Fate of Disciplines," special issue, *Critical Inquiry* 35, no. 4 (Summer 2009).

60 S. Dehaene, *Reading in the Brain: The Science and Evolution of a Human Invention* (New York: Viking, 2009); M. Wolf, *Proust and the Squid: The Story and Science of the Reading Brain* (New York: Harper, 2007); M. Wolf, *Reader, Come Home: The Reading Brain in a Digital World* (New York: HarperCollins, 2018). In this company, Daniel T. Willingham's *The Reading Mind: A Cognitive Approach to Understanding How the Mind Reads* (San Francisco: Jossey-Bass, 2017) seems almost nostalgic in its cognitive approach.

61 G. G. Hruby and U. Goswami, "Neuroscience and Reading: A Review for Reading Education Researchers," *Reading Research Quarterly* 46, no. 2 (2011): 156–72, esp. 168; U. Goswami, "Neuroscience and Education," *British Journal of Educational Psychology* 74 (2004): 1–14, esp. 7; U. Goswami, "Neuroscience and Education: From Research to Practice?" *Nature Reviews Neuroscience* 7, no. 5 (April 2006): 406–11.

62 Wolf, *Reader, Come Home*, 1. There is a large literature on the gains and losses entailed by digital reading: see, for example, D. M. Levy, *Scrolling Forward: Making Sense of Documents in the Digital Age* (New York: Arcade, 2001), N. S. Baron, *Words Onscreen: The Fate of Reading in a Digital World* (Oxford: Oxford University Press, 2015), and N. Carr, *The Shallows: What the Internet is Doing to Our Brains* (New York: W. W. Norton, 2011).

63 Dehaene, *Reading in the Brain*; Wolf, *Proust and the Squid*. See also M. Seidenberg, *Language at the Speed of Sight: How We Read, Why So Many Can't, and What Can Be Done About It* (New York: Basic Books, 2017), a normative book in the tradition of the science wars that nevertheless incorporates neuroscientific perspectives alongside an efficient and judicious survey of the claims and recommendations of the previous century of reading science.

64 Wolf, *Reader, Come Home*, 29–34.

Conclusion

1 US Department of Education, *Data Point: Adult Literacy in the United States* (July 2019), https://nces.ed.gov/pubs2019/2019179.pdf. The National Center for Education Statistics maintains an extraordinarily detailed interactive mapping system for such data—effectively the twenty-first-century successor to Louis Wilson's geographical works of the 1930s—at Program for the International Assessment of Adult Competencies (PIAAC), U.S. Skills Map: State and County Indicators of Adult Literacy and Numeracy, accessed April 9, 2022, https://nces.ed.gov/surveys/piaac/skillsmap/. Its display of inequities is dispiritingly similar to that of Wilson's maps.

2 "The Reading Wars," *The Economist*, June 17, 2021, 23–24. Lest it be thought that its verdict was the last word, one should note that a more recent report by researchers at University College, London, has argued that a phonics-based policy has proved ineffective: D. Wyse and A. Bradbury, "Reading Wars or Reading Reconciliation? A Critical Examination of Robust Research Evidence, Curriculum Policy, and Teach-

ers' Practices for Teaching Phonics and Reading," *Review of Education* 10, 1 (January 2022), 1–53: doi:10.1002/rev3.3314.

3   Compare S. Shapin, "Is There a Crisis of Truth?" *Los Angeles Review of Books,* December 2, 2019, https://www.lareviewofbooks.org/article/is-there-a-crisis-of -truth/.

4   Compare G. Eyal, *The Crisis of Expertise* (Cambridge: Polity, 2019), 43–63.

5   P. Galison, *Image and Logic: A Material Culture of Microphysics* (Chicago: University of Chicago Press, 1997), 784–803.

6   I. Hacking, "Making Up People," *London Review of Books,* August 17, 2006; Hacking, *Rewriting the Soul: Multiple Personality and the Sciences of Memory* (Princeton: Princeton University Press, 1995); Hacking, *The Social Construction of What?* (Cambridge, MA: Harvard University Press, 1999); Hacking, "Kinds of People: Moving Targets," *Proceedings of the British Academy* 151 (2007): 285–318.

7   M. Wolf, *Reader, Come Home: The Reading Brain in a Digital World* (New York: Harper, 2018), 1–4, 11, 80, 107.

8   M. Biagioli and A. Lippman, eds., *Gaming the Metrics: Misconduct and Manipulation in Academic Research* (Cambridge, MA: MIT Press, 2020); J. S. G. Chu and J. Evans, "Too Many Papers? Slowed Canonical Progress in Large Fields of Science," SocArXiv, March 2, 2018, doi:10.31235/osf.io/jk63c.

9   Open Science Collaboration, "Estimating the Reproducibility of Psychological Science," *Science* 349, no. 6251 (August 2015): 943, https://doi.org/10.1126/science .aac4716; M. Serra-Garcia and R. Gneezy, "Nonreplicable Publications are cited more than Replicable Ones," *Science Advances* 7, no. 21 (May 2021): 1–7: https://doi .org/10.1126/sciadv.abd1705.

10   J. C. R. Licklider, "Man-Computer Symbiosis," *Transactions of Human Factors in Electronics* (March 1960): 4–11, esp. 5.

INDEX

Page numbers in italics refer to figures.